Offshore Engineering and Production
Third Edition

Witherby Seamanship International
A Division of Witherby Publishing Group Ltd
4 Dunlop Square, Livingston, Edinburgh, EH54 8SB, Scotland, UK
Tel No: +44(0)1506 463 227 - Fax No: +44(0)1506 468 999
Email: info@emailws.com - Web: www.witherbyseamanship.com

First Published 1995
Second Edition 2000
Third Edition 2011

ISBN: 978-1-905331-98-7
eBook ISBN: 978-1-85609-510-5

British Library Cataloguing in Publication Data
A catalogue record for this book is available from the British Library.

Printed and bound in Great Britain by Bell & Bain Ltd, Glasgow

Published in 2011 by

Witherby Publishing Group Ltd
4 Dunlop Square
Livingston, Edinburgh, EH54 8SB
Scotland, UK

Tel No: +44 (0) 1506 463 227
Fax No: +44 (0) 1506 468 999

Email: info@emailws.com
Web: www.witherbys.com

Offshore Engineering and Production
Third Edition

by

Angus Mather

Tension Leg Platform (TLP), the *'Conoco Hutton'*, which was decommissioned in 2002

Foreword

I certainly felt honoured on being asked to write a foreword for such an impressive and comprehensive manual, covering, as it does, every aspect of today's offshore exploration and production – a section of the industry which hardly existed in the early 50s when my oilfield career commenced. Without a doubt, this will be the essential reference source in the future.

Regarding earlier days, I seem to recall that a well was drilled approximately one mile off the coast of Borneo in 1952, believed to be the first major submarine operation in what, in those days, was known as the British Commonwealth. This particular well had a rather unique distinction: the sole means of transport to and from the shore, for personnel and equipment, was via an aerial ropeway! Some things have changed.

Today, the offshore engineer is required to carry out an enormous range of tasks and a full appreciation of the industry is an essential prerequisite to success. This book should be on the bookshelf, if not a permanent part of every engineer's baggage. Virtually every element of equipment used and each phase of the procedures, various specifications and relevant regulations, etc are covered in understandable terms. It will be as valuable to the new recruit to the industry as to the experienced engineer embarking on a new field of work.

Brendan G. McKeown O.B.E.
Norfolk

Former Amoco Drilling Superintendent, Brendan McKeown remembers September 16th 1969 with great pride and considers himself to be privileged to be present at the beginning of one of the great British success stories of the 20th century, the discovery of North Sea Oil. How did he view it at the time? "It's difficult. You're an oil man, it's your life's business and all of a sudden you are in on a first. It's like winning a cup winner's medal. It was marvellous."

The event Brendan is referring to is the discovery of the first commercial oil discovery in the North Sea, well 28/18–1, now known as the Arbroath field, located some 130 miles off the coast of Scotland.

Born and educated in Belfast, Northern Ireland, Brendan McKeown graduated from Queen's University in 1946 with a B.Sc. in Civil Engineering. Initially, Brendan was employed in Local Government at Stormont as a civil engineer before joining the Kuwait Oil Company. In 1956 Brendan undertook a post graduate diploma in Oil Technology at Imperial College, and then spent the next 10 years *drilling* in Kuwait and Iran. From 1967 until his retirement, Brendan worked for Amoco in the role of Drilling Superintendent, Production Superintendent, Operations Manager and finally as Production Manager. On retiring, Brendan was honoured by Her Majesty the Queen, receiving an OBE for *Services to the Oil Industry.*

The Author

The author, Angus Mather, commenced his career with the Ministry of Defence at Her Majesty's Dockyard, Devonport, in 1970 as a Technician Apprentice. This was followed by 10 years' service as an Engineer Officer in the British Merchant Navy prior to entering the oil and gas industry. The next 25 years were spent employed both in the contracting industry as Engineering Manager of Kvaerner Oil and Gas Services, and as a Classification Society surveyor, involved with oil and gas related projects, both on and offshore, in the UK, the Middle East, the Gulf of Mexico and West Africa.

Preface

Offshore engineering encompasses a considerable number of very specialised and often completely unrelated disciplines. They can be categorised into three core activities: construction, production and reservoir engineering. This book has been written not as a definitive manual, but to provide the reader with a basic explanation of them.

It is hoped that the material contained within this publication will provide the new recruit to the industry with a basic appreciation of what is a relatively complex subject, while at the same time providing the more experienced individual with a fuller appreciation of activities outside of their own particular speciality. The decision on what topics should be included and the depth to which they should be discussed are based largely on my personal experience.

While frequent references are made to the oil and gas industries of the North Sea, the bulk of the text is of a more general nature and so is applicable to offshore engineering on a worldwide basis. Units of measurement are quoted in both imperial and metric, with preference being given to the unit most frequently associated with a particular discipline.

Angus C. Mather
Suffolk
August 2011

Acknowledgements

The author acknowledges with thanks the following companies for their contributions and cooperation in the preparation of this book.

ABB Offshore Technology
Amerada Hess Limited
Baker Service Tools
Bettis Actuators and Controls
BK Kendle Engineering Limited
BP Exploration
Chevron UK Limited
Cooper Oil Tool
Dockwise NV
Framo Engineering AS
Halliburton Energy Services
Hydril Inc.
Hydrovision Limited
Institute of Marine Engineering, Science and Technology
Institute of Petroleum
Kongsberg Offshore As
Lloyd's Register of Shipping
MTL Instruments
Oilfield Publications Limited
RGB Stainless Limited
Schlumberger Oilfield Services
Shaffer Inc.
Shell (UK) Exploration and Production
Slingsby Engineering Limited
SLP Engineering Limited
Statoil
SBM Offshore NV
RWE Dea
Thomas Telford Limited (Offshore Engineer)
Varco International Inc
Weir Pumps Limited

Photographic acknowledgements:
Charles Hodge Photography, Lowestoft
Tony Wilks
David Gold
John Price

Contents

Abbreviations

ABS	American Bureau of Shipping
AC	Alternating current
ACoP	Approved Code of Practice
ACFM	Alternating Current Field Measurement
AIT	Auto ignition temperature
ALARP	As low as reasonably practicable
ANSI	American National Standards Institute
APAU	Accident Prevention Advisory Unit (of HSE)
API	American Petroleum Institute
ARPA	Automatic Radar Plotting Aid
ASME	American Society of Mechanical Engineers
ASNT	American Society for Non-Destructive Testing
ASTM	American Society for Testing and Materials
AWS	American Welding Society
BA	Breathing apparatus
BASEEFA	British Approval Service for Electrical Equipment in Flammable Atmospheres
BHAB	British Helicopter Advisory Board
BOP	Blowout preventer
BPD	Barrels per day
BS	British Standard
BSI	British Standards Institute
BV	Bureau Veritas
CA	Certifying Authority
CAA	Civil Aviation Authority
CAP	Civil Aviation Publication
CALM	Catenary anchor leg mooring
CAPEX	Capital expenditure
CE	Carbon Equivalent
CE	Community European
CHAOS	Consequences of hazards and accidents on offshore structures
CIMAH	Control of Industrial Major Accident Hazards
CoF	Certificate of Fitness
COSHH	Control of Substances Hazardous to Health
CP	Cathodic protection
CRINE	Cost reduction in the new era
CSON	Continental Shelf Operations Notice
CSWIP	Certification scheme for weld inspection personnel
CTOD	Crack tip opening displacement
DC	Direct current
DCR	Design and Construction Regulations
DDC	Deck decompression chamber
DEG	Duoethylene glycol
DEn	Department of Energy
DIN	Deutsches Institut für Normung
DNV	Det Norske Veritas
DoT	Department of Transport
DP	Design pressure
DP	Dynamic positioning
DSM	Diving safety memorandum
DSV	Diving support vessel
DT	Design temperature
EECS	Electrical Equipment Certification Service
EEMUA	Engineering Equipment Material Users Association
EERA	Evacuation, escape and rescue analysis
EOR	Enhanced oil recovery
EPC	Engineer, procure and construct
EPIC	Engineer, procure, install and construct
EPIRB	Emergency position-indicating radio beacon
EPS	Emergency power supply
ESD	Emergency shutdown
ESDV	Emergency shutdown valve

FARSI	Functionality, availability, reliability, survivability and interaction.
FEA	Fire and explosion analysis
FES	Fire and explosion strategy
F&G	Fire and gas
FPS	Floating production system
FPSO	Floating, production, storage and offloading vessel
FPDSO	Floating, production, drilling, storage and offloading vessel
FPV	Floating production vessel
FRAMS	Floating riser and mooring system
FRA	Fire risk analysis
FRC	Fast rescue craft
FSA	Formal safety assessment
FSO	Floating storage and offloading vessel
FSU	Floating storage unit
GA	General alarm
GBS	Gravity base structure
GL	Germanischer Lloyd
HAZ	Heat affected zone
HAZAN	Hazard analysis
HAZOP	Hazard and operability study
HF	High frequency
HGT	High pressure grease tube
HIPS	High integrity protection system
HIC	Hydrogen induced cracking
HP	High pressure
HPHT	High pressure and high temperature
HSC	Health and Safety Commission
HSE	Health and Safety Executive
HSWA	Health and Safety at Work Act
HVAC	Heating, ventilating and air conditioning
ICP	Independent, Competent Persons
IMO	International Maritime Organization
IP	Institute of Petroleum
ISO	International Standards Organisation
ITP	Inspection and Test Plan
JT	Joule Thompson
KO	Knock out
LAT	Lowest atmospheric tide
LCV	Level control valve
LED	Light emitting diode
LEL	Lower explosive limit
LIC	Level indicator controller
LNG	Liquefied natural gas
LOLER	Lifting Operations and Lifting Equipment Regulations
LP	Low pressure
LPG	Liquefied petroleum gas
LRS	Lloyd's Register of Shipping
LSA	Life-saving appliances
MAPD	Major accident prevention document
MAR	Management and Administration Regulations
MCA	Maritime and Coastguard Agency
MEG	Monoethylene Glycol
MHAU	Major Hazards Assessment Unit (HSE)
MHSWR	Management of Health and Safety at Work Regulations
MIG	Metal inert gas, welding
MMSCF	Millions standard cubic feet
MMSCFD	Millions standard cubic feet a day
MLSS	Mudline suspension system
MLSV	Mudline safety valve
MPI	Magnetic particle inspection
MSA	Marine Safety Agency (UK), now the MCA
MSS	Manufacturers Standardization Society of the Valve and Fittings Industry, Inc. (USA)
MWA	Mineral Workings Act
MWP	Maximum working pressure

Multilateral – Multiple boreholes drilled from an existing single bore well.

Multiphase – Practice of flowing unstabilised well fluids (oil with high gas content) in a single pipeline by boosting the pressure to prevent vaporisation of the dissolved gases.

Natural gas – Hydrocarbon gas occurring naturally from underground reservoirs both on and offshore.

NGL – Natural gas liquids, a mixture of hydrocarbon liquids that include butane and ethane obtained from natural gas. May be produced from condensate reservoirs but more probably produced as a by-product of oil production.

Nipple – A section of threaded or socket welded pipe used as an appurtenance that is less than 12 inches in length. Often used to describe any short length of open-ended pipe.

Nozzle – Flanged inlet or outlet connection on a pressure vessel.

Packer – Device for sealing one casing string from another, or from the production tubing.

Pedestal – Large diameter, vertical tube or tub onto which a crane is attached.

Photogrammetric – The use of still photography to capture dimensional information for transposing into drawings.

Pig – Spherical device inserted into a gas subsea pipeline to sweep the line of deposits of rust, scale and condensed liquids. May also be used to clean oil pipelines of wax and may be 'intelligent', ie containing measuring and inspection equipment.

Pig trap – A pressure chamber permitting the entry or removal of equipment into the subsea pipeline, normally pigs.

Pipeline – Piping used to convey fluids between platforms or between a platform and a shore facility.

Pipe spool – Single length of pipe with flanged ends.

Platform, offshore – A fixed offshore structure involved in the production of oil or gas that may be constructed of steel or concrete. Term used frequently by Americans to describe an offshore installation.

Pressure vessel – Container, normally cylindrical, used to contain internal, or occasionally external, pressure.

Produced water – Formation water removed from the oil and gas in the process pressure vessels.

Production separator – Main process vessel used primarily for the separation of gas, oil (and condensate) and water.

Production tubing – Pipe used in wells to conduct fluid from the producing formation into the Christmas tree. Unlike the casing, the tubing is designed to be replaced during the life of the well, if required.

Purge – Maintain gas flow in an over rich or lean concentration to avoid the build–up of oxygen and an explosive mixture.

Quality assurance – A sequence of planned and systematic actions necessary to provide adequate confidence that a product or service will satisfy given requirements of quality.

Quality control – The operational techniques and activities that are used to ensure that a quality product or service will be produced.

Rig – A term normally associated with drilling equipment, ie a drilling rig. Also a slang term used extensively to describe any of the structures and vessels associated with oil and gas exploration and production.

Riser – The vertical portion of a subsea pipeline (including the bottom bend) arriving on or departing from a platform.

Rock dumping – Deposition of rocks onto subsea pipelines to provide protection against anchors and fishermen's nets when burying the pipe is impractical. Rocks and gravel may also be dumped around subsea wellheads and jacket legs to repair scour damage.

Seismic survey – The use of artificially generated sound waves to determine the type of rock formations below the ground or under the seabed by monitoring the reflected sound wave signals.

Scour – Removal of the seabed in the vicinity of a jacket, subsea wellhead or pipeline by tidal action.

Scrubber – Pressure vessel containing equipment designed to remove or scrub liquids from a gas stream.

Shuttle tanker – Moderately sized oil tanker used to transport oil from larger vessels into port.

Skid – Steel framework used to contain equipment, may be transportable.

Stress corrosion cracking – The cracking that results from a combination of stress and corrosion.

Slew ring bearing – Large ball or roller bearing that connects a crane to a pedestal and permits rotation.

Slug – An accumulation of water (may also be sand or condensate) in a gas pipeline.

Spudding – A term used to describe the insertion of the conductor into the seabed when drilling a well. May also be used to describe the process of setting the legs of a jack-up into the seabed.

Stinger – Tubular steel support frame attached to the stern of a pipelay barge to control the bending of the pipe as it enters the water.

Stripping – The removal or replacement of drill pipe or tubing strings from a well under pressure using a stripping BOP.

Stripping gas – Gas, normally process gas, used to assist in the purification of a liquid by reducing the partial pressure of gaseous contaminants to encourage vaporisation.

Swabbing – The lowering of the hydrostatic pressure in the hole due to the upward movement of the drill pipe and/or tools. Also the use of wireline equipment to clean a well by scooping out liquids.

Sulphide stress cracking – Cracking of metallic materials due to exposure to fluids containing hydrogen sulphide.

Telemetry – System for the collection, collation and transmission of information to a remote source using radio, satellite, fibre optics or cable links. Also associated with the remote control of process equipment.

Third party gas – Term used to describe gas sold direct from an oil company or parties other than British Gas, the previous monopoly holders.

Throttle – Regulation of fluid flow by a throttling valve or fixed orifice.

Topsides – Upper part of a fixed installation that sits on top of the jacket and consists of the decks, accommodation and process equipment.

Vam – Trade name for casing thread produced by the Vallourec company of France.

Vent – A pipe or fitting on a vessel that can be opened to the atmosphere.

Vent stack – Open-ended pipe and support framework used to discharge vapours into the atmosphere at a safe location above the installation, without combustion.

Wellhead – Permanent equipment used to secure and seal the casings and production tubing and to provide a mounting place for the christmas trees.

Wireline – Equipment used to introduce tools into the well bore under pressure.

Workover – Re-entry into a completed well for modification or repair work.

Workover rig – Normally a smaller, portable version of the main drilling derrick, which can be used to carry out workover operations on installations that do not have a permanent derrick.

Standards, Guidance Notes and Codes of Practice

Provided below is a list of standards and specifications, most of which are referenced in this book. The list is not exhaustive but identifies the standards and guidance notes which are in most frequent use within the offshore industry, and which provide valuable reference material. A number of the standards referred to have been withdrawn but are still used as reference, particularly in respect of older installations.

British Standards

PD 5500	Specification for unfired fusion welded pressure vessels
BS 5750	Quality systems
BS EN 60079	Electrical apparatus for potentially explosive atmospheres
BS 6755	Part 2. Specification for fire type testing requirements valves (replaced by BS EN 10497:2004)
BS EN 10025-1:2004	Hot rolled products of structural steels. General technical delivery conditions
BS EN 10210	Hot finished structural hollow sections of non-alloy and fine grain steels
BS EN 1011-1:2009	Welding. Recommendations for welding of metallic materials. General guidance for arc welding
BS EN ISO 15607	Specification and qualification of welding procedures for metallic materials - General rules (replaces BS EN 288)
BS EN ISO 9606	Approval testing of welders for fusion welding (previously BS EN 287)
PD 8010	Code of practice for pipelines
BS 4515	Specification for welding of steel pipelines on land and offshore
PD CEN/TR 15135:2005	Welding. Design and non-destructive testing of welds
BS EN ISO 17635:2010	Non-destructive testing of welds. General rules for metallic materials
BS EN ISO 17637:2011	Non-destructive testing of welds. Visual testing of fusion-welded joints
BS EN 1435:1997	Non-destructive examination of welds. Radiographic examination of welded joints
BS EN 583-1:1999	Non-destructive testing. Ultrasonic examination. General principles
BS EN 1714:1998	Non-destructive examination of welded joints. Ultrasonic examination of welded joints
BS EN 571-1:1997	Non-destructive testing. Penetrant testing. General principles
BS EN ISO 17638:2009	Non-destructive testing of welds. Magnetic particle testing

European Equipment Manufacturers and Users Association – EEMUA

EEMUA No. 107	Recommendations for the protection of Diesel Engines in Zone 2 Hazardous Areas
EEMUA No. 150	Steel Specification for Offshore Structures. (Withdrawn. Replaced by EN 10225)
EEMUA No. 158	Construction Specification for Fixed Offshore Structures in the North Sea

American Standards – Various

ANSI/ASME B31.3	Chemical Plant and Petroleum Refinery Piping
ANSI B16.5	Steel Pipe Flanges and Flanged Fittings
ANSI B16.9	Wrought Steel Butt Welded Fittings
ANSI B16.11	Forged Steel Fittings, Socket Welded and Threaded
ASME IX	Boiler and Pressure Vessel Code
AWS D1.1	Structural Welding Code

MSS SP-44	Steel Pipeline Flanges
MSS SP-75	Specification for High Test Wrought Butt Welding Fittings
NACE MR-01-75	Sulphide Stress Cracking Resistant Material for Oilfield Equipment
NACE RP-01-76	Corrosion Control on Steel, Fixed Offshore Platforms Associated with Petroleum Production

American Petroleum Institute (API) Specifications and Recommended Practices

API Spec 5L	Specification for Line Pipe
API Spec 6A	Specification for Wellhead and Christmas Tree Equipment
API Spec 6AF	Specification for Fire Test for Valves
API Spec 14A	Specification for Subsurface Safety Valve Equipment
API Spec 14D	Specification for Wellhead Surface Safety Valves and Underwater Safety Valves for Offshore Installations
API Std 1104	Standard for Welding Pipelines and Related Facilities
API RP 2A	Recommended Practice for Planning, Design and Construction of Fixed Offshore Platforms
API RP 2G	Recommended Practice for Production Facilities on Offshore Structures (now discontinued)
API RP 6AR	Recommended Practice for Repair and Remanufacture of Wellhead and Christmas Tree Equipment
API RP 14B	Recommended Practice for Design, Installation, Repair and Operation of Subsurface Safety Systems
API RP 14C	Recommended Practice for Analysis, Design, Installation and Testing of Basic Surface Safety Systems for Offshore Production Platforms
API RP 14D	Recommended Practice for Wellhead Surface Safety Valves and Underwater Safety Valves for Offshore Service
API RP 14E	Recommended Practice for Design and Installation of Offshore Production Platform Piping Systems
API RP 14F	Recommended Practice for Design and Installation of Electrical Systems for Offshore Production Platforms
API RP 14G	Recommended Practice for Fire Prevention and Control on Open Type Offshore Production Platforms
API RP 14H	Recommended Practice for Installation, Maintenance, and Repair of Surface Safety Valves and Underwater Safety Valves Offshore
API RP 14J	Recommended Practice for Design and Hazards Analysis for Offshore Production Platforms
API RP 16E	Recommended Practice for Design of Control Systems for Drilling Well Control Equipment
API RP 53	Recommended Practice for Blowout Prevention Equipment Systems for Drilling Wells (now discontinued)
API RP 64	Recommended Practice for Diverter Systems Equipment and Operation
API RP 500	Recommended Practice for Classification of Areas for Electrical Installations at Petroleum Facilities Classified as Class 1, Division 1 and Division 2
API RP 520	Recommended Practice for the Design and Installation of Pressure Relieving Systems in Refineries - Parts I and II
API RP 521	Guide for Pressure and Depressuring Systems
API RP 1111	Recommended Practice for Design, Construction, Operation and Maintenance of Offshore Hydrocarbon Pipelines

American Society for Testing and Materials (ASTM) Standards

ASTM A105	Specification for Forgings, Carbon Steel, for Piping Components
ASTM A106	Specification for Seamless Carbon Steel Pipe for High Temperature Service
ASTM A193	Specification for Alloy-Steel and Stainless Steel Bolting Materials for High-Temperature service
ASTM A194	Specification for Carbon and Alloy Steel Nuts for Bolts for High-Pressure and High-Temperature Service
ASTM A234	Specification for Piping Fittings of Wrought Carbon Steel and Alloy Steel for Moderate and Elevated Temperatures
ASTM A312	Specification for Seamless and Welded Austenitic Stainless Steel Pipes
ASTM A320	Specification for Alloy Steel Bolting Materials for Low Temperature Service
ASTM A333	Specification for Seamless and Welded Steel Pipe for Low Temperature Service
ASTM A790	Specification for Seamless and Welded Ferritic/Austenitic Stainless Steel Tube Pipe
ASTM A860	Specification for Wrought High-Strength Ferritic Steel Butt-Welding Fittings

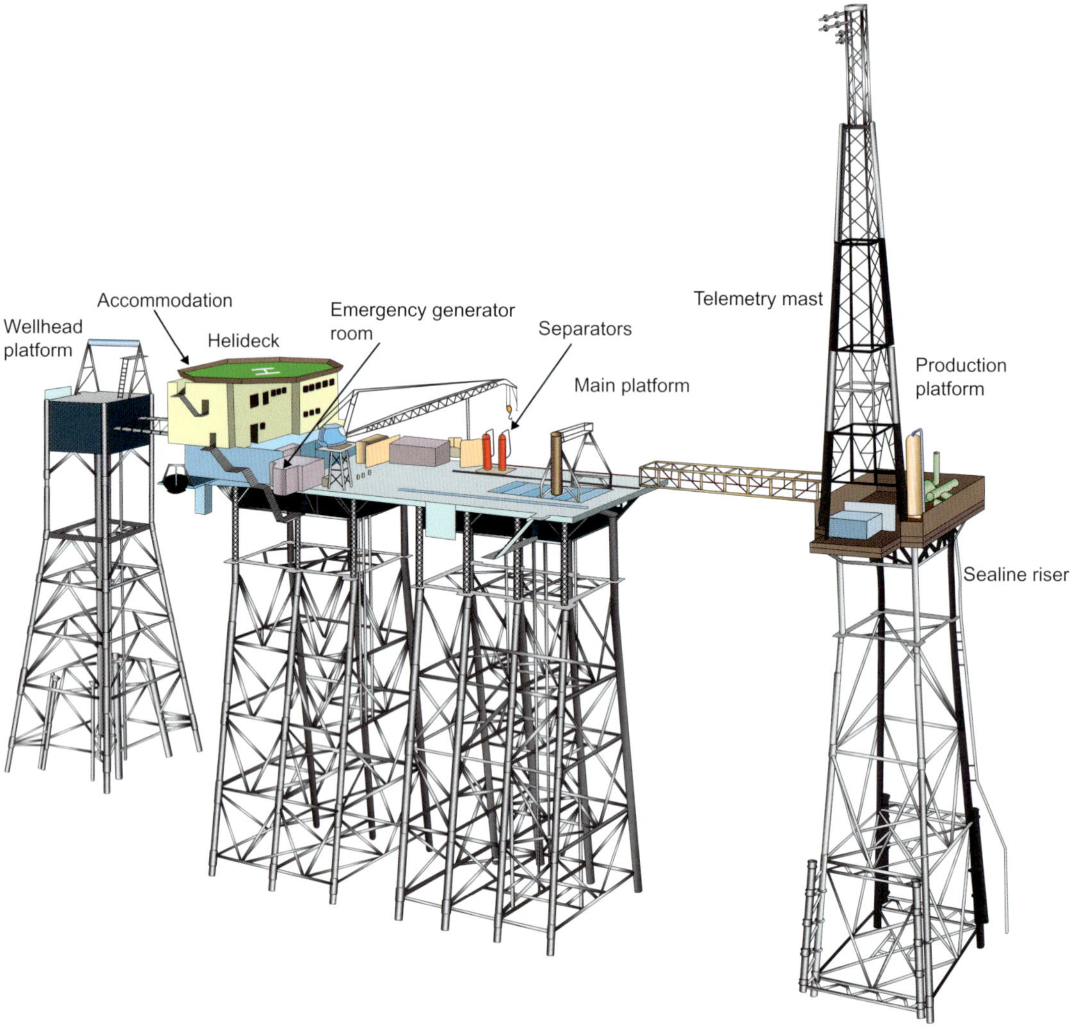

Gas installation

1
Introduction and Background

1 Introduction and Background

Top Ten Oil and Gas Producing Countries Worldwide

	Oil Reserves (1,000 million tonnes)				Gas Reserves (trillion cu ft.)		
	Total World	**205.0**	**100%**		**Total World**	**6,621.2**	100%
1.	Saudi Arabia	36.3	18%	1.	Russian Federation	1,567.1	24%
2.	Iran	18.9	9%	2.	Iran	1,045.7	16%
3.	Iraq	15.5	8%	3.	Qatar	895.8	14%
4.	Venuezuela	24.8	12%	4.	Turkmenestan	286.2	4%
5.	Kuwait	14.0	7%	5.	Saudi Arabia	279.7	4%
6.	UAE	13.0	6%	6.	USA	244.7	4%
7.	Russian Federation	10.2	5%	7.	UAE	227.1	3%
8.	Libya	5.8	3%	8.	Venezuela	200.1	3%
9.	Kazakhstan	5.3	3%	9.	Nigeria	185.4	3%
10.	Canada	5.2	3%	10.	Algeria	159.1	2%
	Others	56.0	27%		Others	1530.3	23%

	Oil Production (million tonnes)				Gas Production (billion cu m.)		
	Total World	**3820.5**	**100%**		**Total World**	**3820.5**	**100%**
1.	Russian Federation	494.2	13%	1.	USA	593.4	16%
2.	Saudi Arabia	459.5	12%	2.	Russian Federation	527.5	14%
3.	USA	325.3	9%	3.	Canada	161.4	4%
4.	Iran	202.4	5%	4.	Iran	131.2	3%
5.	China	189.0	5%	5.	Norway	103.5	3%
6.	Canada	155.7	4%	6.	Qatar	89.3	2%
7.	Mexico	147.5	4%	7.	China	85.2	2%
8.	Venezuela	124.8	3%	8.	Algeria	81.4	2%
9.	Iraq	121.8	3%	9.	Saudi Arabia	77.5	2%
10.	Kuwait	121.3	3%	10.	Indonesia	71.9	2%
	Others	1479.0	39%		Others	1898.2	50%

Source: BP Statistical Review/OGJ

Note: 1 trillion = one thousand billion (10^{12})
1 billion = one thousand million (10^{9})
1 billion cubic metres of gas = 0.83 million tonnes of oil equivalent.
1 tonne = 7.5 barrels of crude oil.

Note: The figures quoted above are for the year ending 2009. While the figures will vary from year to year it is unlikely that table positions will change significantly and they will, therefore, provide a clear indication of the general order of the world's main hydrocarbon producers.

	Oil Consumption (million tonnes)		
	Total World	**3820.5**	**100%**
1.	USA	842.9	22%
2.	China	404.6	11%
3.	Japan	197.6	5%
4.	India	148.5	4%
5.	Russian Federation	124.9	3%
6.	Germany	113.9	3%
7.	Brazil	104.3	3%
8.	Saudi Arabia	121.8	3%
9.	South Korea	104.3	3%
10.	Brazil	104.3	3%
	Others	1553.4	41%

Source: BP Statistical Review/OGJ

1.1 Worldwide Oil and Gas Production – Perspective

The tables showing worldwide oil production, gas production and oil consumption make interesting reading. While most people are fully aware of the significance of OPEC's contribution, particularly Saudi Arabia, many will not appreciate the significance that the world's two major superpowers play in both the supply and consumption of oil and gas. It is also interesting to note that in the last ten years, Norway, small country that it is, has moved from 8th place to 5th in the list of top gas producers. Ten years ago, the UK featured in the top ten tables for both oil and gas production but, with heavily depleted reserves, now appears in neither. It is also interesting to note that Japan consumes more oil

North Sea

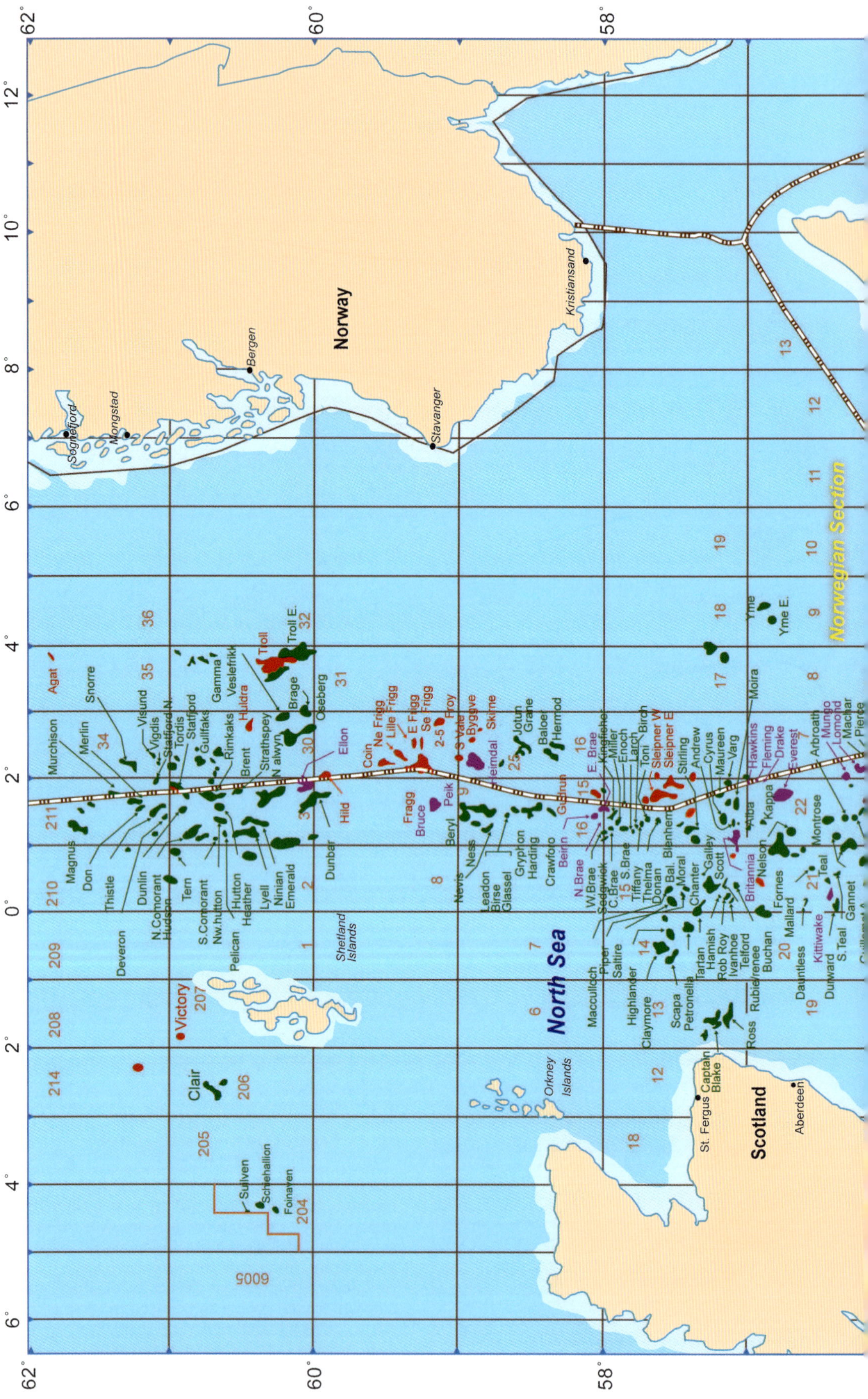

Figure 1.1 – The North Sea UK oil and gas fields

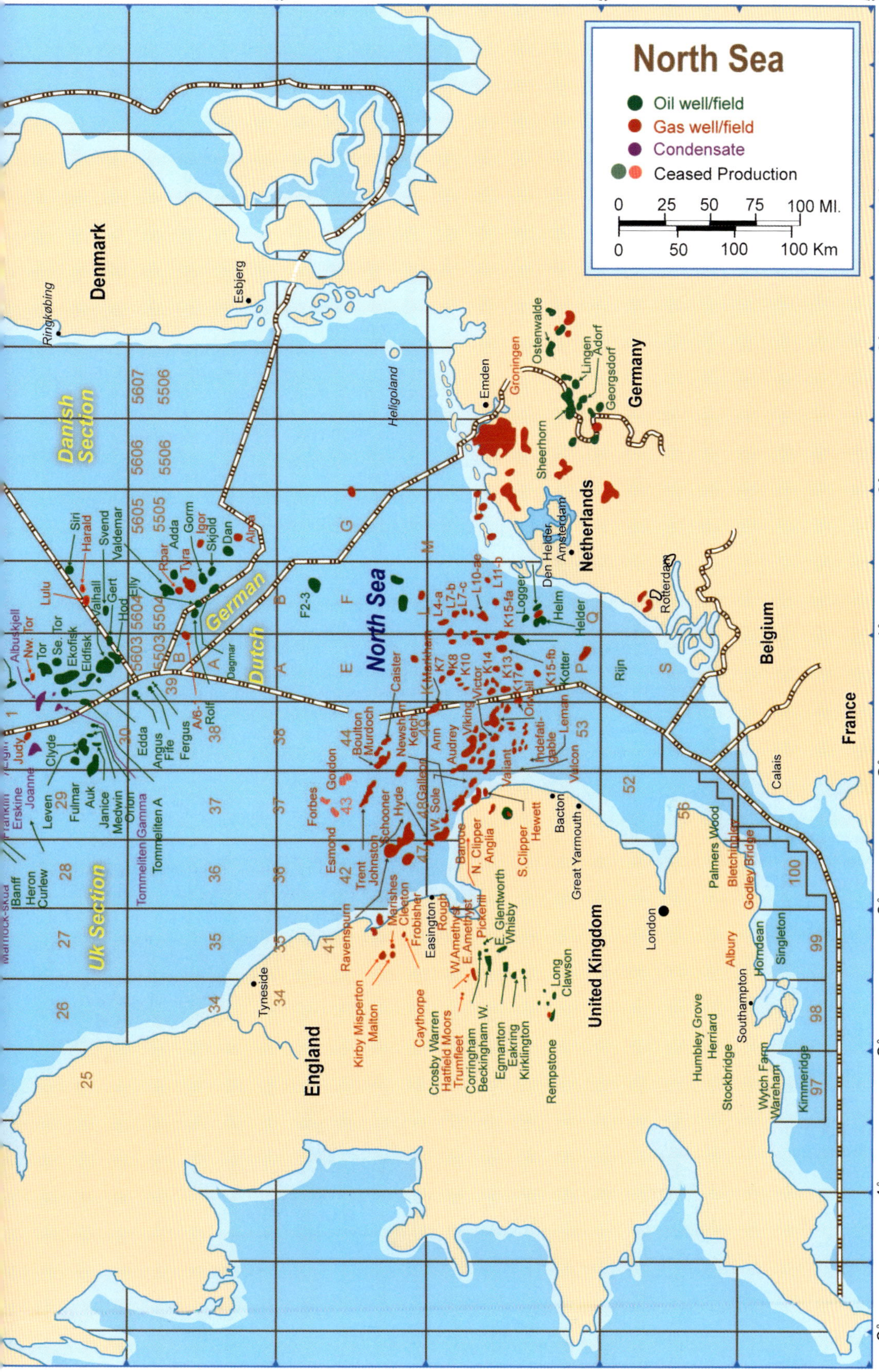
North Sea
Oil well/field
Gas well/field
Condensate
Ceased Production
0 25 50 75 100 MI.
0 50 100 100 Km
Denmark
Germany
Netherlands
Belgium
France
United Kingdom
England
Danish Section
German
Dutch
Uk Section
North Sea
Ringkøbing
Esbjerg
Heligoland
Emden
Groningen
Ostenwalde
Lingen
Adorf
Georgsdorf
Sheerhorn
Den Helder
Amsterdam
Rotterdam
Logger
Helm
Helder
Kotter
Calais
London
Great Yarmouth
Bacton
Tyneside
Southampton
Easington
Ekofisk
Eldfisk
Tor
Valhall
Hod
Gert
Lulu
Siri
Harald
Svend
Valdemar
Elly
Roar
Adda
Tyra
Gorm
Igor
Skjold
Dan
Alma
Dagmar
Rolf
Albuskjell
Nw. Tor
Se. Tor
Judy
Joanne
Clyde
Erskine
Leven
Fulmar
Auk
Janice
Medwin
Orion
Edda
Angus
Fife
Fergus
Tommeliten Gamma
Tommeliten A
Banff
Heron
Curlew
F2-3
Caister
Boulton
Murdoch
Newsham
Ketch
Gordon
Forbes
Esmond
Schooner
Hyde
Galleon
Ann
Audrey
W. Sole
Barque
N. Clipper
Anglia
S.Clipper
Hewett
Valiant
Vulcan
Indefatigable
Leman
Viking
Victor
Trent
Johnston
Ravenspurn
Marishes
Cleeton
Frobisher
Rough
W.Amethyst
E.Amethyst
Pickerill
Glentworth
Whisby
Long Clawson
Kirby Misperton
Malton
Caythorpe
Crosby Warren
Hatfield Moors
Trumfleet
Corringham
Beckingham W.
Egmanton
Eakring
Kirklington
Rempstone
Humbley Grove
Herriard
Stockbridge
Wytch Farm
Wareham
Kimmeridge
Palmers Wood
Bletchingley
Godley Bridge
Albury
Horndean
Singleton
K7
K8
K10
K14
K13
K17
K15-fb
K15-fa
L4-a
L7-b
L7-c
L10-ae
L11-b
54°
52°
50°
12°
10°
8°
6°
4°
2°
0°

than both the Russian Federation and India and that Iran, which holds gas reserves second only to Russia, is the fourth largest producer and exports nothing.

Whilst the UK does not possess reserves which enable it to compete with the Russian Federation and the OPEC nations, what it can provide is the technology and expertise developed from 40 years experience in the North Sea, technology and expertise that has recently tamed the Atlantic Ocean frontier provinces, West of Shetland.

As the North Sea is now considered to be a mature province, the majority of oil companies are looking for development opportunities elsewhere in the world such as Nigeria, the Gulf of Mexico, Brazil and even Russia. The Caspian Sea area of the former Soviet Union is also a prime target. For years their oil and gas fields have been in decline due to lack of investment and it is to the well-established international oil companies that the new governments have been looking to for assistance in reviving their oil and gas industries.

The four main countries in the Caspian region are Azerbaijan, Kazakhstan, Turkmenistan and Uzbeckistan and oil majors such as Agip, Amoco, BP, British Gas, Chevron, Statoil and Texaco have taken a large stake in the various consortiums which have been formed to share the costs, profits and risks associated with these new ventures. The risks are great because the ownership of the oil is often in doubt as the various countries are in dispute over their territories, particularly offshore where there are problems over the establishment of maritime boundaries.

Transportation is another problem area with potentially troublesome countries bordering the new states that have, on occasions, tempered the enthusiasm of western oil companies. However, the rewards can be considerable, with the *Tengiz* field in Kazakhstan having estimated reserves of between 6 billion and 9 billion barrels of oil and the *Karachaganak* field having estimated reserves of 2.4 billion barrels of oil and 16 trillion cubic feet of gas.

1.2 Field Developments – Field Size

Oil field size is defined in terms of millions of barrels (mn/bbls) of recoverable oil. Daily production figures are quoted in thousands of barrels of oil a day (bbls/d) and give an indication as to the size of the platform required to exploit a particular reservoir. The oil and gas fields currently under development tend to be much smaller than those that represent the golden years of the offshore industry. A realistic picture of current development trends within the UK sector may be obtained by considering the reserves contained within PanCanadian's *Buzzard* field which came on-stream in 2007. The *Buzzard* development is one of the largest of the decade and will eventually provide nearly 10% of the UK's oil. It contains estimated oil reserves of 500 million barrels which equates to an output of nearly 200,000 barrels of oil a day.

The fields currently under development in the UK compare favourably with field sizes being developed elsewhere in the world. For instance, the *Mars* field was until recently the largest find in the Gulf of Mexico for 20 years and at 500 million barrels is comparable with the *Buzzard* development. In terms of oilfield size, an oil field with reserves of 400 to 500 million barrels is still regarded as being quite a respectable development proposition, with the vast majority of new finds now ranging from 20 to 70 million barrels. As far as production rates are concerned, there are relatively few fields which produce in excess of 100,000 barrels of oil a day, 15,000 to 50,000 barrels being more the norm.

Gas field reserves are generally defined in billions of cubic metres (bcm), although they are still regularly quoted in units of standard cubic feet (scf). Daily output figures are measured either in millions of cubic metres a day (mcm/d), or millions of standard cubic feet a day (mscf/d). As far as gas is concerned, a field containing reserves of 70 to 80 billion cubic metres is regarded as being a sizeable find, 15 to 20 billion cubic metres currently representing the average size discovery, whilst fields of 1.5 to 2 billion cubic metres are still considered to be worth developing. Whilst there are some prodigious gas producers, the vast majority of gas fields produce less that 100 mscf day (2.8 mcm) once the initial pressure peak has subsided.

1.3 Reserves

Whilst every effort has been taken to ensure the accuracy of the figures reproduced in this chapter, and indeed the book as a whole, it should be noted that figures quoted by the oil and gas industries are notoriously inconsistent and vary by as much as 400%. As an example, in spite of the sophisticated measuring equipment available the UK, proven oil reserves are ambiguously quoted as being somewhere between 610 and 2015 million tonnes with the undiscovered reserves estimated at between 285 and 2,680 million tonnes. Gas reserves are similarly quoted as being somewhere between 630 and 1985 billion cubic metres for proven reserves and 500 to 1,700 billion cubic metres for estimated undiscovered reserves. However, these results will provide a basis for general comparisons to be made with the global picture.

1.4 Oil and Gas Prices

Crude Oil Prices (Average per year in $US per barrel)											
1979	18.0	1984	28.6	1989	18.6	1994	14.9	1999	17.6	2004	33.9
1980	34.0	1985	27.5	1990	23.7	1995	16.6	2000	26.9	2005	50.6
1981	35.8	1986	14.2	1991	19.6	1996	20.8	2001	20.5	2006	60.4
1982	32.8	1987	18.5	1992	19.4	1997	18.5	2002	22.8	2007	68.7
1983	29.5	1988	15.2	1993	16.9	1998	12.9	2003	27.3	2008	96.5
										2009	75.0

Oil production is a particularly lucrative occupation as consideration of the production figures of a large FPSO or fixed platform will show. A daily oil production rate of 100,000 barrels, a healthy but not exceptional figure, and an oil price of $75 a barrel will bring in revenues of $7.5 million a day. This equates to $2.74 billion a year which could well cover the entire development cost of the project.

Gas Prices

For years, gas prices did not fluctuate dramatically because of the long-term contracts signed between the major operators and distributors. The operators received approximately 20 pence per Therm for the gas, a figure which equates to £70,000 per million cubic metres, or £2,000 per million cubic feet of gas. Nowadays, the price of gas fluctuates wildly with spot market prices at periods of peak demand rising to as much as 70 pence a Therm. In this market, small fields command a premium price, there being a wider and more competitive market for small volumes than for large fields. Whilst a 50 billion cubic feet (1,400 million cubic metres) gas field is considered small, it still represents sales revenues of £100 million, even when sold at 20 pence a Therm. Further changes are sure to occur as the oil majors become gas suppliers to the public.

Facts and Figures Pertaining to the Oil and Gas Industry

1 tonne = 7.5 barrels (crude oil).
1 cubic metre = 35.31 cubic feet.
1 tonne of oil equivalent is defined as having a calorific value of 397 Therms
Current world oil demand – 82 million barrels a day.
Average UK oil production – 1.37 million barrels a day.
Total UK oil production – 77,380,000 tonnes a year.
UK gas production – 205 million cubic metres a day.
Total UK gas production – 75.2 billion cubic metres a year.
Overall average cost (development and production, North Sea) per Therm of gas – 13 pence.
Average oil production costs from North Sea fields – $4 to $6 barrel.
Combined overall average cost (development and production, North Sea) per barrel of oil – $16.
New field, overall average cost (development and production, North Sea) per barrel of oil – $12.
Average oil production costs from Gulf of Mexico FPSO – $4 to $8 barrel.
Average oil production costs from Gulf of Mexico TLP – $6 to $16 barrel.
Average oil production costs for Saudi Arabian crude – $1.50 barrel.
Typical transportation cost by sea – $2.5 a barrel.

2 Offshore Structures and Support Vessels

2 Offshore Structures and Support Vessels

2.1 Offshore Structures

Figure 2.1 – Fixed steel installation

The structure shown is instantly recognisable as an offshore rig. This type of structure, more correctly described as a fixed steel offshore installation or platform, forms the backbone of the offshore industry and there are in excess of 7,500 such structures dotted about the oceans of the world.

Not so familiar are the structures and vessels shown elsewhere in this chapter, which provide assistance to, or are in competition with, the fixed steel structure. They could all be loosely described as oil rigs (or gas rigs), which provides an indication as to the ambiguity of the expression and hints at the complexity of the offshore industry.

2.1.1 Fixed Steel Structures

The traditional offshore installation, shown in Figure 2.2, utilises a welded steel, tubular framework or jacket to support the topside facilities, and this arrangement is referred to as a fixed steel structure. The topside facilities will vary slightly depending on whether it is an oil or gas producing installation, but they will include hydrocarbon process equipment, power generation, a helideck, and accommodation and hotel services designed to cater to the needs of the personnel employed in the operation and maintenance of the installation.

The single jacket installation is typical of the rigs found in deepwater environments such as the

Figure 2.2 – Gas platform, Mobile, Alabama

Figure 2.3 – Oil platform

Figure 2.4 – The '*Tethys*' installation. A good example of the sea harvester design, built by SLP in Lowestoft

northern sector of the North Sea. It should be noted that the helideck and accommodation facilities are situated as far from the potentially dangerous hydrocarbon processing area as is physically possible.

An installation may consist of any number of bridge linked jackets, with modern designs tending to favour a separate jacket to house the accommodation and helideck. These multi-jacket installations tend to be restricted to shallow water developments where the construction costs are considerably less than their deepwater counterparts.

While the fixed steel structure is likely to remain as the industry's first choice of installation design based on operational requirements, its future use may well be restricted to the development of large fields located in intermediate water depths where a substantial return on the capital sums invested can be guaranteed. The ever changing demands of the offshore industry has spawned a number of competitors more ideally suited to specific applications.

The concrete structure has emerged as a viable alternative to the steel jacket, but they both tend to suffer from the same inherent disadvantages when used to develop deepwater reserves. The scantling requirements of structures, that must be capable of withstanding the extreme weather conditions associated with the North Sea, are considerable and this is reflected in the construction costs. Generally speaking, the tension leg platform (TLP) or the floating production system (FPS) will be preferred for the development of deepwater reserves. The FPS, used in conjunction with subsea wells, also represents the most cost effective solution for the exploitation of smaller, marginal fields. The alternatives to the fixed steel structure will now be discussed and further information on fixed steel structures can be found in Part 4.

Figure 2.5 – Concrete gravity base

2.1.2 Concrete Gravity Base Structure (GBS)

While the vast majority of fixed offshore platforms employ a tubular steel jacket to support the topside facilities, a number of installations have been constructed using a base manufactured from reinforced concrete, as shown in Figure 2.5. This type

Figure 2.6 – Topside modules being towed into position on a GBS

of installation was pioneered by the Norwegians, who had experienced difficulties in producing the enormous quantities of high quality steel products and the large skilled workforce that are essential ingredients to the success of steel fabrication projects. Also, the presence of the Norwegian Trough, a large undersea valley, made the laying of subsea pipelines an expensive and challenging operation, so the void spaces that feature prominently in the design of a GBS provide valuable storage space for crude oil prior to discharge into oil tankers via a single buoy mooring (SBM).

The first concrete structure to be installed in the North Sea was constructed by the Norwegians in 1973 and used to develop the *Ekofisk* field. Since 1973, they have installed a steady string of concrete structures and it came as no surprise when the Norwegian government elected to develop the *Troll* field with a concrete structure. The '*Troll Alpha*' platform which stands in 350 metres (1,150 feet) of water, is currently the largest offshore oil and gas field in Europe and at 1,270,000 tonnes is the largest concrete platform in the world.

Figure 2.7 – '*Troll Alpha*' platform under tow

The construction of a concrete installation base normally begins in a dry-dock. The design of the base includes void spaces or caissons, suitably dimensioned to provide the structure with a natural buoyancy that will enable it to be floated clear of the dry-dock for finishing off prior to being towed to its final destination. Once on location, the void spaces are flooded to facilitate positioning of the base on the seabed while the topside modules are lifted into place. The void spaces may then be pumped dry and used as storage compartments for crude oil, or filled with permanent iron ore ballast. The colossal weight of the concrete structures removes the need to install foundation piles, so giving the name gravity base structure.

In Europe, technological developments and improved construction techniques have led to the use of concrete for the building of tension leg platforms, floating production systems and semi-submersible vessels. Further afield, the concrete structure has been selected in preference to other designs because of its ability to withstand impact from icebergs, a feature of vital importance for installations operating in the more northerly waters of Europe, Russia and Canada. In addition to the threat from icebergs, these remote locations frequently prohibit the laying of a subsea pipeline and, as was found in Norway, the facility to store the oil within the structure prior to transfer to an oil tanker has proven a major factor in determining the viability of the development of the fields.

It is worth noting that not all gravity base structures are made of concrete. In 1982 the Phillips Petroleum Company installed a steel gravity base structure over the *Maureen* field and, since then, at least two further steel based structures have been installed in the North Sea.

The *'Maureen'* platform sat on three steel storage legs or tanks and under normal operating conditions the tanks were filled with varying quantities of crude oil and water, the water being used to displace the oil without the aid of pumps. Up to 50,000 barrels were stored in this way, prior to being transferred into tankers.

As the *Maureen* field reached the end of its commercial life, Phillips actively sought a buyer for the platform with a view to simply re-floating her and towing her to a new destination. Unfortunately, an interested buyer never materialised so the installation was eventually floated away and scrapped.

2.1.3 Tension Leg Platform (TLP)

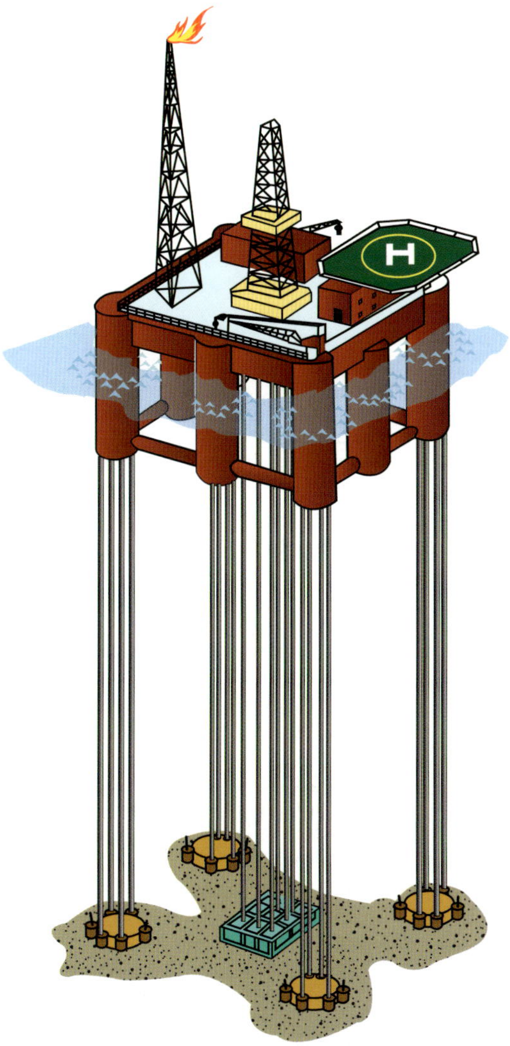

Figure 2.8 – Tension Leg Platform (TLP)

The tension leg platform concept was devised by the Conoco Oil Company as an alternative to the fixed steel structure and floating production system (FPS) for the development of deepwater oil and gas fields. The world's first TLP was built by Conoco in 1984 and used to develop the *Hutton* field in the North Sea. Located in 148 metres (500 feet) of water, the *Hutton* field could have been developed using a traditional fixed steel installation, but the harsh environment associated with the North Sea was judged by Conoco to provide the ideal test bed for the TLP design prior to venturing into deeper waters.

The TLP fulfils a role midway between the fixed offshore installation and the floating production system. It combines the initial cost saving benefits associated with floating production systems with the operational benefits attributed to the fixed offshore installation.

The TLP consists of a floating production facility that is tethered to the seabed by a number of tensioned legs. The legs are typically constructed from large diameter (250 mm to 750 mm) steel pipes or wire ropes (125 mm diameter). One end of the leg is secured in a foundation template piled into the seabed while the other end is connected to a tensioning winch within the floating superstructure. While the legs remain in tension, the platform can resist vertical wave-induced motion.

The TLP foundation design may appear elaborate when compared to the traditional ship type anchorage systems employed by floating production systems, but the advantages are considerable. The restriction on vertical movement permits the use of fixed installation type wellhead equipment and rigid steel conductor-riser assemblies. In contrast, the floating production systems must rely on flexible riser pipes emanating from subsea wellheads. These incur a considerable penalty in terms of installation costs and maintenance availability.

In 1995, Conoco constructed the world's first concrete TLP to develop the *Heidrun* field. Located in 325 metres (1,100 feet) of water, the field is one of the largest currently under development in the North Sea and demanded a structure of truly epic proportions. The four concrete foundation units that secure the *'Heidrun's'* tethers each weigh 21,000 tonnes, are 48 metres (160 feet) in diameter and are 26 metres (85 feet) high.

With TLPs already in use in water depths of 4,000 ft plus, continuing refinement and design will see their use at ever increasing depths in the future.

2.1.4 Semi-Submersible Vessel (Semi-Sub)

A semi-sub is an enormous box section barge supported on twin hulls. It has a colossal capacity to consume water ballast and, with minimal water plane area and displacement of up to 170,000 tonnes, presents an immovable object to wave action in anything but the most severe weather conditions.

Figure 2.9 – Semi-submersible drill ship

The design of a semi-sub does not permit the installation of an effective main propulsion plant. It is, therefore, heavily dependant on the assistance of support ships for towage to its destination and for the deployment of anchors (typically eight). However, some vessels are fitted with computer controlled azimuthing thruster units that provide an accurate manoeuvring capability, particularly when operated through a satellite navigation system. This facility enables the semi-sub to hold station in water depths in excess of 650 metres (2,200 feet), where the laying of anchors may prove impractical.

Figure 2.10 – Semi-submersible anchoring arrangement

(a) Anchors on the deck of a drilling rig

(b) Anchor cable

(c) Mooring windlass for anchor and cable

Figure 2.11 – Anchoring arrangements of the semi-sub

The semi-sub is an ideal choice of vessel for performing operations where accurate station keeping and exceptional stability are prerequisites to success. Depending on the equipment fitted to the main deck, the vessel may perform any one of six roles:

i) Heavy lift

The semi-sub will most frequently be encountered in its role as a heavy lift crane barge during the installation of offshore platforms. Over the years, the lifting capacity of the cranes has steadily increased from 2,000 tonnes to 7,000 tonnes, and the more modern vessels are fitted with tandem cranes that are capable of lifts of up to 14,000 tonnes. Sophisticated rapid acting ballast systems counteract the huge weight transfers that occur as structures are lifted and manoeuvred into position.

ii) Accommodation

Some of the older, smaller, heavy lift semi-subs are now used for modification or repair projects where the facility to provide extra accommodation and storage space takes precedence over the lifting capacity of the crane (Figure 2.12). They have been joined by purpose-built accommodation vessels, or flotels with up to 800 beds, that can remain in close attendance to fixed structures for months at a time. Transfer of personnel may be by bridge or by helicopter and the inherent stability of the semi-sub ensures that the workforce sleep soundly, even during a winter in the North Sea.

iii) Drilling exploration vessel

Depending on water depth, the drilling of exploration wells will be carried out by either a self-elevating jack-up, a semi-submersible vessel or a monohull drill ship. The jack-up is limited to operations in water depths of approximately 120 metres (400 feet), but no such restrictions apply to the semi-sub. The current world water depth drilling record stands at over 10,000 feet, at a field in the Gulf of Mexico.

Figure 2.12 – Semi-sub utilised for accommodation

The combination of satellite positioning, stability and an abundance of deck space for the storage of drill pipe and well test equipment make the semi-sub the first choice of vessel for deepwater exploration projects.

iv) Pipelay barge

The sketch and pictures in Figures 2.13 to 2.15 show pipelay barges of the type that has installed nearly 10,000 kilometres of subsea pipeline in the North Sea during the past 25 years.

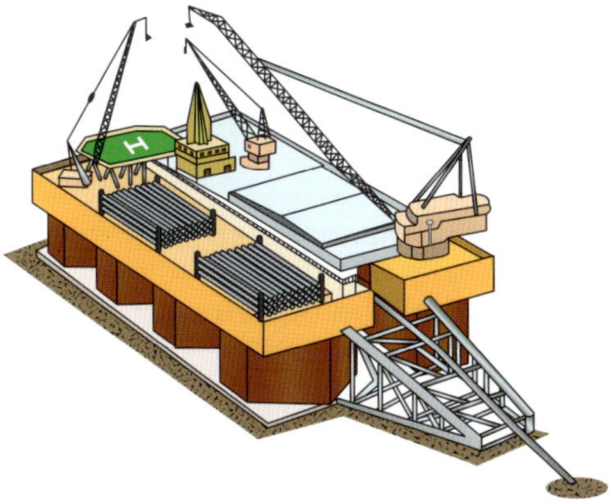

Figure 2.13 – Semi-submersible pipelay barge

Once started, the laying of a subsea pipeline becomes a non-stop operation as the lay barge slowly winches its way forward on its anchors. A modern vessel can lay 2 to 4 kilometres of pipeline a day, azimuthing thruster units occasionally replacing the anchor spread in very deepwater locations.

Subsea pipelines vary in diameter from 3″ to 42″, with 30″ and 36″ lines being favoured for the main transmission routes. The pipe is supplied to the barge in 12 metre lengths, ready painted to provide corrosion protection, gas pipelines being coated in concrete to provide the negative buoyancy required to keep them on the seabed.

On the barge the lengths of pipe are double jointed, ie two lengths of pipe are welded together prior to being lowered into the firing line, the inside of the barge where the double joint lengths are welded to the pipeline as it enters the sea. The pipeline is welded using a part manual, part automatic welding process. The welds are then X-rayed or ultrasonically scanned and, if free from defects, are given a protective coating of bitumen before being launched into the sea from the stinger framework situated on the stern of the vessel. The stinger prevents buckling of the pipe as it enters the water.

The completed pipeline will be subjected to a hydrostatic pressure test to prove its structural integrity, and may then be trenched (buried) or rock dumped (covered in rocks) to protect it from hazards such as fishermen's nets and ships' anchors. Alternatively, it may simply be left to rest on the seabed.

v) Pipe reeling

The cost of laying subsea pipelines is considerable and there is a niche market in laying infield pipelines that are generally of a diameter less than the optimum 16″ diameter, which can be cost effectively laid by a conventional pipelay barge. The reel lay vessels, often diving support vessels, are particularly suitable for this type of work because they can lay pipe in DP (dynamically positioned) mode and so do not require a complex anchor spread, a particularly desirable feature when there are several vessels working in the field or where the water depth is beyond that in which anchors can be used.

Figure 2.14 – The '*Castoro Sei*' semi-submersible pipelay vessel. The pipeline is guided into the sea from the 'stinger' framework attached to the vessel's stern

Figure 2.15 – Pipereel vessel

Laying flexible pipes from a reel is a relatively easy thing to do, but it is also possible to lay conventional rigid steel pipes using the reeling process, and this is not quite so easy to imagine. The pipe, which is wound onto large diameter reels, up to 25 metres in diameter, is basically the same type of pipe that would be laid by a conventional lay barge. Pipe ranging from 3" diameter and 8 mm wall thickness up to 16" diameter and 28 mm wall thickness can be reeled. While nominally any basic API grade of pipe can be reeled, up to API 5L X65, it must be precisely specified at the time of purchase, with very specific restrictions on both the chemical properties, the ultimate tensile strength and the yield strength to ensure a uniformity in properties from one pipe joint to the next as it is being reeled.

Figure 2.16 – Pipelaying barge about to start laying pipe

vi) Floating production facility

Figure 2.17 – Floating semi-submersible offshore oil production platform

2.1.5 Floating Production Systems (FPS)

Floating production systems (FPS) have been with us for many years with the first oil field in UK waters, the *Argyll* field, developed using a floating production system in 1975. There has been a considerable growth in the number of floating production systems constructed for service in the North Sea, West Africa, Australia, the Gulf of Mexico and offshore Brazil.

A floating production system is a floating oil rig (as shown in Figure 2.18). It contains all the equipment associated with a fixed installation and is used in conjunction with subsea wellheads to exploit moderate to deepwater oil fields.

Figure 2.18 – Floating Production Systems (FPS)

Floating production systems are selected for field developments for one of three reasons:

i) The field is small and can only be developed economically by a reusable asset such as an FPS.
ii) The field is isolated and an established pipeline infrastructure does not exist.
iii) The field is located in very deep water where it would not be possible to install a conventional fixed platform.

A major advantage of floating production systems lies in the fact that they can lift anchors and depart to new locations when oil production reaches a commercially unprofitable level.

Figure 2.19 – Monohull FPSO *'Uisge Gorm'* in the North Sea

Figure 2.20 – Floating production system, Gulf of Mexico

2.1.5.1 Vessel types

Conventional ship-shaped monohull vessels and semi-submersible vessels may both be used as the basis for floating production systems and there are advantages and disadvantages associated with each. The monohull designs are more suited to coping with strong winds and tides than semi-submersibles, have ample onboard storage capacity for the stabilised crude oil and are well suited to off-loading duties into shuttle tankers. While semi-submersible production units may require a support vessel for oil storage and to facilitate off-loading (FSU or FSO), they are inherently more stable and offer a greater production capacity than a monohulled vessel, being capable of accommodating more than double the number of risers, up to 100.

In the North Sea and the seas west of Shetland, monohull floating production systems have been the preferred option, while the Brazilians, who are the undisputed leaders in deepwater floating production technology, have tended to favour semi-submersible vessels. As far back as 1977, Petrobras broke the 1,000 metre (3,250 feet) water depth barrier with the *Marlim Sul* field development and they currently operate 23 FPSs with a further 17 FPSs and FPSOs scheduled to enter service by the year 2020, predominantly in the 800–1,400 metre (2,660–4,660 feet) water depth range. They are also currently developing the giant *Roncodor* field, which at 2,600 metres (8,660 feet) was, until recently, a world record depth.

2.1.5.2 Vessel configurations

While the FPS exists as a production installation in its own right, the term floating production system is frequently used to describe any vessel engaged in the oil production or oil storage process, and includes the FSU, FSO and FPSO. A brief description of each type is provided below but it is worth noting that there are many variations on these basic themes. It is the most frequently used arrangements that have been described.

FSO
The floating, storage and off-loading unit (FSO or FSU), is a storage tanker into which processed oil is pumped from a fixed platform (or semi-submersible FPS) that has no storage capacity of its own and is not connected to a subsea pipeline. The oil is subsequently off-loaded from the FSU into shuttle tankers for transportation to a refinery.

FPS
In most cases, the term 'floating production system' is used to describe a production vessel that is connected to a subsea pipeline, rather than one that has the capability to discharge oil into shuttle tankers. Normally, it is a semi-submersible vessel and the *Rob Roy* field development illustrated in Chapter 6 is an example of this type.

FPSO
Like the FPS, the floating, production, storage and off-loading vessel normally replaces a conventional platform in its entirety. The FPSO differs from the FPS because it is not connected to the beach by a subsea pipeline. The oil, which is again extracted via subsea wells, is processed and stored onboard the vessel, normally a monohull design, prior to being off-loaded into shuttle tankers, as illustrated in Figure 2.21.

FPDSO
Currently, no monohull FPSOs are fitted with drilling derricks so drilling and workovers have to be carried out by drill ships. The floating, production, drilling, storage and off-loading vessel (FPDSO) is, therefore, the latest development in the FPS family. Still at the conceptual stage, the FPDSO will only provide advantages for compact field developments, the wells being too far apart for drilling to be carried out from a single location in many other cases.

PSV
Pioneered by BP with the *'Seillean',* the production storage vessel design is unique. It is a purpose-built vessel designed for the exploitation of small fields that are located in moderate water depths. It is a floating production system and shuttle tanker in one, with a dynamic positioning capability that enables the vessel to remain on station over the wells without the need to deploy anchors while producing oil. Once the vessel's storage tanks are full (typically 20 days' production), the vessel disengages from the wellhead and proceeds to a shore terminal to discharge the cargo.

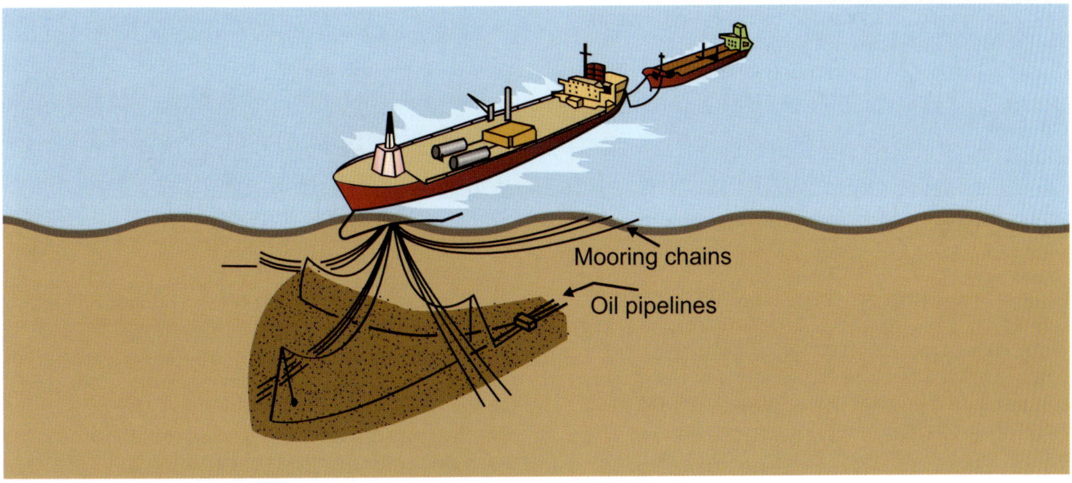

Figure 2.21 – Floating production storage and off-loading system

Figure 2.22 – The FPSO *'Petrojarl 1'* discharging into a shuttle tanker

2.1.5.3 Principal features – monohull FPSO

The floating production system is a floating oil platform that may be equipped with gas compression and water injection systems, in addition to oil processing equipment. After processing, the stabilised crude oil is stored in the ship's cargo tanks prior to being transferred into shuttle tankers via a mooring hawser and hose, reeled from the stern of the vessel. Gas produced as a by-product of oil production is normally used as fuel for the gas turbine powered electricity generators or, where significant volumes are produced and an export pipeline is not available, it may be re-injected into the formation or flared when production exceeds demand.

There are some significant design challenges that have had to be overcome to make the FPSO a viable proposition for long-term field development goals. While these relate primarily to the mooring system, turret and the flexible risers, the basic hull and the topside processing equipment, all must be designed to a higher specification than those of a conventional tanker or offshore structure. These items will now be described briefly as they relate to a typical monohulled vessel, permanently moored on station via an internal turret (Figure 2.23).

Hull

A number of existing tankers have been converted for service as FPSOs and, in some cases, they have

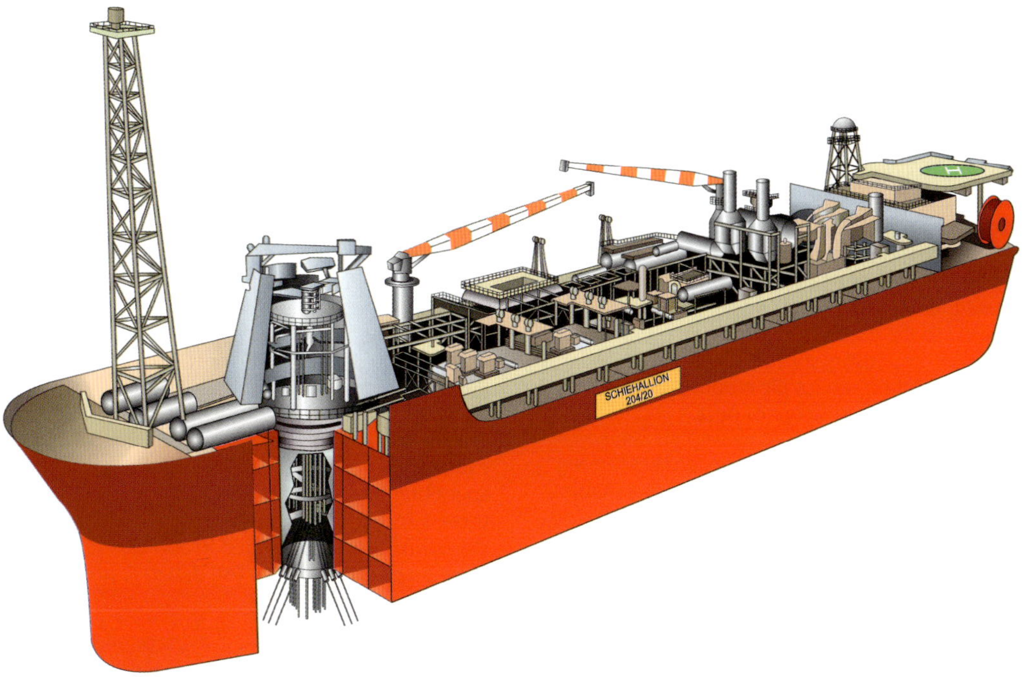

Figure 2.23 – BP *'Schiehallion'*

retained their original propulsion system. Where operators have opted for a new purpose-built vessel, main propulsion systems are generally omitted, the vessel relying on the electrically powered thruster units to assist in stationkeeping. The hulls are built in accordance with rules provided by the ship Classification Societies and are designed to remain on station for a period of 10–20 years without dry-docking.

Topsides

The topside oil and gas processing equipment is designed and constructed in accordance with fixed offshore platform and refinery standards, specifications and recommendations, such as those produced by the American Petroleum Institute (API), American Society of Mechanical Engineers (ASME) and British Standards Institute (BSI). However, the codes and standards must be adapted to take account of vessel motion. The additional loadings due to wave action can be considerable and have a serious effect on the fatigue life of the equipment, particularly the process pressure vessels and pipework.

The process equipment is normally pre-assembled into skids, the skids being mounted on a framework or pillars 3 metres above the main deck of the vessel, the air gap providing some protection against green seas, which are large waves breaking on the deck, and to provide adequate hazardous area boundaries.

Turret

The turret consists of a large diameter (16–32 metres), vertical cylinder that sits within the hull of the vessel, mounted on heavy duty roller bearings. The mooring wires and the flexible subsea risers are attached to the turret, which prevents it from rotating. The vessel is free to weathervane (rotate through an angle of typically 270°) around the fixed turret under the influence of wind, waves and currents, the extent of rotation being controlled by the vessel's azimuth thruster units.

Mooring

A spread of 8 – 14 anchors ensures that the FPSO remains on location while producing oil, the anchor spread typically consisting of a combination of wires and chains that are tensioned by winches within the turret. Conventional anchors, suction anchors or piles may be used to make the connection with the sea bed. For the ultra-deep waters of the *Campus Basin* in Brazil, Petrobras is experimenting with fibre rope mooring systems and twelve, 711 tonne breaking strength, Superline Polyester ropes are used to secure an FPSO in 1,400 metres of water at the *Marlim South* field.

Risers

To permit the FPSO to move both vertically and laterally, the connections to the subsea wells are usually made through flexible, steel reinforced risers. The risers are designed to absorb any wave-induced motion that might affect the position of the vessel, often being draped over submerged midline buoys that help in reducing the loads on both the risers and the turret, the oil being transferred from the turret to the process equipment through a swivel stack.

It is the swivels and, to a lesser extent, the risers that are currently the limiting factors in the development of ultra-deep water FPSOs, gas export lines in particular being difficult to install successfully and limited to a diameter of less than 16 inches (0.406 metres) at operating pressures of 340 bar (5,000 psi). Another problem that increases with water depth is the risk of hydrostatic collapse of the risers, a typical maximum water depth for a steel-reinforced riser of 8-10 inch (0.2-0.25 metres) diameter being approximately 800-1,000 metres (2,660-3,330 feet). However, risers made from conventional steel pipe have been used successfully in very deep water locations where the installations are not subject to extreme weather conditions.

2.1.5.4 Shuttle tankers

There are two types of tanker equipped for offshore loading operations. Crude oil carriers, which transport oil over long distances, are provided with fairly basic loading and mooring equipment because the time spent loading is minimal when compared to the duration of the subsequent sea voyage. In comparison, shuttle tankers spend a large proportion of their service life loading, discharging and manoeuvring in and out of port.

Prior to the boom in FPSOs, shuttle tankers tended to be converted product tankers to which suitable mooring equipment was added. However, the shortage of suitable conversion tonnage has lead to a ship building boom and most major shipyards now have a shuttle tanker within their design portfolio. Approximately 20% of the UK's oil production and 60% of Norway's oil is transported by shuttle tanker rather than by pipeline.

The advanced specification of the new shuttle tankers tends to reflect the industry's awareness of the hazards associated with oil spills caused by grounding or collisions. A far greater proportion of the vessel's life will be spent loading and discharging than would be the case for a traditional VLCC (very large crude carrier). Consequently, the latest designs tend to specify redundancy in steering, manoeuvring and dynamic positioning equipment. These ships tend to be in the 100,000 – 125,000 dwt range, approximately half the size of a VLCC.

The manoeuvrability of a shuttle tanker is extremely important because of the long periods spent holding station while loading oil. Typically, three 1,750 kW bow thrusters and two 1,750 kW stern thrusters are provided in addition to the main propulsion system. The dynamic positioning

Figure 2.24 – Shuttle tanker loading at Draugen loading buoy

and mooring systems are designed to cope with maximum wave heights of 6.0 metres and mean wind speeds of up to 72 knots, although typical loading restrictions are applied at wave heights of 4.0 – 4.5 metres for connecting operations, with oil transfer ceasing when wave height exceeds 4.5 metres.

The next generation of multi-purpose shuttle tankers (MST) are likely to be designed to facilitate drilling, as well as for operation in the shuttle tanker mode, and for conversion to FSO or FPSOs with the minimum of work should the need arise.

2.1.6 Self-Elevating Jack-Up

The origin of self-elevating jack-up dates back over 70 years, when it could be found searching for oil in the muddy swamps of Louisiana.

The self-elevating jack-up consists of a triangular shaped (sometimes rectangular), box section barge fitted with three (sometimes four) moveable legs that enable the vessel to stand on the seabed in water depths of up to approximately 120 metres (400 feet) (Figure 2.25).

Figure 2.25 – Self-elevating jack-up rig under construction

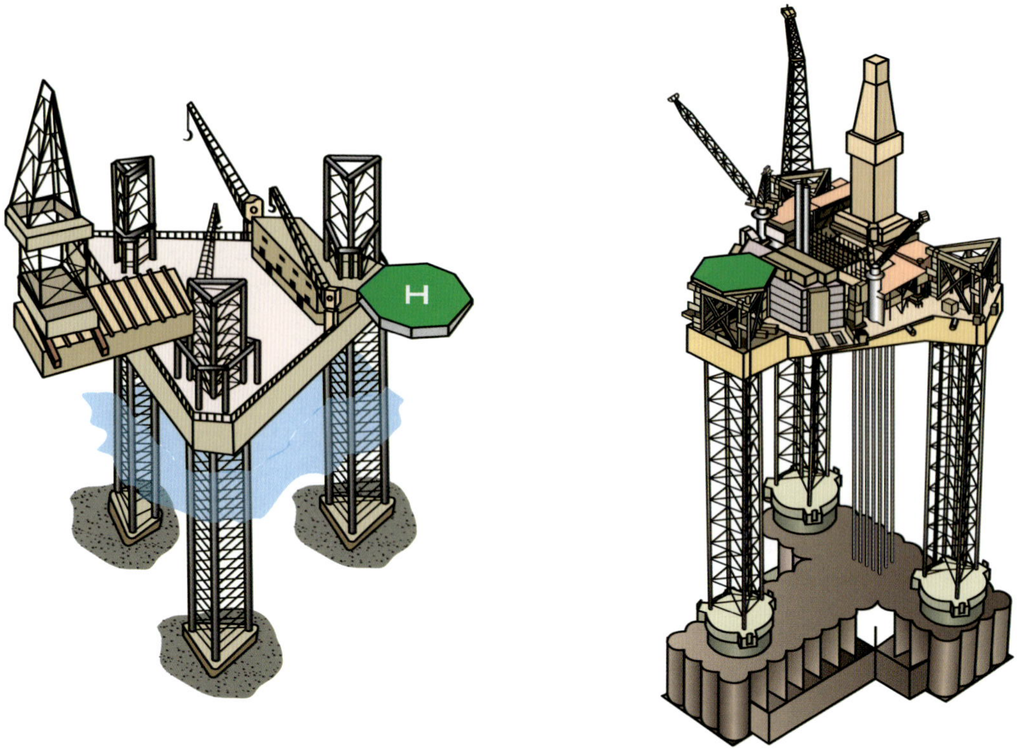

Figure 2.26 – Self-elevating jack-up

The shape of the vessel imposes design restrictions similar to those encountered with semi-submersible vessels, which prevent the fitting of a main propulsion unit. Transportation over short distances is under the tow of tugs or anchoring handling vessels, while it is both quicker and safer to accommodate a jack-up on the back of a submersible heavy lift ship to undertake long sea passages (Figure 2.27).

Figure 2.27 – Drilling programme complete, the *'Arch Rowan'* is lowered to the water and towed to its new destination. The *'Wrestler'* is taking the tow

Once on location, thruster units enable the jack-up to maintain position while the legs are lowered to the seabed and the hull is jacked into position, clear of wave action. The base of each leg is fitted with a spud can, which consists of a plate or dish designed to spread the load and prevent over-penetration of the leg into the seabed. High pressure jets of water or compressed air may be used to remove loose debris in the vicinity of the spud cans while the legs are manoeuvred into position, this process being referred to as spudding in. The legs are raised and lowered by means of a rack and pinion arrangement, the racks being attached to the corners of each leg, running from top to bottom, while the pinions are driven by electric motors via reduction gearboxes, typically 12 motors for each leg. The operation of the electric motors is synchronised to ensure that the hull of the jack-up is raised and lowered on an even keel.

Jack-ups are primarily used for drilling operations, but a number have been constructed to act specifically as accommodation support vessels to provide assistance to fixed installations during construction, modification or repair programmes. The jack-ups used for drilling operations fulfil two completely different roles. The majority of jack-ups are used for exploration (wildcat drilling) purposes.

The other role is in the drilling of wells for permanent installations, particularly gas producing platforms such as those located in the southern sector of the North Sea. The gas production process does not require the same level of ongoing drilling activity and well modification work that is associated with the production of oil. Consequently, derricks are not normally fitted to gas producing installations and all well modifications and drilling operations

Figure 2.28 – Jack-up during self-elevating process

Figure 2.29 – The *'Ocean Benarmin'* jack-up with its drilling derrick cantilevered over the Mobil *'Camelot 53/1a'* unmanned gas platform. The *'Benarmin'* is completing the wells following installation of the topside structure.

must be carried out by a jack-up sited alongside. The drilling derrick is fitted to rails so that it can be cantilevered into a position that provides direct access to the wellhead area.

The future of the self-elevating jack-up looks assured with old rigs being replaced with designs capable of operating in ever deepening, harsh water environments. In fact, jack-ups look set to increase their share of the offshore market with a new generation of rigs designed to operate as process facilities for the development of marginal oil fields using both conventional and subsea wells. The BP *Harding* development provides a good example of these new trends and was a particularly innovative concept.

2.1.7 Single Point Mooring (SPM)

The single point mooring concept originated in the 1960s as a solution to the problem of transferring crude oil from an onshore reception facility or refinery into VLCCs that were physically too big to enter port. The tankers are moored to a large buoy located at some considerable distance from the coast, often several miles, the buoy itself being secured by a spread of anchors. The oil is transferred from a subsea pipeline, through a swivel connector in the buoy into floating loading hoses attached to the tanker, the tanker being free to weathervane around the buoy, independent of both wind and tide.

For single point moorings of the tethered buoy and flexible riser type, the catenary anchor leg mooring (CALM) buoy is used extensively and a number of more sophisticated designs such as the buoyant or submerged turret loading (STL) system have been developed to meet the needs of the industry.

Figure 2.30 (b) – CALM buoy

There are a number of variations on the SPM, (Figure 2.31) which employ a rigid structure or riser column to convey crude oil from the seabed to the surface, such as the single anchor leg mooring (SALM) and the articulated loading platform (ALP). Regardless of the design, they all attempt to achieve the same end result, which is to provide a safe mooring while the tankers load oil in wind speeds of up to 80 km/hour (50 mph) and wave heights of up to 5.5 metres (18 feet).

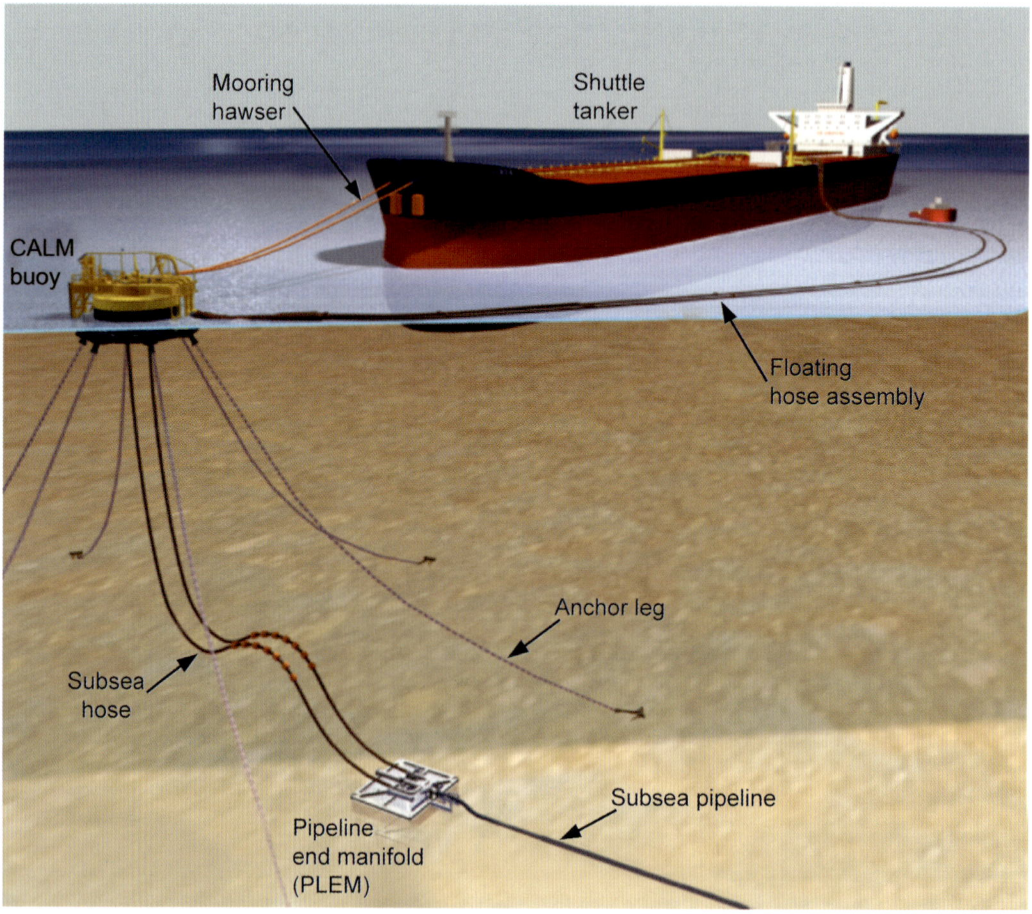

Figure 2.30 (a) – CALM buoy

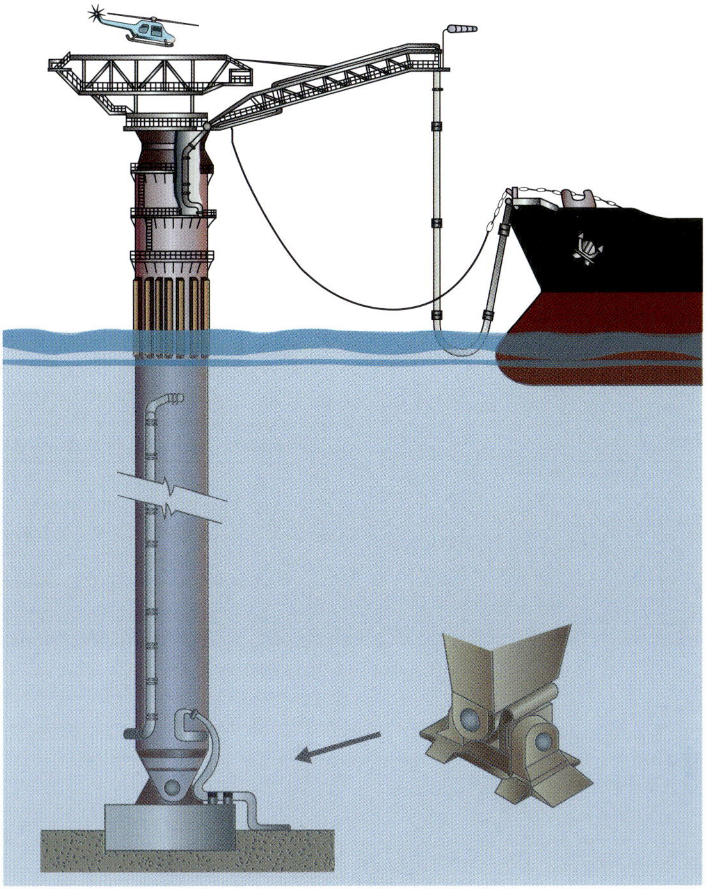

(a) SPM – ***Statfjord B***

(b) Loading platform, Sakhalin Island

(c) Shell '*Sea Eagle*' and articulated loading platform

(d) **Articulated loading platform**

Figure 2.31 – SPM variations

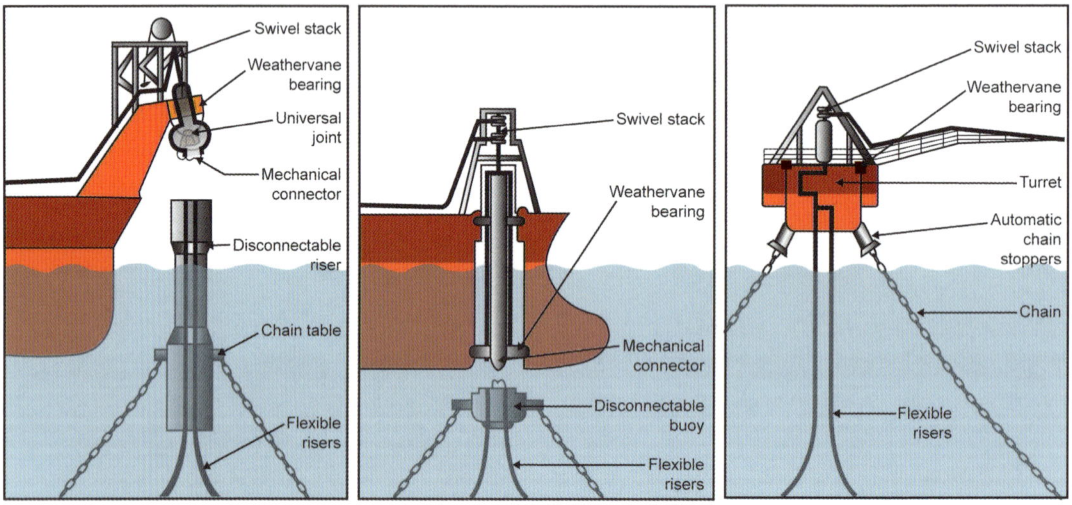

Figure 2.32 – Disconnectable riser turret mooring | **Disconnectable buoyant turret mooring** | **Permanent turret mooring**

The following brief account of a fixed column SPM is representative of the type used as a loading terminal by Statoil for their *Statfjord B* development, Figure 2.31a.

This particular SPM (or SALM) consists of three main components:

Gravity base

The gravity base shown measures 20 × 20 × 8 metres (65 × 68 × 27 feet), weighs 950 tonnes and contains 4,500 tonnes of iron ore ballast to assist the forces of gravity in keeping the column firmly rooted to the seabed.

Column

The steel column measures 170 metres (560 feet) in height, 9 metres (30 feet) in diameter and weighs 4,000 tonnes. To assist emplacement, the column is filled with 2,400 tonnes of iron ore in addition to the permanent water ballast. A Cardan type articulated joint provides an attachment that also permits a degree of lateral movement. The crude oil is transferred from the manifold at the gravity base up to the rotating head by two 36 inch diameter (750 mm) pipelines routed up the outside of the column.

Rotating head

The rotating head contains the quick release mooring and loading attachments, temporary accommodation and a helideck, and is designed to act as a moving pivot for a ship to rotate around under the influence of wind and tide. The oil is transferred from the column into the loading hoses through a swivel coupling at discharge rates of up to 57,000 barrels an hour (8,000 tonnes), and this particular SPM has the facility to load ships ranging in size from 80,000 to 150,000 tonnes.

Riser column SPMs have two main applications when used in support of fixed installations.

Connected to a subsea pipeline, the riser towers provide the ideal way of loading as the SPM enables tankers to load oil destined for global export without the inconvenience and expense of entering port. They are also used in support of fixed installations located in isolated environments where the laying of a subsea pipeline would be impractical or prohibitively expensive. Processed oil from the platform is pumped into tankers via the riser tower, which is installed at a safe distance from the installation, connected by a short subsea pipeline.

The other area where the SPM has grown in popularity in recent years is in the subsea market. When a field is developed exclusively through subsea wells, the oil is normally processed by a floating production system (FPS) prior to export. The wells may be connected to the floating production system by individual flexible risers attached to a turret within the hull of the vessel or, alternatively, they may be manifolded on the seabed and connected to the FPS by a detachable single point mooring tower or buoy. This is often a more suitable arrangement in locations where hurricanes and typhoons occur.

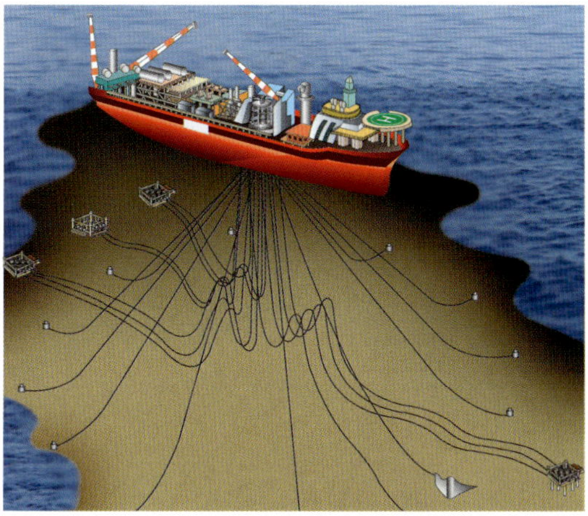

Figure 2.33 – Floating production system (FPS)

2.2 Support Vessels

Support vessels can loosely be divided into construction or support, where the larger vessels tend to be employed on the installation, maintenance and repair of subsea pipelines and fixed platforms. The smaller vessels consist primarily of diving support (DSV), survey, supply and standby boats.

2.2.1 The Offshore Supply Boat

The distinctive profile of the offshore supply boat cannot be confused with any other vessel. The high bow and forward accommodation are designed to withstand the severest of weather conditions and permit 360° of unrestricted vision from the wheelhouse, while the long flat afterdeck provides an ideal platform for the stowage of containers, drill pipe and a 20 tonne anchor. Below deck, a refrigerated cargo hold facilitates the transportation of perishable food stuffs and potable water, diesel fuel, cement and barytes are carried in purpose-built tanks.

'Barytes' or 'barites' is the naturally occurring mineral form of barium sulphate. It has a high SG of 4.5 and is used in oil drilling operations.

Figure 2.34 – Supply boat

Figure 2.35 – Supply boat in support of offshore operations

Figure 2.36 – Supply boat with anchor handling

The supply boat has great power and manoeuvrability. A twin engine, twin propeller arrangement provides propulsive power and a degree of mechanical redundancy, while twin rudders and bow thrusters provide the manoeuvrability. The entire operation is controlled from a wheelhouse that contains duplicate controls facing forward and aft. The helmsman has fingertip control over the vessel to enable him to maintain close attendance to an offshore installation while the stores are plucked from the afterdeck by the platform crane.

Figure 2.37 – View of the rig and crane from the deck of a supply vessel

To ensure regular employment in a fiercely competitive market, the supply boat has had to develop into a general use vessel and, in addition to the basic function of delivering stores, the more modern vessels will be encountered towing rigs, handling the anchors of semi-submersible vessels and providing fixed installations with fire-fighting cover during major repair programmes.

Figure 2.38 – View of the supply boat from the rig

2.2.2 Standby Vessels

Under Statutory Instrument No. 1542, *'Offshore Installations (Emergency Procedures)* 1976', now replaced with the PFEER regulations, *'Offshore Installations (Prevention of Fire and Explosion, and Emergency Response) Regulations, SI 1995/743S* 743', all manned offshore installations located in the UK sector of the North Sea had to have a standby vessel in attendance. This vessel was to be capable of accommodating the entire complement of the installation and of providing first aid facilities.

Figure 2.39 – Standby boat

Figure 2.40 – Standby boat – in a typical North Sea environment

The Regulations permitted one standby vessel to cover any number of installations located within a five mile radius. In practice, the operators of offshore installations tended to employ a number of vessels considerably in excess of this minimum requirement, particularly when engaged in operations such as underdeck scaffolding, painting and inspection programmes where there was a possibility that individuals may fall into the sea.

Prior to the *'Piper Alpha'* disaster in 1988, the majority of the 160 strong standby fleet operating within the UK sector of the North Sea consisted of converted trawlers. However, the disaster highlighted deficiencies in their operational capabilities and they were severely criticised during the Cullen Inquiry.

In response to the Cullen Report, the Department of Transport (DoT) and the Health and Safety Executive (HSE) formulated a code of practice for standby vessels aimed primarily at improving their manoeuvrability and speed of response. New vessels must be at least 11 metres in length, capable of a speed of 10 knots, and provide 360° unrestricted vision of the surrounding seas from the wheelhouse. They must be of either twin screw (propeller), or single screw assisted by a 360° thruster powered propulsion unit, and be fitted with a bow thruster. They must also carry

Figure 2.41 – The *'Putford Artemis'* standby boat and the *'DB101'* semi-submersible crane vessel (SSCV) at work installing a topside module. Inset – A fast rescue craft

two fast rescue craft available for immediate deployment.

There was initially some flexibility in the acceptance of older vessels that did not comply fully with the new recommendations, but the days of the converted trawler were numbered and there are few left in service. Under the PFEER regulations, the requirements for standby vessels will be determined on an installation by installation basis, governed primarily by the results of the assessment required by regulation 5, and the requirement under regulation 17 to provide *"effective arrangements for recovery and rescue"*, which only *"a suitable vessel standing by will provide"*. The purpose of the assessment is to identify the events that could give rise to a major accident involving fire or explosion, or the need to evacuate the installation.

Further information

i) Assessment of The Suitability of Standby Vessels Attending Offshore Installations, published by HMSO.

ii) Guidelines for The Management of Emergency Response for Offshore Installations, published by UKOOA.

Figure 2.42 – Offshore support vessels

1 Multi-functional support vessels (MSVs)
2 Accommodation units
3 Fallpipe dumping vessels
4 Pipeline bury barges
5 Seismic survey vessels
6 Flexible pipelay vessels
7 Derrick lay barges
8 Pipelay ships
9 Derrick and crane barges
10 Heavy transportation vessels
11 Anchor handling, tug, supply vessels
12 Diving support vessels
13 Well service vessels (see also 18)
14 Multi-purpose vessels (see also 1)
15 Reel pipelay vessels
16 Pipelay barges
17 Standby vessels
18 Well stimulation vessels

2.3 Offshore Installation Descriptions

Radio Mast
A steel tower that accommodates communication components such as satellite and telemetry dishes.

Telemetry Dishes

Flare Boom – (oil producing installations only)
The flare stack provides a safe, remote location for the disposal by burning of unwanted gaseous hydrocarbon by-products.

Process Area
The process area contains the pressure vessels and associated equipment required to remove impurities and by-products from the oil or gas prior to their discharge into the subsea pipeline.

Cellar Deck
Lowermost deck in process area.

Spider Deck
Walkway located above the high waterline that facilitates inspection and maintenance of the jacket structure. It also provides an emergency escape route to the sea

Drilling Derrick – (normally oil producing installations only)
The drilling derrick is used throughout the life of the installation to drill new wells, to drill wells for enhanced oil recovery, and to modify and repair existing wells.

Wellhead Area
The wellhead area contains the christmas trees that regulate the flow of hydrocarbon products from individual wells to the process equipment.

Pedestal Cranes
Used to assist in maintenance operations and to facilitate the unloading and loading of stores from supply boats.

Helideck
Helicopters provide the means by which personnel are transported to and from offshore installations and they are the primary method of evacuation in the event of an emergency. In UK waters, helicopters and helidecks come under the jurisdiction of the Civil Aviation Authority (CAA).

Accommodation
The accommodation block provides a full range of hotel services designed to cater to the needs of the personnel employed in the operation and maintenance of the installation.

On new installations the accommodation may be designated as the Temporary Refuge (TR), and will be designed to provide maximum protection to personnel during an emergency situation.

Lifeboats
In the event of an emergency that necessitates abandonment of the installation, the lifeboats provide a way of escape to the sea in the absence of helicopter assistance.

Power Generation
The majority of offshore installations are located a considerable distance from the coast and, as such, must be self-sufficient in all aspects, including the generation of electricity. The alternators may be driven by reciprocating diesel or gas fuelled engines, or by gas turbines

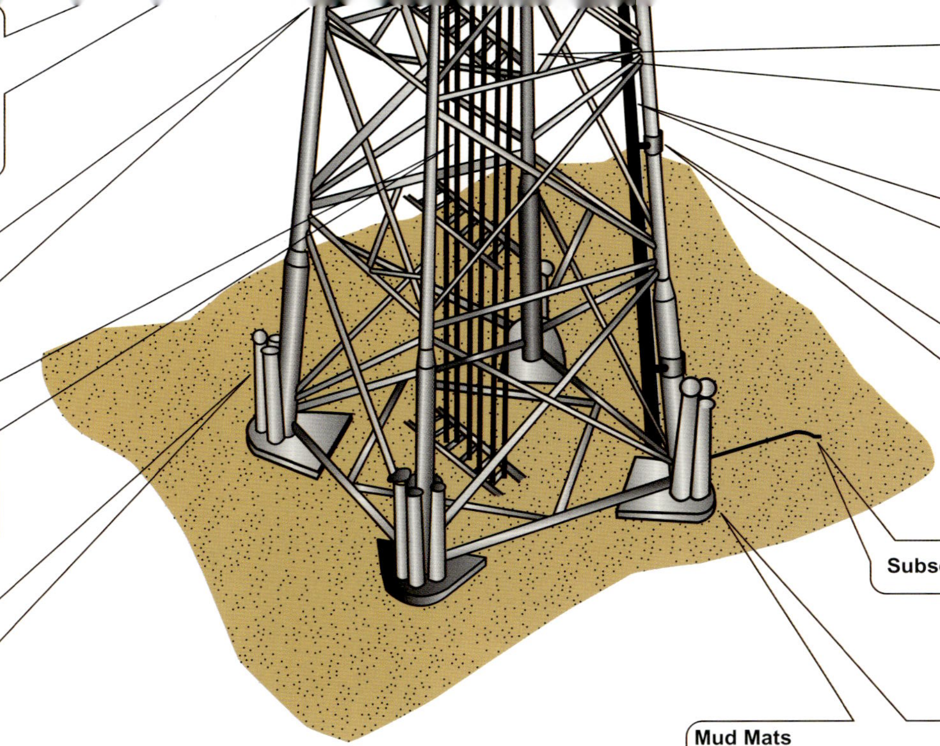

Not shown

Vent Stack – (gas producing installations only)

Open-ended discharge pipe through which process gas may be expelled to the atmosphere to de-pressure and make safe the gas process equipment.

'J' Tubes

Open-ended 'J' shaped pipe attached to the jacket structure and extending from cellar deck to seabed. Provides protection for flexible flowlines and umbilicals emanating from subsea wells.

Figure 2.43 – Typical fixed steel structure

2.4 Fixed Steel Structures – Installation

The North Sea is one of the most hostile marine environments in the world and a considerable amount of specialised technology has been developed over the years to cope with the extremes of mother nature. The bulk of the offshore engineering prior to 1965 was carried out in the relatively calm waters of the Gulf of Mexico and it was Gulf technology that was used to construct the early North Sea installations. Some of the first process modules were built in the USA and shipped over to the UK. Since those early days the technology developed to cope with the North Sea's extremes of weather has elevated British offshore engineering into a position as a world leader.

This section deals with fixed steel structures, which represent the vast majority of offshore installations. They vary considerably in size and on the UKCS (UK Continental Shelf) there is a very definite North-South divide created by the tremendous differences in water depths, weather conditions and platform complexity. The southern sector installations stand in relatively shallow water depths of 12 – 40 metres (40 – 135 feet) and the basic nature of the gas processing equipment permits the construction of small, lightweight structures. The support structures, or jackets, weigh in the region of 250 – 1,500 tonnes and the topsides or superstructures from 500 – 10,000 tonnes.

In the northern sector, jacket weights tend to be in the region of 5,000–20,000 tonnes, although the giant rigs installed in the boom years were much bigger. These colossal structures must support topside weights of up to 40,000 tonnes in water depths of 100 – 160 metres (325 – 550 feet) and reflect the considerable quantity of process and drilling equipment required to produce and process oil.

The jacket and topside dimensions quoted for the North Sea are representative of the size of installations found at all geographic locations where water depths are similar.

Before embarking on a description of the construction, transportation and installation of offshore structures, it is worth mentioning the subsea template because it is often the template that is the first part of the structure to be installed. The template is piled into the seabed in a location considered to be the most favourable for reaching the hydrocarbon deposits. The main objective of the template is to provide a guide frame through which wells can be drilled prior to the arrival of the jacket. Drilling wells takes a considerable amount of time and it is advantageous to enlist the services of a drilling vessel to pre-drill some or all of the wells while the jacket is under construction on the mainland. In addition to well guides, guides are provided in the template to assist in the accurate positioning of the jacket over the template. Subsea templates are unlikely to be necessary on small installations situated in shallow waters.

Figure 2.44 – The installation of a drilling template. The template is being installed on the end of a drill string from a jack-up. The template has slots for nine wells and there are two pile guides. Inset – subsea wellhead protection frame loaded onto the deck of an offshore supply boat for transportation to its final resting place.

2.4.1 Construction

The construction process for all jacket types is essentially the same, regardless of the jacket size. The jacket legs are formed by welding a series of rolled steel cans end to end. Typically, the first two legs will be laid out horizontally while horizontal and diagonal braces are added, the bracings are normally supplied in the form of structural steel pipes. The remaining two legs are then set in place while further braces are added. Once the basic jacket has been welded the attachments are added. These include the mud mats, lifting pad eyes, sacrificial anodes and spider deck attachments. Depending on the design, pile clusters may be required and these will then be welded in place prior to painting the jacket. On large jackets, only the area in the vicinity of the splash zone is painted while the tendency on small jackets is to paint the entire jacket. The transportation photos show two jacket styles, a conventional jacket and what is referred to as a 'Sea Harvester' style jacket, which has a central core comprising three or four tubulars supported at the perimeter by skirt piles. While this type of design has become very popular, the conventional four leg, pile through the leg style of design is still widely used. There are advantages and disadvantages inherent in both designs.

Very large jackets are constructed in a broadly similar manner to the jackets just described. However, the intersection points of tubulars, referred to as nodes, may be prefabricated or even cast in order to save production time. Due to the sheer size of the component parts, three sides are normally fabricated in the horizontal position and these are then rolled up and welded into position before adding the final bracings.

Once the jacket framework is complete, the pre-fabricated pile clusters and mud mats are installed. On a large jacket, such as *'Britannia'*, the pile clusters are massive fabrications and it is through these that the foundation piles will eventually be installed. The mud mats prevent over-penetration of the jacket legs into the seabed during installation and help keep the structure level while piling operations are proceeding.

Figure 2.45 – The roll-up of the outer legs of the *'Britannia'* jacket at the fabrication yard in Dragodos, Spain. For ease of fabrication, the jacket frames are welded in the horizontal position and, when finished, are rolled up into the vertical by using a combination of cranes and winches. The frames are then restrained by guide wires while the infill bracings are installed.

Figure 2.46 – *'Sea Harvester'* jacket awaiting transportation to site

Figure 2.47 – The *'Cavendish'* gas platform jacket with piles pre-installed

2.4.2 Load Out and Transportation

The size of the jacket will ultimately dictate the way it will be taken to the offshore site. Except in the case of exceptionally large jackets that are too big, the vast majority of jackets are transported to the offshore site on a barge. In most cases, the jackets will be placed on the barge in the horizontal position, although shallow water jackets can be transported in the vertical position. Exceptionally large jackets, those too big to fit on a barge, may have to be fitted with flotation devices and towed to the site.

Figure 2.48 – The massive 20,000 tonne *'Britannia'* jacket being guided on to the transportion barge

Figure 2.49 – Barge transportation of shallow water jacket

2.4.3 Installation

Small jackets that have been placed on the barge in the vertical position can be lifted from the transportation barge by a large crane, often with the piles pre-installed.

If the jackets have been transported in the horizontal position a delicate up-ending process must be followed, normally because the installation crane has insufficient headroom for a direct lift. Once the jacket has been installed the piling operation commences. The photos show the installation of a *Sea Harvester* platform.

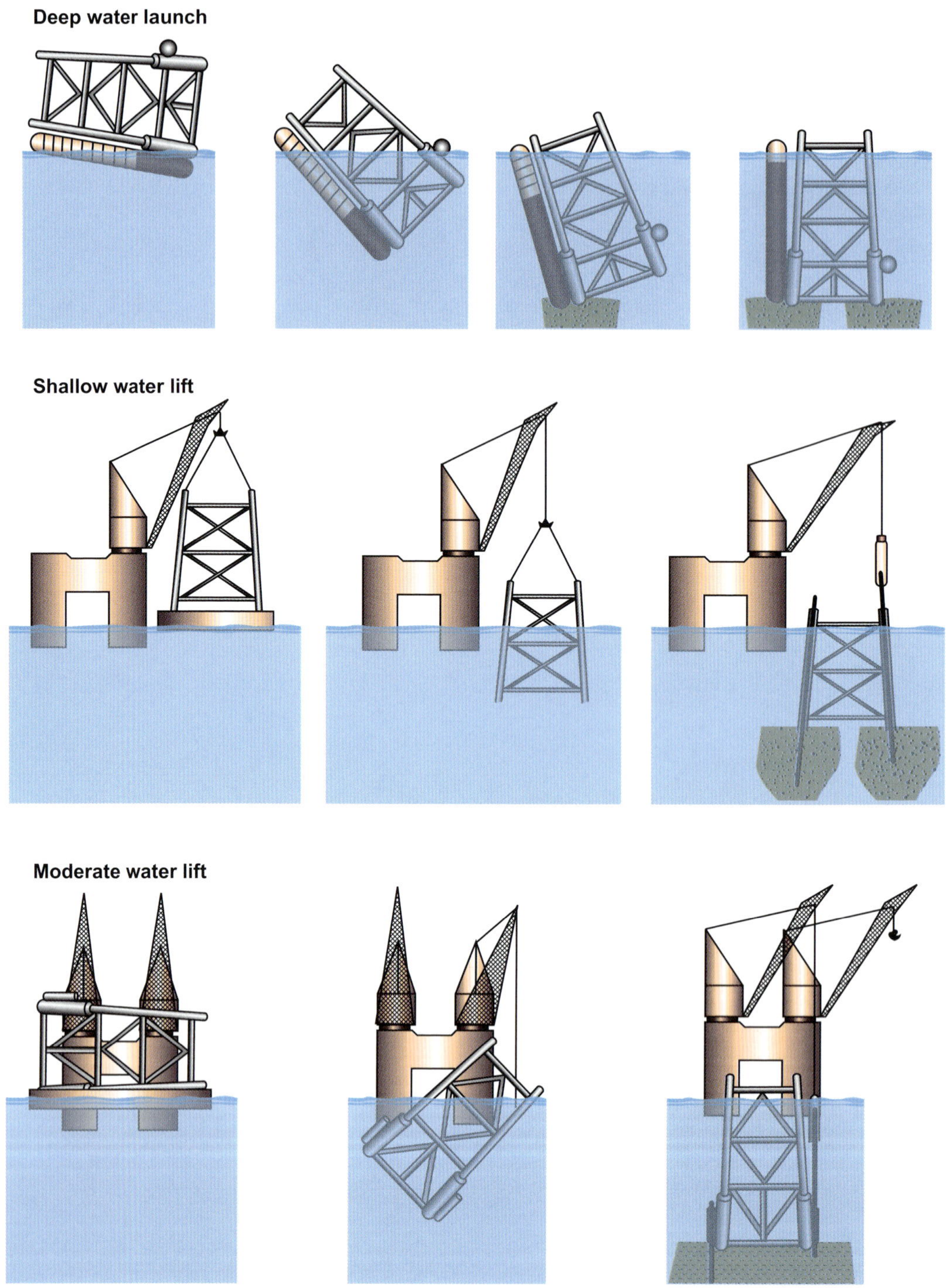

Figure 2.50 – Jacket installation

Figure 2.51(a) – Commencement of jacket up-ending process

Figure 2.51(b) – Hydraulic release mechanism for installation cable

Figure 2.52 – Grouting lines and grout line cutting mechanism on the pile guide

Jackets that are too large to be installed with a crane are normally installed using a barge that can be ballasted to such an angle that the jacket can be slid into the water. Flotation devices are used to position the jacket over the installation site and these are then sequentially flooded to enable the jacket to sink slowly onto its final resting place.

Figure 2.53 – Barge launch of deepwater jacket

2.4.4 Piling Operations

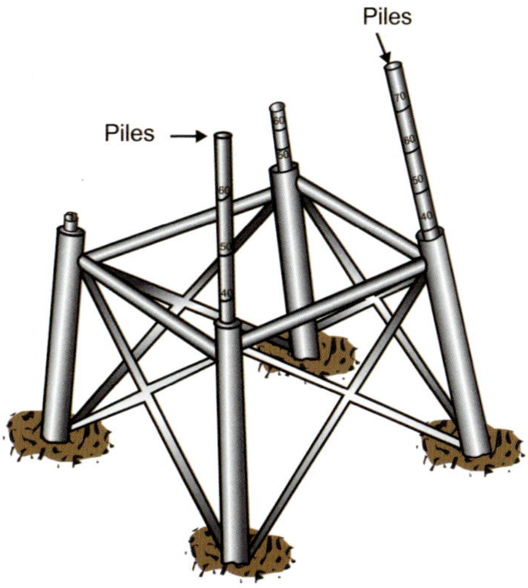

Figure 2.54(a) – Piling – through jacket

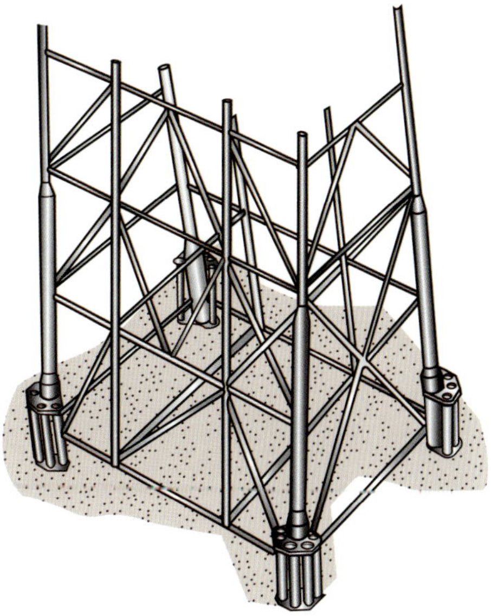

Figure 2.54(b) – Pile clusters – deep water

Jacket size dictates the method of pile installation, as can be seen from the sketches and photos. On the smaller jackets, tubular piles are driven through the legs to a pre-determined depth of 30–50 metres (100–165 feet), or inserted into pre-drilled holes and grouted (cemented) when the formation resists conventional piling techniques. The piles are fitted with spacers to ensure that they locate centrally within the jacket leg and, on completion of piling, steel shims (referred to as crown plates) are inserted between the top of the pile and the jacket leg and welded. The pile is then cut level, just above the top of the jacket, and grout may then be pumped into the base of each jacket leg until it flows from a vent hole cut in the top of the leg. Filling the annulus space formed by the jacket leg and pile in this manner ensures an extremely rigid structure is produced. This arrangement has been used to secure jackets in water depths of up to 100 metres (335 feet).

Figure 2.55 – Grouting operations, the grout being clearly visible as it pours from the vent holes. Grout is a mixture of cement and water with no sand. This is an old picture as can be seen from the relaxed attire of the construction crew.

The immense size of the deepwater jackets makes piling through the legs impractical, so a pile cluster is fitted to the base of each leg. The piles can then be driven through the cluster guides to the required depth of 60 – 90 metres (200 – 300 feet) before being grouted into position. Over 6,000 tonnes of piles were used to secure the BP *'Magnus'* jacket.

Skirt piles, as they are often referred to, are also used on small jackets to give the jacket a larger and more stable footprint on the seabed.

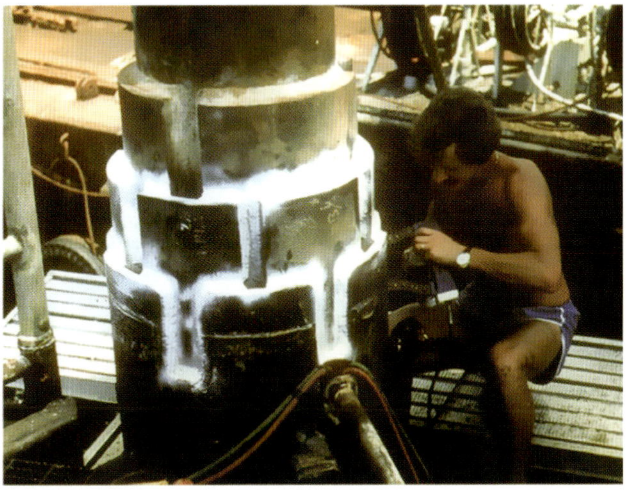

Figure 2.56 – Magnetic particle inspection of the crown plate welding. Note the white contrast paint used to enhance the inspection process.

Figure 2.57 – Piling of skirt piles

Figure 2.58 – The offshore industry works around the clock, 365 days a year. This night-time shot shows the steam hammer of the *'Heerema Balder'* in action driving the foundation piles through the legs of the Shell/Esso *49/26 'Foxtrot'* jacket. Inset – one of the Shell/Esso *48/19a 'Clipper'* jackets and foundation piles during transportation. Note the comparative size of the men on the barge, the unpainted lower steel structure and the depth graduations on the piles.

An alternative to piling was developed by Statoil and it was used to secure the *'Europipe'* riser jacket. The jacket legs sit within skirted mud mats, essentially large inverted buckets, 12 metres (40 feet) in diameter and 6 metres (20 feet) deep, which penetrate the seabed under the weight of the jacket, the jacket legs being filled with water ballast during installation to assist the penetration process. Suction provides further penetration prior to the addition of 2,200 tonnes of permanent grout ballast that is poured into the top of the buckets. This offered considerable savings on installation time and a similar process is often used in the form of suction anchors to secure FPSOs.

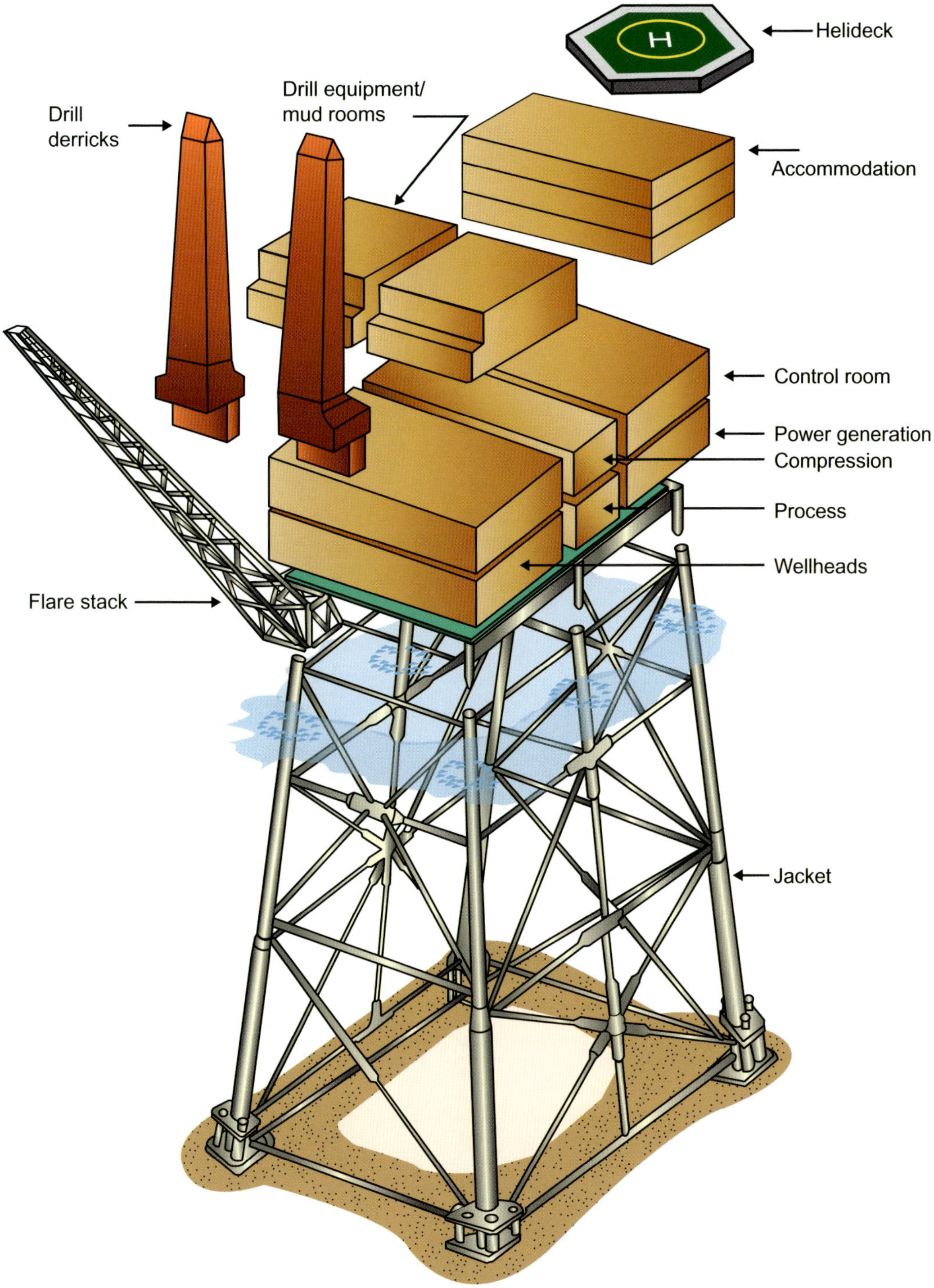

Figure 2.59 – Lifting plan

2.4.5 Topside Installation

The topside weight of an installation may be as low as 500 tonnes for an unmanned installation in the southern North Sea, or as high as 35,000 tonnes for the northern sector. While a floating crane can quite easily lift 500 tonnes onto a jacket, a single lift of 35,000 tonnes is beyond the capacity of any crane. Consequently, larger platforms were designed within the constraints of the lifting facilities available at the time and the photo shows the installation, module by module, of the *'North West Hutton'* platform (now decommissioned). Over the years, the lifting capacity of crane barges has increased considerably and vessels are available that can lift up to 14,000 tonnes. This has facilitated the construction of what are referred to as 'single integrated decks', enormous decks containing a number of modules that can be lifted as one unit.

The heavy lift cranes used for installing rigs are mounted on semi-submersible barges that have been purpose built for offshore work. These enormous catamaran type vessels, with deck capacities of up to 11,500 tonnes, often accommodate the smaller topsides modules as deck cargo for transportation to the site. Where this is not practical, the modules are loaded onto flat top barges and towed to their destination.

Figure 2.60 – Absolutely ideal conditions to lift 3,000 tonnes and demonstrate the capabilities of a semi-submersible heavy lift crane vessel (SSCV). The platform is the *'North West Hutton'*, physically one of the largest oil installations in the North Sea and the crane vessel is the *'Hereema Balder'* (1981). This platform has since been dismantled in the same manner as it was assembled.

Figure 2.61 – At 11,000 tonnes, the single integrated deck of the Conoco/Chevron *'Britannia'* was, at the time, the heaviest offshore lift recorded. The strain was taken by the Heeremac heavy lift crane barge the *'DB102'* (1998).

To assist installation, the modules and jackets are fitted with male and female type location devices, referred to as 'stab ins' or 'bucket guides', examples of which can be seen in the installation photographs. On shallow water jackets with through-leg piles, the topside structure is welded to the top of the pile rather than to the jacket.

Assembly of an offshore installation in this manner creates tremendous competition within the module fabrication industry. Until recently, a module destined for service in the UK sector of the North Sea would have been constructed almost entirely in the UK. However, the advent of the common market has seen the emergence of the 'Euro rig', which can contain a collection of modules fabricated in the UK, Spain, France, Italy and Portugal. The *'Britannia'* installation is a good example of this type of practice.

The *'Britannia'* is the largest gas platform to be installed in the UK waters of the North Sea and supplies 8% of the UK's gas. The jacket weighs 21,000 tonnes and was constructed in Dragados in Spain, while the topside structures, which weigh 28,000 tonnes, were built in the UK. The jacket is secured by the longest piles in the North Sea, 123 metres (400 feet) in length and 2.74 metres (9 feet) in diameter.

2.4.6 Abandonment

Decommissioning and abandonment of offshore installations is rapidly becoming a growth industry, particular in locations such as the Gulf of Mexico where there are currently in excess of 4,000 installations. In Europe there are considerably fewer installations but they tend to be much larger, there being approximate 250 structures in the UK sector of the North Sea and about 70 structures in the Norwegian sector. Further afield, there are nearly 100 offshore installations in Thailand, 200 in Malaysia, over 400 installations in Indonesia and 700 in the Middle East, all of which will one day require removal.

There are three main options available for the removal of offshore installations that are considered to be commercially unviable propositions. In all cases, the topside equipment and decks are removed, which leaves the jacket structure to be disposed of.

The first option involves removing the jacket in its entirety, which is the favoured option for jackets located in shallow waters, the jacket simply being cut at the seabed and lifted onto a barge. Alternatively, the upper jacket structure may be removed with the lower section left in place, cut at a depth below the surface where it will not prove a hazard to shipping. The third option is to use explosives to topple the jacket onto the seabed where it will be left to form a natural reef.

Oil companies are keen to adopt what is perceived by the general public to be the correct environmental option when decommissioning offshore installations and considerable thought is given to the removal and recycling processes. However, some of the very large structures are simply too big to permit their safe removal and an artificial reef is likely to prove the only viable solution to the problem.

They are assembled in building brick fashion, as can be seen from the sketches. The various stages of installation will now be explained.

2.4.7 RWE Dea UK *'Cavendish'* Installation

Figure 2.62 – Arrival of topsides and jacket

Figure 2.63 – Jacket and piles

Figure 2.64 – Lifting preparation

Figure 2.65 – Installation of the RWE *'Cavendish'* jacket and piles. A conventional four leg jacket – note the unpainted steelwork, the sacrificial anodes, 'J' tubes and gas risers.

Figure 2.66(a) – Piling operations

Figure 2.66(b) – Installing the next section of pile, which will be aligned and then welded to the previous section

Figure 2.67(a) – Crown plates prior to welding

Figure 2.67(b) – Welding

Figure 2.67(c) – Pile cutting

Figure 2.67(d) – Crown plates, welding, NDT and painting complete

Figure 2.68 – Installation of the topsides by the heavy lift ship the *'Stanislav Yudin'*

Figure 2.69 – Installation of the topsides

Figure 2.70 – The RWE *'Cavendish'* minimum facilities installation newly installed and awaiting the arrival of the drilling jack-up

Figure 2.71 – Gas installations are not provided with drilling derricks so a jack-up must be mobilised when drilling operations are required

Figure 2.72 – The RWE *'Cavendish'* when commissioned

3

Safety Systems

3 Safety Systems

(a) Lifeboat of the gravity davit type that is lowered to the water on wire rope falls

(b) Free-fall lifeboat

Figure 3.1 – Lifeboat systems on offshore installations

Safety Case Overview

In the UK, the requirement to have a Statutory Operations Manual for every installation has been superseded by the requirement for a 'Safety Case' (See Appendix VI). It shares many similarities with an Operations Manual and uses quantitative risk assessment as part of the Safety Management System. The information includes:

1. General information
   The name and address of the operator of the installation.

2. Structure – layout and notation
   A description, with scale diagrams of the main and secondary structures and plant, the connections to pipelines and any wells connected to the installation.

3. Plans and environmental criteria
   A scale plan of the location of the installation and of anything connected to it with particulars of the meteorological and oceanographic conditions the installation may be subjected to and the properties of the seabed and subsoil at its location.

4. Operational parameters
   Particulars of the types of operation, and activities in connection with an operation, which the installation is capable of performing.

5. Accommodation arrangements
   The maximum number of persons expected on the installation and for whom accommodation is to be provided.

6. Well control arrangements
   Particulars of the plant and arrangements for the control of operations on a well, including how to control well pressures, how to prevent the uncontrolled release of hazardous substances and how to minimise the effects of damage to subsea equipment by drilling equipment.

7. Pipelines
   A description of any pipeline with the potential to cause a major accident which includes its dimensions and layout, maximum operating conditions and safety devices.

8. Fire and gas detection
   Particulars of plant and arrangements for the detection of toxic or flammable gas, and for the detection, prevention and mitigation of fires.

9. Temporary refuge
   A description of the arrangements made to protect persons from the hazards of explosion, fire, heat, smoke, toxic gas or fumes following an accident and for enabling personnel to be evacuated. This description should include details of the temporary refuge (including how incidents can be monitored from within), escape routes and evacuation arrangements.

10. Performance standards
    A statement of performance standards which have been established in relation to the arrangements required for compliance with paragraph 9 which should include a description of how long a period of time the temporary refuge is designed to provide protection for.

11. Quantitative risk assessments
    A demonstration by reference to the results of a quantitative risk assessment that the measures taken in relation to the hazards referred to in paragraph 9 will reduce risks to health and safety to the lowest reasonably practical level.

12. Specifications and design codes
    Particulars of the main requirements in the specification for the design of the installation and its plant, including any limits for safe operation.

13. Design criteria
Sufficient particulars to demonstrate that the design of the installation, its plant and pipelines are such that the risks from a major accident are at a level that is as low as is reasonably practical.

14. Modifications and repairs
Particulars concerning any remedial work to be carried out to the installation or plant and the time in which it must be carried out.

Public Address and Alarm Indications			
Function	**Status Lights**	**Tone**	**Action**
Installation Safe	Green	'Bell' followed by announcement	• Work as normal
General Alarm • Fire detector • gas detector • manual initiation	Yellow	Intermittent Signal Constant Frequency	• Stop work • make area safe • muster
Prepare to abandon installation	Yellow	Continuous Signal Variable Frequency	• Prepare lifeboats • await instructions
Toxic Gas	Red	Continuous Signal Constant Frequency	• Muster

Think *Safety* at all times
Study the location of safety equipment
Know your muster station

IF IN DOUBT – ASK

Fire

1. Raise alarm by activating break glass unit
2. Announce *FIRE* over public address system
3. Try to extinguish a small fire after raising alarm
4. Retreat to muster area

Man Overboard

1. Throw man a lifebuoy
2. Keep man in sight at all times
3. Raise alarm by public address system

Survival Equipment

Lifejackets, immersion suits, smoke hoods, gloves and torches are located:

- In cabins
- at muster points
- in boxes as shown on Life Saving Plan

Muster Points

- Accommodation Locker Room
- Lifeboat station

Lifeboats

- AP Platform – West Side
- AD Platform – East Side

A Sample Platform Safety Notice

Figure 3.2 – Status lights

3.1 Fire-fighting Equipment

Legislation requires operators to provide a combination of active and passive fire-fighting protection measures to provide protection for personnel against the hazards of fire and the release of gas on offshore installations. These measures include fire and gas detection systems, remote control safety devices, fire alarms, fire mains, water deluge systems, automatic sprinkler systems, fixed and portable fire-fighting equipment, and the protection of strategic locations with passive fire protection materials.

3.1.1 Equipment Approval

Each and every piece of fire-fighting equipment must be approved by a National Administration or a National Standards Organisation such as the Maritime and Coastguard Agency (MCA) in the UK and the United States Coast Guard in the USA. For installations registered in countries that are member states of the European Union, the equipment must also be certified in accordance with the requirements of the Marine Equipment Directive (MED).

3.1.2 Guidance Notes and Standards

3.1.2.1 International Convention for the Safety of Life at Sea (SOLAS)

While the SOLAS regulations are primarily aimed at the marine industry, most of the requirements pertaining to lifesaving equipment and fire-fighting arrangements on offshore installations are based on them.

3.1.2.2 EN ISO 13072:1999 - Petroleum and Natural Gas Industries - Control and Mitigation of Fires and Explosions on Offshore Production Installations - Requirements and Guidelines

This standard, published by the International Standards Organisation (ISO), provides extensive guidance on fire-fighting systems and is applicable to all installations located in the waters of the European Union.

3.1.2.3 Prevention of Fire and Explosions and Emergency Response (PFEER) Regulations SI 1995/743

Before 1995, the requirements for active fire-fighting equipment on offshore installations located in the territorial waters of the United Kingdom were governed by Statutory Instrument No. 1978/611, Offshore Installations (Fire-fighting Equipment). In 1995, SI 611 was revoked and replaced by the Offshore Installations (Prevention of Fire and Explosion, and Emergency Response) Regulations, SI 1995/743, the Regulations being referred to as PFEER. The Health and Safety Executive (HSE) publish an Approved Code of Practice and Guidance to assist with the interpretation of the Regulations.

3.1.2.4 National Fire Prevention Association (NFPA)

In the USA, the NFPA publish a range of standards that cover all manner of fire protection equipment. They are referred to in Europe as well as the USA, particularly for guidance on sprinkler systems and water spray systems.

A brief description of the essential component parts of active and passive fire protection systems will now be given. Where performance standards are quoted, such as for fire hoses and nozzles, these relate to what has historically been accepted by the industry and they are not mandatory requirements. Reference to the regulations and specific design specifications pertaining to a particular installation should be made to ensure the requirements of the duty holder and administration are met in each case.

3.1.3 Active Fire Protection

The various regulations require the provision of fire and gas detection systems, remote control safety devices, fire extinguishers and fireman's equipment on all installations (manned or un-manned), and fire alarms, fire mains, hydrants and hoses, water deluge systems or monitors, and automatic sprinkler systems on all installations that are normally manned.

Figure 3.3 – Fire drill

3.1.3.1 Fire and gas detection

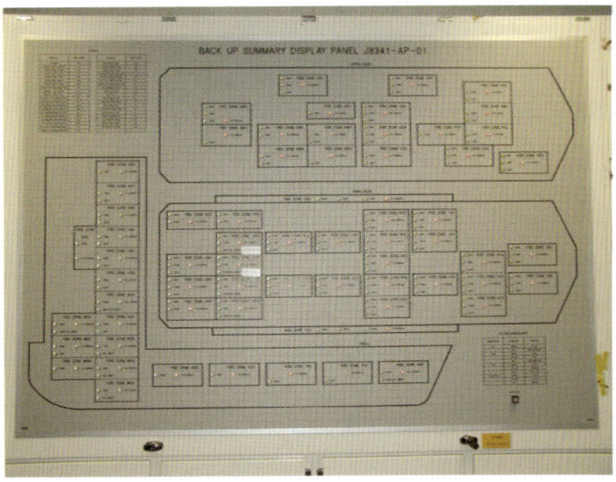

Figure 3.4 – Automatic fire detection display panel

Automatic fire detection systems should be fitted throughout an installation and the equipment used should be of a self-monitoring design so that if a fault develops a warning alarm will be initiated. Similarly, gas detection should be provided on all parts of the installation where flammable gas may accidentally accumulate.

Manually activated alarm points should be provided at strategic positions in the accommodation and working areas.

Figure 3.5 – Manually activated alarm point

Detectors sensitive to smoke, heat, ultraviolet light and gas may be used individually or in combination, depending on the hazard most likely to occur in any particular location.

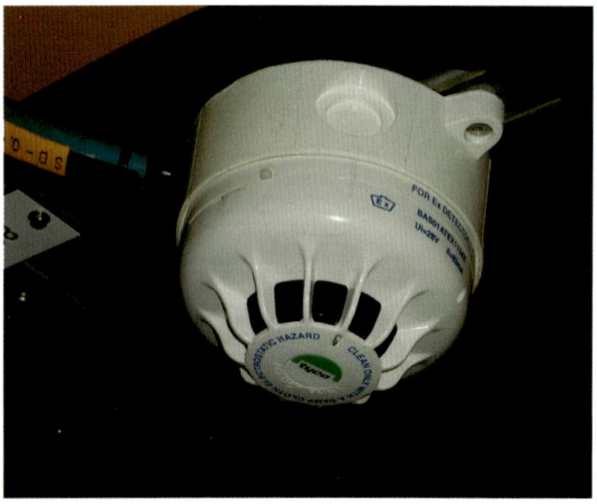

Figure 3.6 – Smoke detector

The main control room is usually dedicated for use as the fire control station because the operation of production equipment and communications can most easily be coordinated at that point. The main control room is normally manned 24 hours per day, at least on the larger installations (where the control room is not normally manned, a repeater alarm must sound in a manned location or cabin).

If a detector should sense a hazard or a manual alarm point has been activated, an audible and visual alarm should be initiated at the hazard and in the main control room. There is no requirement for the alarm to sound throughout the installation or for the automatic operation of the fire-fighting equipment. The location of the hazard must be identified in the main control station and the operations staff can then decide what action is to be taken.

3.1.3.2 Fire-fighting equipment

The active fire-fighting system centres around a ring main that is pressurised by at least two fire pumps, as shown in Figure 3.7. The system is normally maintained at a constant pressure of approximately 10 bars (150 psi) by a seawater service pump and pressure tank. The fire pumps may be manually activated from strategic locations such as the main control room, helideck and process areas, or automatically by a significant drop in ring main pressure, as would occur when a fire hydrant is opened or the deluge system activated.

Fire pumps

The fire pumps consist of a centrifugal deepwell pump that is normally powered by a self-contained diesel engine via a gearbox. Electric submersible pumps are occasionally used, self-contained units with a dedicated diesel engine and generator to provide the necessary electrical power being preferred.

Figure 3.7 – Fire pump

The number of fire pumps will be determined from the fire and explosion analysis, but normally at least two independently powered fire pumps will be found on an offshore installation. The number of pumps installed should reflect the possibility of the unavailability of equipment due to breakdown or maintenance requirements. They should be located remotely from each other in areas of minimal risk and be capable of operating for a clearly defined period of time, normally at least 12 hours unattended.

An engine protection system should be provided but it should not be capable of automatically shutting the engine down under fault conditions. The protection system should initiate alarms under fault conditions, which must then be addressed by platform staff and suitable corrective action taken. The only exception to this rule is the ingestion of gas at the air inlet manifold, which will shut the engine down prior to an overspeed situation developing.

Each pump should be capable of supplying adequate water to operate the largest section of deluge equipment in addition to maintaining a pressure of 3.5 kg/cm^2 at two fire hydrants, which are fitted with nozzles of 19 mm diameter.

The fire main should be constructed from corrosion resistant materials such as galvanised mild steel, stainless steel or cunifer (copper/nickel alloy), and must be protected against the effects of both over pressure and freezing.

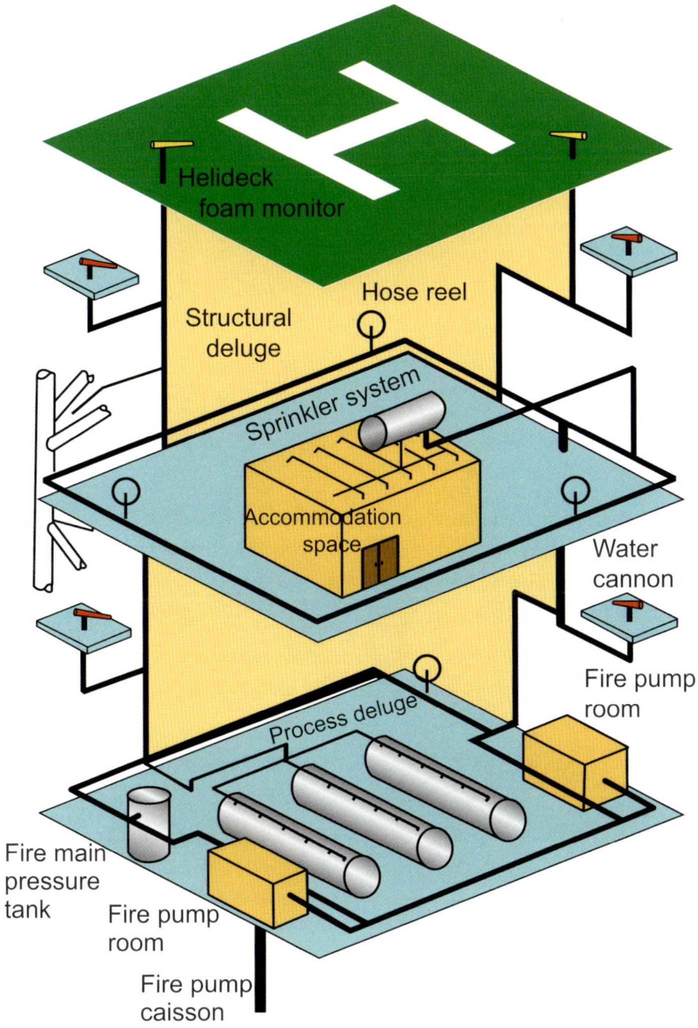

Figure 3.8 – Fire main

Wherever possible, the fire main should be routed clear of hazardous areas, with full use made of any protective shielding afforded by structural members. Isolation valves should be installed so that a damaged section of fire main will not adversely affect the operation of the remaining system.

Figure 3.9 – Section of fire main

In some parts of the world, glass reinforced plastic (GRP) fire mains have been installed on offshore installations but at the time of writing, they are not approved for use in the UK due to their failure to meet the material fire test requirements. At the present moment in time, they are only capable of passing fire tests under flowing conditions, failure occurring when the flow of water ceases. However, with the dramatic improvement in material properties, it is only a matter of time before full approval is obtained.

Hose reels

The fire hydrants and hose reels provide a back-up to the fixed fire-fighting systems and permit boundary cooling to be applied to spaces in which an outbreak of fire has occurred. Hydrants should be located in readily accessible positions, particularly in way of access points, and a sufficient number of hose reels stations must be provided so that water can be brought to a fire in any location from at least two directions. At least one of the required jets must be supplied from a single standard length of hose.

Standard fire hose is regarded as being of 64 mm bore, 18 metres in length and made from unlined canvas. However, smaller diameter hoses are permitted, such as the 45 mm bore lined varieties, provided they have a similar throughput to the larger hoses. Non-collapsible hoses of 25 mm diameter may also be used but only where deluge cover is provided as the primary means of fire protection for the space concerned. The 25 mm hoses may also be used in accommodation spaces, but again, only as an addition to the larger hoses.

All hydrants, nozzles and hose couplings should be of a standard type throughout the installation. The nozzles should be of robust construction and capable of providing a water jet of 12 metres length and a minimum discharge rate of 26 m^3/hour, while a pressure of not less than 3.5 bar is being maintained in the fire main. The nozzles should also be capable of providing a fine water spray curtain of approximately 4.1 metres diameter at a distance of 2.1 metres.

Figure 3.10 – Hose reel

Deluge

Figure 3.11 – Deluge system test

The deluge system pipework is similar in design to a sprinkler system, but on a much larger scale and with the exception that frangible bulbs are not fitted, the pipework remaining open and dry until the deluge control valves on the fire ring main are activated. The outlet nozzles are of the open type and are fitted with deflector plates designed to create an even water droplet spray pattern.

The primary function of the deluge system is to minimise the effect of heat arising from a major fire involving hydrocarbons. The deluge water is intended to cool the process plant and reduce the escalation of a fire while safety systems, designed to isolate or vent the supply of hydrocarbons, take effect and any ignited hydrocarbons are consumed by the fire.

Figure 3.12 – Deluge system protecting equipment

The deluge system will cover the wellheads, gas process pressure vessels and pipework, process plant and storage vessels in their entirety, and may be extended to cover the main structural steelwork to prevent it from weakening if exposed to fire. The system may be manual or automatic in operation, with the preference being for automatically activated systems.

The water coverage provided by the deluge system will be determined by the size of the potential hazards within the area to be protected. A minimum water coverage figure of 12.2 litres per minute over the reference area is specified for all areas, while this increases to 20 litres per minute where there are high pressure hydrocarbon inventories present (and for structures when required). The figure rises to 400 litres a minute for wellhead areas.

Water monitors

A monitor is a permanently fixed device for directing water or foam in jet or spray form onto a fire, in effect, a water cannon.

Figure 3.13 – Water monitor

The monitors are designed for the application of water in open deck areas and should be capable of providing the desired cover irrespective of weather conditions. In theory, it is acceptable to use water monitors in place of a deluge system to protect process areas, but in practice the monitors are normally fitted to supplement the deluge systems.

Water monitors are also used to protect the helideck on normally manned installations. They must be capable of discharging a sufficient quantity of low expansion foam to completely cover the helideck, even in adverse weather conditions. A tool kit containing equipment to assist in releasing the occupants from helicopter wreckage will also be found in the immediate vicinity of the water monitors.

Sprinkler system

It is recommended, although no longer mandatory in the UK, that every accommodation space be protected by an automatic sprinkler system. An accommodation space is defined as any room used for eating, sleeping, cooking or recreation, or as an office, sick bay, laundry room or locker room, any corridor giving access to any of these rooms, and any store-room in the vicinity of any of these rooms.

The sprinkler system must be supplied from a permanently pressurised water source, located remotely from the accommodation, which is capable of operating for a period of at least four hours. Sprinkler systems are normally charged with freshwater via a pressure tank, with changeover arrangements provided to admit seawater from the fire main once activated. However, statically charging the system with seawater is acceptable provided that it can be demonstrated that the efficiency of the system will not be impaired by corrosion, marine growth or sediment.

The sprinkler system's frangible bulbs are designed to burst at between 68°C and 70°C in the accommodation spaces, while in the galley higher settings may be used, up to 30°C above the maximum deck head temperature. The heads should provide a water supply of 5 L/m^2 per minute in spray form. In cooking areas where cooking oils and fats are used, deflector plates should be positioned to prevent water impingement.

Operation of the sprinkler system should initiate an audible and visual alarm in the main control station and indicate the location of the hazard. The automatic activation of a general fire alarm is not required, further action being left to the discretion of the control room staff who can manually activate further alarms or alert personnel by means of the public address system, if required.

Arrangements are to be provided to enable the sprinkler system to be tested without detracting from the operational efficiency or disrupting the routine operation of the installation.

Figure 3.14 – Sprinkler

Water spray

Water spray, or fog systems, share many similarities with sprinkler systems and are simple in design, having no moving parts and so requiring minimal maintenance. Unlike a sprinkler system, water spray systems can also be used to protect machinery spaces and are suitable for fighting both oil and hydrocarbon fires.

In most cases, the systems comprise of dedicated air (or nitrogen) and water storage vessels, which come complete with associated controls. The principle of operation is that stored air, or nitrogen pressure, is used as both the propellant for the water supply and for the atomisation of the water at the nozzles. The extinguishing mechanism is one of heat removal by the water, the microscopic water droplets or fog turning to steam that starves the fire of oxygen to a level at which combustion cannot be sustained.

Water fog systems are normally supplied in modular form as dedicated stand alone units for protecting specific areas. They are designed as a one-shot system and consume very small volumes of water with no further back-up being required. They operate at low pressures (3–5 bar) so the water can be distributed to the nozzles via small bore tubing or flexible hoses (which remain dry until activated), making installation particularly easy. Water spray systems are rapidly gaining in popularity, not least because they are seen as the most environmentally acceptable medium.

Figure 3.15 – Pressurised storage vessels

3.1.3.3 Fixed fire fighting systems

Fire-fighting equipment that is installed to protect a specific compartment or area is generally described as a fixed fire-fighting system. An inert gas is predominately used as the extinguishant, although dry powder systems are used extensively. Sprinkler systems and water fog systems could also be described as fixed fire-fighting systems, but have been dealt with in the previous section because water is used as the extinguishant and this is more in keeping with the wet and dry philosophy of the mediums concerned.

Figure 3.16 – Water fog system

3.1.3.4 Essential features

The system chosen should function with minimal operator intervention. Automatic or manual initiation may be provided, with automatic systems being provided with a manual release facility.

It is a requirement that certain locations, such as control rooms and machinery spaces, are protected by fixed fire extinguishing systems. There are a number of requirements that must be addressed when using gaseous extinguishants such as CO_2.

The ventilation systems must be shut down and visual and audible alarms activated in the affected area at least 15 seconds before the release of the gas. Status lights must be fitted outside all protected spaces to indicate whether the system is in the automatic, manual or the discharged condition.

Initiation of the fixed fire-fighting equipment may be automatic but manual operation is often preferred for normally manned locations. Where systems are arranged for automatic release, warning notices should be provided at compartment entrance points. The systems must also be capable of being activated manually and full operating instructions should be provided at the control point.

Any electrical circuits essential to the operation of a fixed fire-fighting system should be run in heat-resistant cable, eg mineral insulated cable (MIC).

3.1.3.5 Storage

The gas storage cylinders associated with fixed fire-fighting systems should be located outside the space to be protected in a well ventilated, gas tight storage room, preferably accessed from the open air. Pressure relief devices fitted to CO_2 cylinders, tanks or manifolds should be arranged so that when operated there will be no danger to personnel from the resultant discharge of the gas. Continuous monitoring of the levels within the gas cylinders is to be provided and automatic over-pressure protection which, in the event of the container being exposed to a fire, will release the gas and protect the space.

At a period agreed by the authorities, but at least after 20 years, the gas storage cylinders should be hydraulically tested to confirm their soundness. After the 20 year test, 10% of the cylinders should be tested every 5 years.

3.1.3.6 Fire-fighting mediums

The various mediums are:

Halon

Halon is a lighter than air gas that relies on a chemical reaction to break the combustion chain and extinguish the fire. It is virtually instantaneous in operation, leaves no mess and is easily dispersed once the area has been made safe. Halon can be stored at low pressures (25 bar) and requires only a 5-7% concentration to be effective which makes the equipment light and compact.

Halon may still be found on some of the older installations providing protection to control rooms, plant rooms, machinery spaces and any locations housing delicate equipment. It is the ideal fire-fighting gas, but, because of the effect it is reported to have on the thinning of the ozone layer, its use has been gradually phased out by all countries who are signatories to the Montreal Protocol. Since July 1992, the use of halons on new installations was restricted to those locations designated as essential, ie normally manned spaces such as main control rooms.

Carbon dioxide (CO_2)

Carbon dioxide was used almost universally prior to the introduction of halon and has made a return with halon being phased out. While carbon dioxide is termed a greenhouse gas and is as environmentally unacceptable as halon, it is produced as a by-product of chemical manufacturing processes and would normally be vented to atmosphere anyway.

Carbon dioxide is particularly suitable in locations such as vent stacks, where the heavier than air smothering effect that characterises the gas can be put to the most effective use. The disadvantages with CO_2 are that it is a high pressure gas (60 bar) and a 35% concentration is required to be effective, so the equipment is both heavy and occupies more space than equivalent mediums. It is also an asphyxiant and great care must, therefore, be taken in both the storage arrangements and use in manned locations.

Dry powder
Like halon, dry powder relies on a chemical reaction to break the combustion chain and while it is very effective it leaves a tremendous mess. It is also extremely effective in fighting vent stack fires and may be used as an alternative to CO_2.

Alternative gases
Chemical manufacturers have produced alternatives to halon that have been approved by the various statutory bodies. Two such gases are inergen and argonite, inergen being the more widely used of the two.

Inergen
Inergen is a mixture of nitrogen, argon and CO_2 (50/42/8%), which has many of the benefits of halon and can be used as a straight replacement with relatively few modifications. The main disadvantage with inergen is the volume required, approximately 50% concentration by volume, so a considerable number of storage cylinders are required.

Argonite
Argonite is a mixture of naturally occurring nitrogen and argon and, like inergen, has many of the benefits of halon, including the use of similar equipment. When argonite is injected into a room, it reduces oxygen to levels that will not sustain combustion, yet will permit essential safety personnel to operate. It is of a similar molecular weight to air and lingers longer than halon, so the protected spaces do not need to be hermetically sealed, a considerable cost saving advantage.

3.1.4 Portable Fire-fighting Equipment

A considerable number of portable fire-fighting extinguishers are located throughout offshore installations. Extinguishers using water, foam, dry powder and CO_2 will be installed at vantage points as specified on the installation fire-fighting plan.

Foam concentrate in hand-held extinguishers should be replaced at intervals not exceeding 10 years and, in the case of extinguishers of 45 litres capacity or greater, the charge should be withdrawn

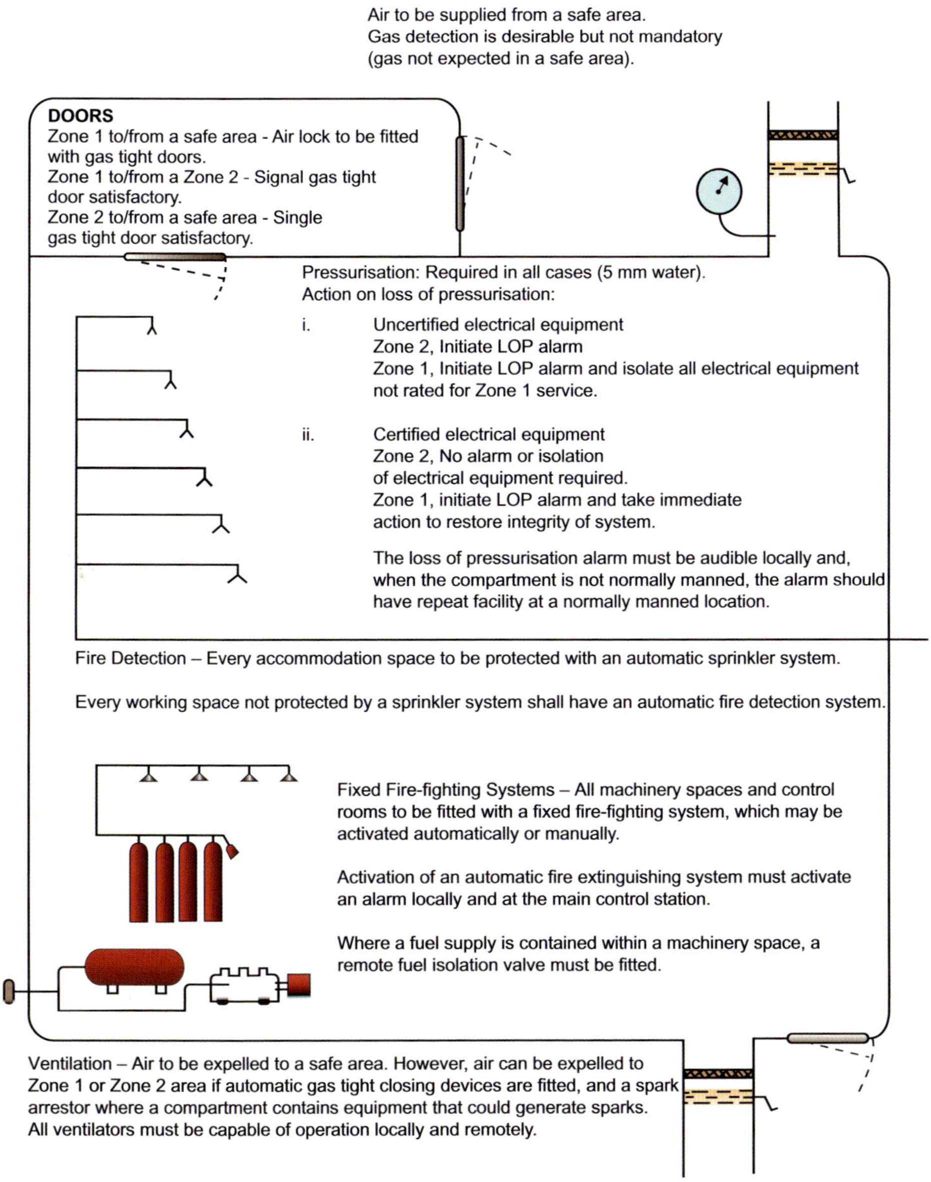

Figure 3.17 – Fire and gas protection plan

at least once in four years and the internal condition of the extinguisher verified.

Portable CO_2 extinguishers should be recharged if the weight of gas falls to less than 90% of the stated capacity. If, when the cylinders are emptied, four years have elapsed since the date of the last hydraulic test, the cylinders should be tested to the bursting pressure of the safety disc before being recharged. Portable fire extinguishers other than CO_2 should be tested to 21 bar every four years, while dry powder extinguishers should be tested to 1.5 × working pressure or 21 bar, whichever is the greater, every four years.

3.1.5 Helideck Fire-fighting Equipment

Figure 3.18 – Helideck

Adequate fire-fighting facilities are to be provided to deal with any unplanned helicopter incidents. This will include a combination of fixed and portable equipment.

Fixed fire-fighting equipment
Foam monitors are to be provided that can cover the landing area with a layer of foam, regardless of weather conditions, the delivery of foam commencing within 30 seconds of the start of the incident. Sufficient foam is to be provided to deal with more than one incident and to permit training exercises to be undertaken. Foam does not generally have a shelf life but it should be tested annually to determine that it is still fit for purpose. In addition to fixed foam monitors, it is recommended that at least two portable foam branches are provided, with a capacity of at least 250 litres/min per hose.

Figure 3.19 – Helideck foam monitor

Portable extinguishers
While foam is considered to be the principal media for dealing with helideck fires, at least those where burning kerosene is involved, potentially less serious fires must also be considered, such as an engine fire or hydraulic oil fire. Carbon dioxide and dry powder extinguishers are more suited to these types of fire so it is a recommendation that two 45 kg dry powder extinguishers are provided and an 18 kg carbon dioxide extinguisher, normally wheeled for ease of use.

Fireman's outfits
All personnel assigned to fire-fighting duties must be provided with suitable personal protective equipment. This will normally include a fireman's protective suit, including a helmet and gloves, and this equipment should be worn throughout helicopter landing operations. Self-contained breathing apparatus should also be provided and a tool box containing equipment that can be used in the event of a helicopter crashing onto the deck. The following items are required:

- Adjustable spanner
- rescue axe
- bolt cutters
- crowbar
- boat hook
- heavy duty hacksaw and six spare blades
- fire resistant blanket
- ladder
- lifeline
- tin snips
- set of assorted screwdrivers
- knife
- fire resistant gloves
- breathing apparatus, two sets.

(a) – Fire extinguishers

(b) – Rescue tool

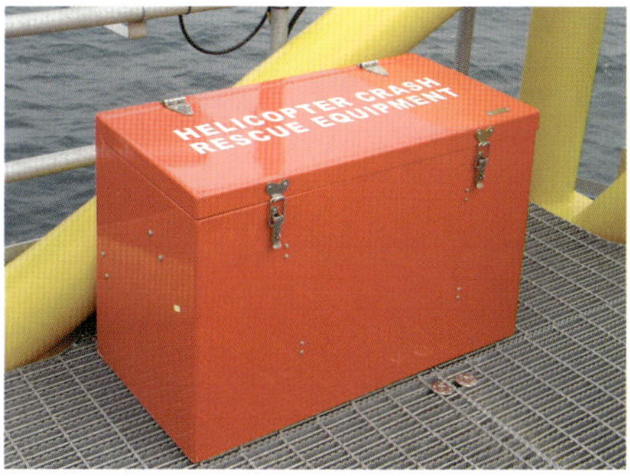

(c) – Helideck tool box

(d) – Breathing apparatus

Figure 3.20 – Helideck fire-fighting equipment

3.1.6 Remote Stops and Closing Devices

In the UK, to minimise the spread of fire and maximise the efficiency of the fixed fire-fighting mediums, the following mandatory requirements are applicable:

3.1.6.1 Ventilation systems

Remote stops for ventilation fans and remote closing devices for ventilator flaps should be provided. Modern installations may employ a centralised ventilation control system that can be operated from the main control room. This facilitates the isolation of specific areas, that can be returned to service as and when required, to assist in the dispersion of smoke.

Figure 3.21 – HVAC - Ventilation damper

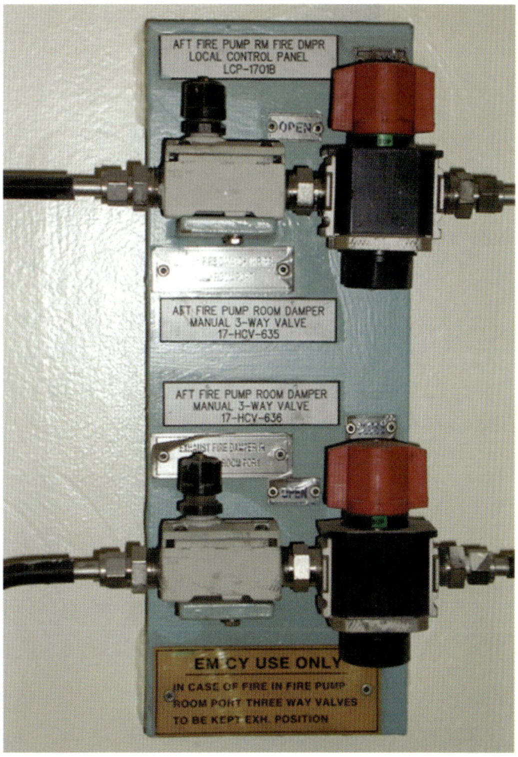

Figure 3.22 – Remote stops for ventilation systems

3.1.6.2 Fuel systems

Compartments containing gas or diesel powered machinery must be provided with fuel isolation valves located externally to the machinery space.

3.2 Passive Fire Protection

UK requirements for passive fire protection are outlined in the *Department of Energy Guidance Notes on the Design, Construction and Certification of Offshore Installations* (withdrawn) and are generally in accordance with the recommendation of the *International Convention For The Safety Of Life At Sea* (SOLAS).

Passive fire protection involves the application of non-combustible heat resisting materials to insulate strategic locations against the effects of fire and smoke. The materials used are generally mineral wools, fibre partitioning boards, cementitious and intumescent coatings, and all must be certified as having had their effectiveness proved by independent prototype testing within a controlled furnace atmosphere. On satisfactory completion of testing, the materials are issued with a fire test certificate, which is generally valid for a period of five years. It is important that passive fire protection materials are installed strictly in accordance with the approved arrangements if the desired protection is to be achieved. The manufacturers of these materials will readily provide literature to show how their products must be installed, and copies of the fire test certification.

Figure 3.23 – Fireproofed vessel and support

Figure 3.24 – Application of cementitious coating

3.2.1 Fire Divisions – Designations

Various designations are used to specify the structural fire protection requirements and A, B and H-class divisions will be terms frequently encountered. A time interval is normally included within the fire rating so that if, for example, an A-60 barrier is specified then this will maintain the integrity required of an A-class division for a period of 60 minutes.

The most commonly encountered divisions are as follows:

3.2.1.1 A-class divisions

These are constructed from steel, or an equivalent material suitably stiffened so as to maintain their structural integrity, and are capable of preventing the passage of smoke and flame to the end of the 60 minute standard fire test.

The standard or *cellulosic* fire test, as defined by ISO 834, involves heating one side of the insulated material and its support structure to a furnace temperature in excess of 925°C. The average temperature on the unexposed side of the division should not increase by more than 139°C above the initial temperature, nor should the temperature at any one point, including any joint, rise by more than 180°C above the initial temperature within the duration of the test, normally 60 minutes. Barriers that pass this test are classified as A-60.

As previously stated, an A-60 barrier should remain intact for a period of 60 minutes and this most arduous of specifications is used to protect locations such as the main control room, emergency sources of power, communication systems, and all accommodation boundaries that face a hazardous area on offshore installations. A-0 designated barriers will also be encountered, and these will withstand the standard fire test and so prevent the passage of heat and smoke for a period of 60 minutes, but they are not insulated. The A-0 barriers are used to classify items such as galley serving hatch shutters.

Typical A-60 arrangements are shown in Figures 3.25 and 3.26, a steel bulkhead providing structural integrity while a dense mineral wool provides the thermal insulation.

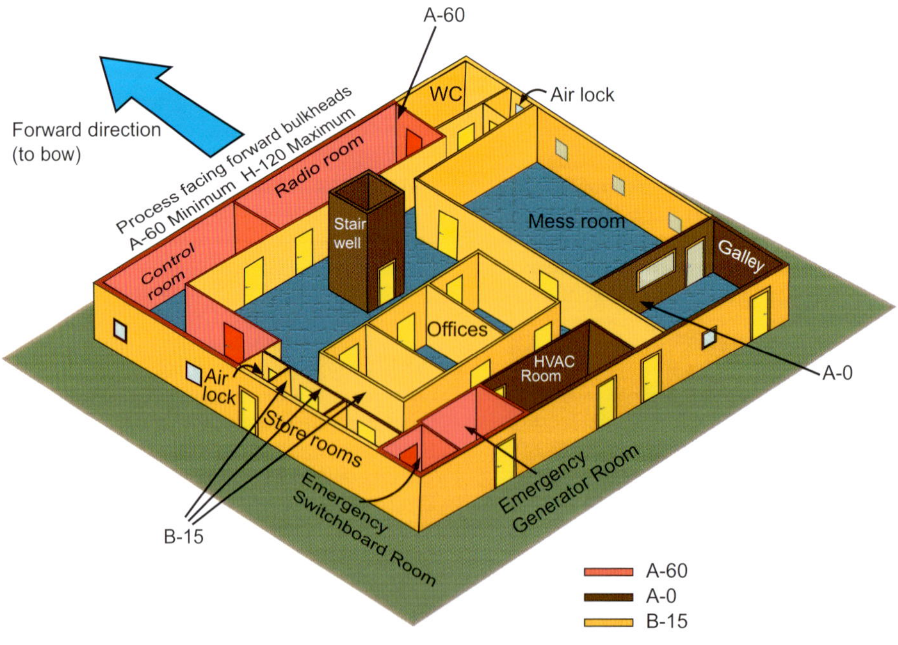

Figure 3.25 – Passive fire protection on a typical FPSO

3.2.1.2 B-class divisions

These divisions are constructed from non-combustible materials (not necessarily steel because they are not required to be load bearing) and for testing are exposed to a standard fire test in a similar manner to A-class divisions. They must be capable of preventing the passage of flame (but not smoke) towards the end of the first 30 minutes of the fire test but the accredited notation is typically only 15 minutes, eg B-15. The temperature rise restrictions on the unexposed face are the same as for A-class divisions.

The B-15 fire rated fibre board partitioning panels are used extensively for the construction of bulkheads, ceilings and corridors in the vicinity of non-critical areas such as accommodation rooms located in safe areas.

3.2.1.3 H-class divisions

These are constructed in a similar manner to A-class divisions, but additional layers of mineral wool or improved ceramic fibre based insulations are applied to withstand the more rigorous demands of a hydrocarbon fire test. The furnace temperature specified for a hydrocarbon fire test is in excess of 1100°C, the temperature rise restrictions on the unexposed face being the same as for A and B-class divisions. The tests are carried out for periods ranging from 30 to 120 minutes, the latter giving an H-120 rated division.

The H-120 specification is recommended for use on accommodation boundaries that face locations where a hydrocarbon fire could occur, such as wellheads and process areas. It may also be specified for the protection of subsea riser pipes in the vicinity of the emergency shutdown valve (ESDV).

3.2.1.4 J-class protection

In addition to the standard and hydrocarbon fire tests, there is a third fire test that is used to determine the suitability of materials to resist jet fires. The test attempts to simulate the effects of a fracture in a pipeline riser or flow line, which is characterised by a particularly fierce and intense fire with greater heat flux and higher levels of turbulence of the flame, and with the added danger of erosion of the insulating material.

International Standards for jet fires, with input from oil companies, manufacturers of fire protection materials and certifying authorities, are published as the Offshore Technology Report-OTI 95 634 - Jet-Fire Resistance Test of Passive Fire Protection Materials and ISO/CD 22899-2.

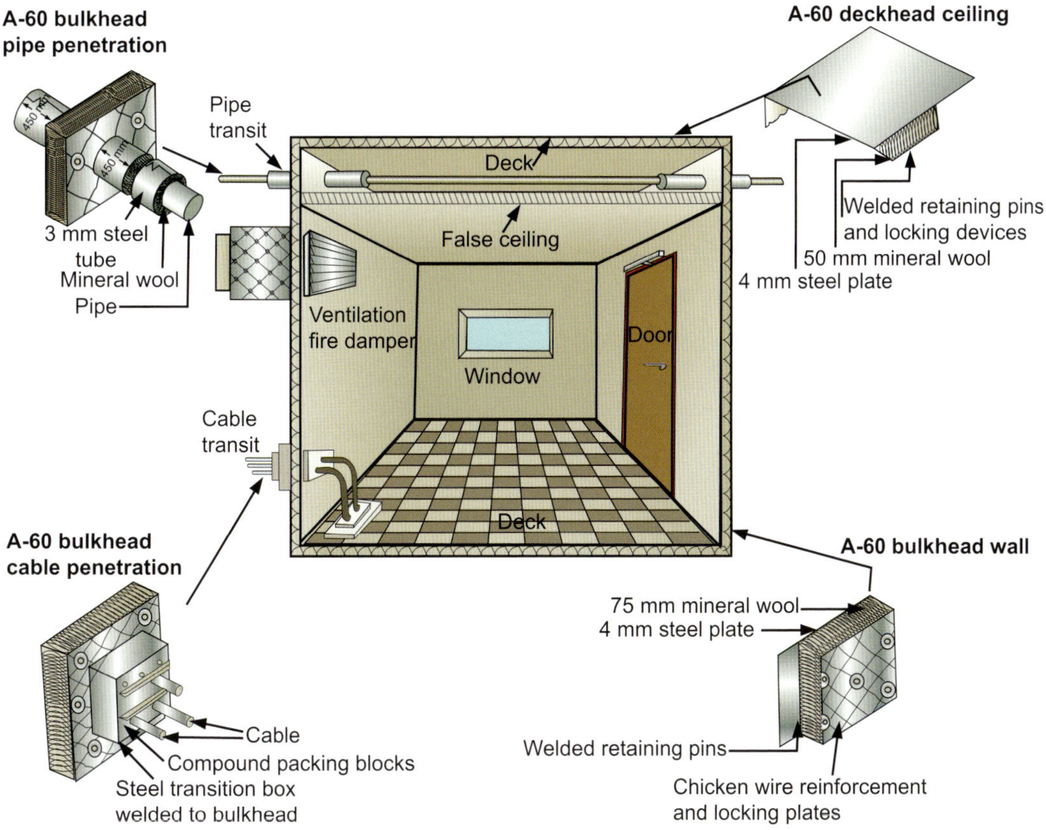

Figure 3.26 – Typical passive fire protection arrangements

A 'J' rating is used to identify jet fire tested materials and the period of effectiveness ranges from 30 to 120 minutes, the basic requirement being to keep the core temperature of the protected material to below 400°C. However, jet fire conditions and ratings are considerably more complex than the A-class and H-class barriers and each case must be considered separately and account taken of the particular hazards determined for the intended application.

3.2.2 Fire Divisions – Penetrations

All items that penetrate fire rated divisions are to be certified to the same standards as the division itself. This means all doors, windows, ventilators and ducting, pipe and cable transits must be approved as H-120, A-60 or B-15, or to whatever fire rating the design drawings specify.

Figure 3.27 – Passive fire protection on primary steelwork

All passive fire protection materials and fittings must be installed strictly in accordance with manufacturers' recommendations to achieve the desired protection and to comply with the fire test approval certificates. This is particularly important where doors are concerned because they are approved as a complete assembly that includes the frame, hinges and locks.

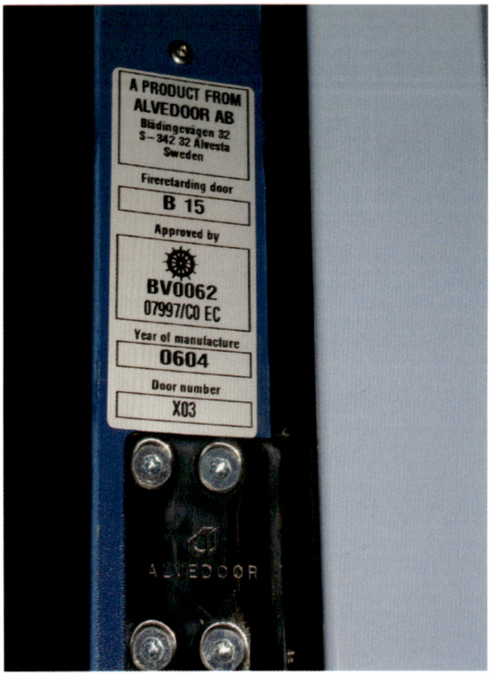

Figure 3.28 – Certification panel, B-15 fire rated accommodation door

Electrical cables are another area where careful consideration is required because they can be supplied as either fireproof, fire retarding or low flame spread.

3.2.3 Draught Stops

Draught stops should be fitted in void spaces behind ceilings, bulkheads and stair wells, both to prevent the passage of smoke and fire and to prevent cold air from feeding a fire. They should be installed at intervals of 14 metres and may take the form of a fixed partition or a curtain made from mineral wool.

3.2.4 Furnishings

In addition to structural fire protection requirements, it is recommended that all furnishings and internal decor is constructed to comply with the requirements of the various parts of BS 476. This defines standards for the resistance, non-combustibility and non-flammability of building material and structures. This includes primary deck coverings such as self-levelling compounds, which are used extensively on ships and offshore installations.

3.2.5 New Developments

As a general rule, all materials used in the construction of an offshore installation, and this includes major equipment items, should be composed of materials that are consistent with their function and should be incombustible. Consequently, any items constructed from new materials, such as GRP, should only be used if they have been proved as being incombustible by fire testing prototype samples. This includes such items as fuel tanks, fire mains and deck gratings.

Reference standards

1. Department of Energy Guidance Notes on the Design, Construction and Certification of Offshore Installations (withdrawn).

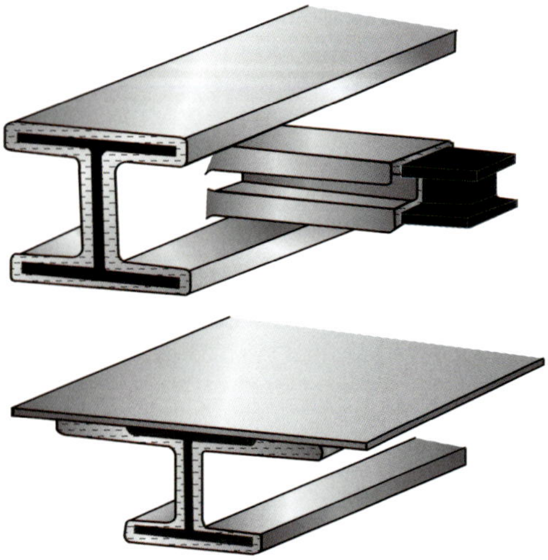

Figure 3.29 – Structural steelwork cementitious coatings

2. Department of Energy Offshore Installations: Guidance on Fire Fighting Equipment.
3. Offshore Installations (Prevention of Fire and Explosion, and Emergency Response) Regulations, SI 1995/743, (PFEER) and Approved Code of Practice and Guidance.

4. SOLAS (Safety Of Life At Sea) rules.
5. BS 476 Fire tests on building materials and structures.
6. Offshore Technology Report-OTO 95 634 – Jet Fire Resistance Test of Passive Fire Protection Materials (Health and Safety Executive).
7. Determination of the resistance to jet fires of passive fire protection materials – Part 2: Validation, interpretation and classification methods ISO/CD 22899- 2.
8. EN ISO 13702:1999 - Petroleum and Natural Gas Industries - Control and Mitigation of Fires and Explosions on Offshore Production Installations - Requirements and Guidelines.

3.3 Life-Saving Appliances

Introduction

Life-saving appliances are normally provided in accordance with standards laid down by the national administration of the country concerned, such as the United States Coast Guard (USCG) in the USA and the Maritime and Coastguard Agency (MCA) in the UK. These requirements are normally based on The International Life-Saving Appliance Code, a mandatory marine code under the terms of the International Convention For The Safety Of Life At Sea 1974 (SOLAS), a code developed by the International Maritime Organization. While SOLAS is mandatory only for ships operating under the national flag of signatory nations, the majority of oil and gas producing countries are signatories to SOLAS and use the Code as the basis for specifying the minimum requirements for life-saving appliances on offshore installations.

Approval of equipment

The International Life-Saving Appliance Code specifies requirements for the manufacture and testing of life-saving equipment and, generally speaking, all equipment must be manufactured in accordance with a type approval certificate issued by the National Administration or Flag State.

Type approval certificates are issued following confirmation by the national administration that the equipment provides safety standards at least equivalent to the SOLAS recommendations and that successful prototype tests have been completed. National administrations such as the United States

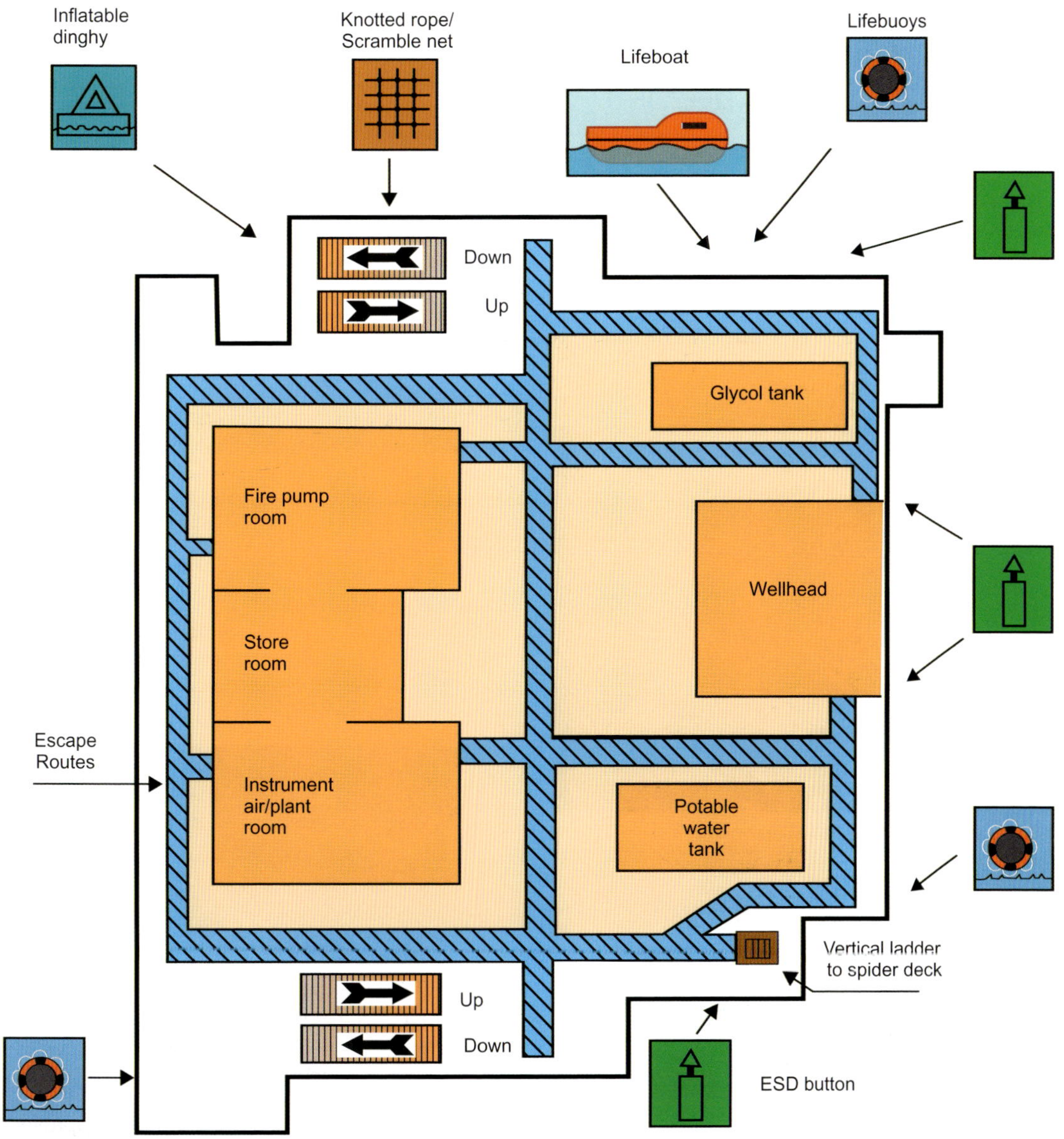

Figure 3.30 – Life-saving appliance (LSA) plan on Cellar Deck

Coast Guard and the Maritime and Coastguard Agency (MCA), previously the Department of Transport (DoT), frequently choose to delegate the approval process to organisations such as the major ship Classification Societies.

For vessels registered in countries that are member states of the European Union, the equipment must also be certified in accordance with the requirements of the Marine Equipment Directive (MED).

Manufacture and certification
Large items of equipment such as lifeboats and davits will be built under the survey of representatives of the local administration, or one of the ship classification societies, while smaller items may be supplied with a type approval certificate issued by the manufacturer. Some of the equipment that is not specifically for marine use, such as torches, gloves and smoke hoods, may be manufactured in accordance with a national standard and be accompanied with a certificate of conformity issued by the manufacturer against appropriate company standards. It may also be supplied against a CE number, which shows that it has been manufactured in accordance with the relevant European regulations.

3.3.1 Life-Saving Equipment

3.3.1.1 Life-saving appliance plan

All life-saving appliances should be installed in accordance with an approved Life-saving Appliance Plan, (Figure 3.30), copies of which should be provided in the Statutory Operations Manual or the Safety Case. The plan should be clearly displayed in the accommodation and at an appropriate working space such as the main control room. The plan should also show escape routes and muster areas and, normally, the locations of fire-fighting equipment.

Figure 3.31 – Emergency escape route

3.3.1.2 Alarm and public address system

A general alarm must be provided and a public address system that permits aural communication, a talk-back facility, which is clearly audible at all locations on the installation.

Where the alarm and public address systems are located in noisy surroundings, a visual form of indication must be provided. Both systems must be supplied with two separate sources of electrical power, one of which is designed to function in an emergency situation.

Figure 3.32 – Alarm and public address system

3.3.1.3 Temporary refuge and means of escape

A temporary refuge or safe haven must be provided, where personnel can congregate should an emergency situation, involving fire and explosion or a release of gas, develop. Clear lines of access should be provided for personnel entering the refuge and for egress to the evacuation stations. Life support systems are to be provided that will protect the occupants for a period of time determined from the results of the fire and explosion analysis (FEA) and evacuation, escape and rescue analysis (EERA). The temporary refuge must provide a way of monitoring the emergency situation and facilities for communication with the emergency services.

There must be a suitable method for personnel to reach the helideck and the sea by at least two routes. Escape to the sea may be effected using fixed ladders, knotted ropes or scramble nets. Personnel descent devices, a form of abseiling equipment, may also be installed, often referred to as 'donuts'.

All escape routes are to be clearly signposted, kept clear of obstructions and provided with adequate emergency lighting arrangements.

3.3.2 Lifeboats

General requirements
Every normally manned installation must be provided with a totally enclosed motor propelled survival craft (TEMPSC), capable of containing all persons onboard (POB).

Figure 3.33 – Lifeboats

Clear instructions that will enable one person to launch and operate the lifeboat are to be provided. The lifeboat must be self-righting, constructed of fire retarding materials and protected against heat by an external sprinkler system. There must be sufficient fuel onboard to propel the boat for 12 hours and the engine to run for at least ten minutes, and a compressed air supply that enables the occupants to breathe. A radio that can transmit distress signals must also be included in the survival equipment.

Figure 3.34 – Gravity launched lifeboat

Testing and commissioning

Lifeboats and davits are to be tested in accordance with IMO Assembly Resolution A.689(17), Testing of Life-Saving Appliances. The requirements are quite onerous, particularly for new lifeboats and davits.

The lifeboat must be load tested by installing test weights that represent 1.1 × SWL (safe working load). The brakes should be capable of holding this load and they should be tested at some point while the load is lowered to the water. It must be confirmed that the lifeboat can be released from the falls when fully waterborne in the 110% overload condition, and in the light condition. While in the water, the action of the hydrostatic release mechanism can be confirmed (if fitted) and that there are three turns of wire remaining on the winch drums, allowances being made for the lowest expected tide.

On completion of load testing, the speed of lowering the lifeboat should be confirmed as being not less than 0.45 metres per second (90 feet per minute) and not more than 0.9 metres per second (180 feet per minute).

Davits, winches and loose gear

Lifeboat davits and winch structural members are to be designed with a factor of safety of 4.5 and on completion of manufacture should be subjected to a static proof load test of 2.2 × SWL. The SWL for each davit is half the weight of the lifeboat, fully loaded with equipment and weights representing the maximum number of people, based on a weight of 75 kilograms per person. Each winch should be capable of lowering and holding a test load of 1.5 × SWL.

Attachments that support the sheave blocks, the blocks themselves and the attachments of the rope terminations are to be tested to 2.5 × SWL. The breaking tensile strength of the wire rope falls, chains and links should be at least 6 × SWL. Each lifeboat release mechanism must be statically proof tested to 2.5 × SWL by an approved test house, and be permanently marked with:

- Manufacturer's name
- type and serial number
- date of manufacture
- SWL
- test certificate number.

Figure 3.35 – Lifeboat release hook and maintenance pennant

Once in service, the wire rope falls should be changed, end for end, at intervals of 30 months and should be replaced every five years unless they are made of stainless steel wire, in which case a longer interval may be agreed with the authorities. The davits should also be load tested every five years.

In service examination

Once installed, the lifeboats should be examined on a regular basis and this should include an in water test, preferably every three months but at least once a year. It is only by manoeuvring the lifeboat that the equipment can be fully tested and particular attention should be given to the lifeboat release mechanisms. Some are designed to release the boat from the wire rope falls only when it is off-load, ie when the boat is fully waterborne. Other systems are designed to release on-load, ie before the boat is fully buoyant. Some designs are launched from a cradle and free-fall into the sea. Only personnel fully trained in the operation of the release mechanisms should carry out these tests because there have been numerous fatalities over the years that were associated with the routine testing of lifeboats.

3.3.3 Liferafts

The requirements for liferaft depend on the capacity of the lifeboats. If one lifeboat can accommodate the entire POB then additional capacity equal to the POB must be provided by liferafts. The combined capacity of lifeboats and liferafts equals 2 × POB.

The alternative is to provide two lifeboats whose combined capacity equals 1.5 × POB, in which case no liferafts need be fitted. In practice, inflatable liferafts are positioned at strategic locations regardless of lifeboat capacity. These inflatable liferafts should be serviced annually. Liferafts are often referred to generically as RFDs, a trade name of one of the many manufacturers.

Figure 3.36(a) – Liferafts

The lanyard lengths should be checked on liferafts, along with the drop lengths of knotted ropes and scramble nets to ensure that they are adequate and will reach the water.

In general, liferafts are designed to withstand launching from a height of 18 metres above sea level and if they are to be stowed at a greater height than this then a stronger casing is required. The liferafts are to be secured to the vessel by a painter, the length of which is determined by taking the stowage height if the raft above sea level in the light ship condition, plus 10 meters. Unless the liferaft is designed to be davit launched, it should not weigh more than 185 kg.

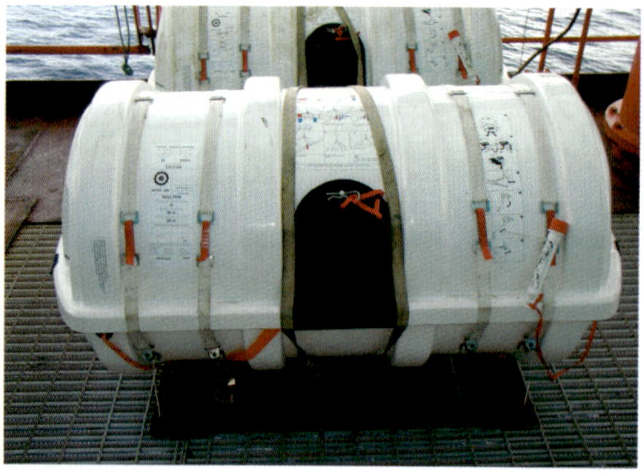

Figure 3.36(b) – Liferafts

All survival craft must contain a first aid kit, an adequate supply of drinking water and a waterproof electric hand lamp suitable for signalling purposes.

3.3.4 Descent Devices

At the end of the 1980s, most operators in the southern sector of the North Sea embarked on a programme to de-man the majority of their older installations and to ensure that, where practicable, all their new installations were designed for un-manned operation. In fact, un-manned installations, or not normally attended as they may be referred to, outnumber the manned installations by a ratio in the region of 10 to 1. Not all operators have elected to install lifeboats on their un-manned installations, mainly because of the high maintenance requirements of the boats and davits. Some have installed alternative arrangements, such as the Sea-scape system, while others have simply provided personnel descent devices and liferafts.

Figure 3.37 – Un-manned platform *'Tethys'*

The Sea-scape system is a net tower with liferafts attached that is lowered to the sea in the event of an emergency. Personnel enter the tower and descend through a number of 'floors', eventually ending up on a platform from which they can enter an inflatable liferaft.

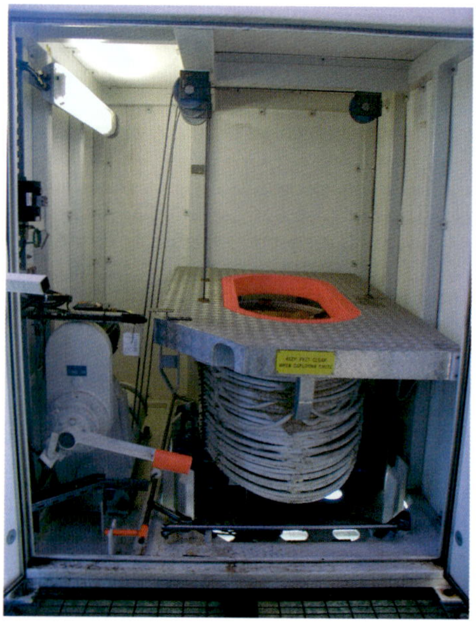

Figure 3.38 – Sea-scape system

The personal descent devices, or donuts, provide an alternative way of escape and may be supplied in addition to a Sea-scape system and lifeboats. They comprise a 'diaper' arrangement attached to an abseiling system. Once donned, the free end is made fast to the superstructure allowing the individual to descend in a controlled manner to the sea. They can either tie off above the waves and await the arrival of a fast rescue craft, or cut themselves free and enter one of the installation's inflatable liferafts. Regardless of which system is installed, a vertical ladder to sea level must always be provided.

Figure 3.39(a) – Personal descent device (DONUT)

Figure 3.39(b) – 'DONUT' system in use

3.3.5 Lifejackets and Personal Survival Equipment

On each installation there should be a quantity of lifejackets equal to 1.5 × the POB. One must be stored by each bed, with the remainder being located at embarkation or muster points. Donning instructions must be provided wherever the jackets are stored. Each jacket should be capable of turning the wearer so that they float, face up, within five seconds of entering the water, and subsequently keep them afloat for a period of 24 hours.

The jackets are to be manufactured from low flammability, rot proof, high visibility materials and should be provided with a non-metallic whistle and, preferably, a light. A suitable ring or loop should also be fitted to the jacket to facilitate rescue.

(a) Lifejacket

(b) Smoke hood, heat proof gloves and portable light source

Figure 3.40 – Lifejacket and personal survival equipment

In the UK, legislation requires the provision of personal protective equipment designed to increase the chances of survival of an individual in an emergency. The equipment should provide protection against the effects of fire, heat, smoke, fumes or toxic gas and immersion in the sea. A set of protective equipment should be located in the accommodation for every individual and should include a smoke hood, portable light source and heat proof gloves. A survival suit should also be provided in addition to a lifejacket.

3.3.6 Lifebuoys

The number of lifebuoys required will depend on the number of persons onboard, but at least eight must be positioned at locations readily available

Figure 3.41 – Lifebuoy

to assist a person who has fallen into the sea. The lifebuoy must be fitted with a light that will operate automatically when immersed for a period of at least 45 minutes. The light should have a luminosity of 2 candelas. Lights need not be fitted to lifebuoys that are installed in addition to the minimum requirement of eight.

Lifeboats and lifebuoys must be highly visible and bear the name of the installation.

3.3.7 Maintenance

In the UK, the maintenance of life-saving appliances is frequently delegated to specialist companies who are authorised by the equipment manufacturers to service their products. Similar practices are followed to those on ships with most items of equipment, including the inflatable liferafts being serviced annually.

Figure 3.42 – Liferaft maintenance

3.4 Navigational Aids – Fixed Offshore Installations

To provide protection against shipping, every offshore installation located within the territorial waters of the UK must be fitted with visual and audible navigational aids in accordance with the requirements of a schedule provided by the *Department of Transport* under the *Coast Protection Act* 1949 as amended by Coast Protection Act (CPA) 1949, section 34 (as extended by the *Continental Shelf Act* (CSA) 1964 section 4(1)).

3.4.1 Visual Navigation Aids

3.4.1.1 White lights

Figure 3.43 – White navigational lights

It is a requirement that a flashing white light can be seen, whatever direction an installation is approached from.

A light intensity of 12,000 candelas will give a range of approximately 15 miles. Should the bulb filament fail, a secondary bulb with a range of 10 miles should illuminate automatically and initiate a bulb failure alarm.

The lights, which are normally mains powered, should be provided with an independent secondary source of power (normally batteries) capable of ensuring uninterrupted operation for a period of four days (96 hours). In the event of a mains power failure, the change over to the secondary power supply should take place automatically, initiating an alarm at the main control station.

3.4.1.2 Red lights

Figure 3.44 – Red navigation light

The horizontal extremities of an installation should be illuminated with flashing red subsidiary lights.

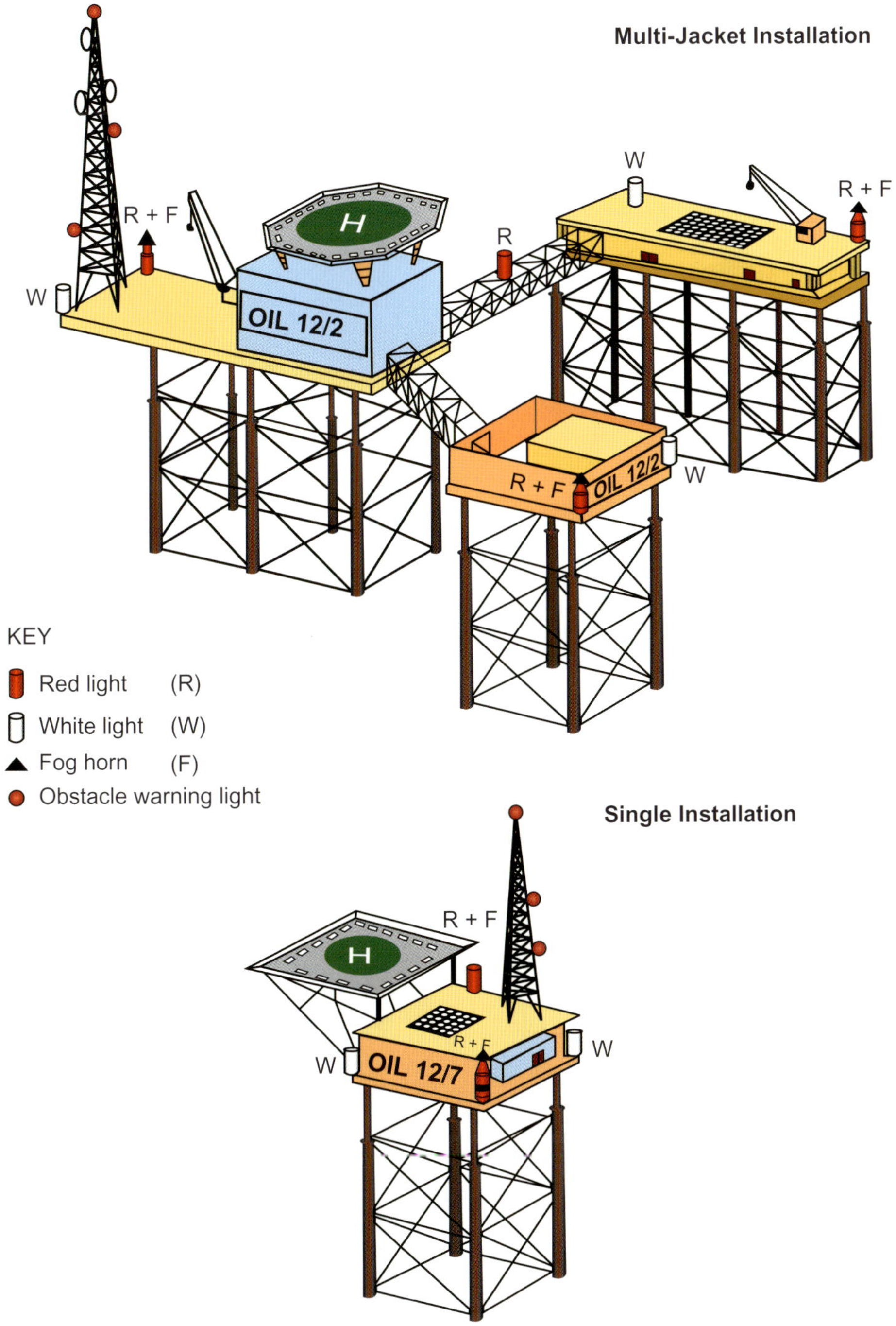

Figure 3.45 – Navigational aids

A red light need not be fitted to an extremity where a white light already exists.

A light intensity of 1,200 candelas will give a range of approximately three miles.

3.4.1.3 Lighting – general

The sketches in Figure 3.45 show typical arrangements. Two white and two red lights will usually be sufficient when two or more jackets are bridge linked and in line with each other.

The white lights must flash in unison. The red lights must also flash in unison, but they need not flash in unison with the white lights.

The lights should spell the letter 'U' in Morse Code every 15 seconds. They should be mounted at a position between 12 and 30 metres (40 and 100 feet) above sea level so that maximum visibility is obtained.

The lights should be exhibited from 15 minutes before sunset until sunrise, and at all times when meteorological visibility is less than two miles.

3.4.2 Audible Navigational Aids

Figure 3.46 – Audible navigational aid

Fog signals should be activated whenever meteorological visibility falls below four miles, and deactivated only after visibility improves to five miles.

The fog signal should spell the letter 'U' in Morse Code over a period of 30 seconds, and have a range of at least two miles.

The primary fog signal, which may be mains powered, should be provided with a secondary fog signal that will commence operation automatically on failure of the primary system and initiate an alarm at the main control station. The secondary system must have a range of at least half a mile.

The main and secondary fog signals should each be capable of operating continuously at full power for at least four days (96 hours), from a power source independent of the mains supply (normally batteries).

The secondary power supplies for navigation aids are to be located within compartments structurally protected against fire to at least A-60 requirements.

3.4.3 Navigation Aids Failure

The appropriate hydrographic office must be informed of any failure of an installation's aids to navigation if:

- It is likely that it will not be possible to show main or subsidiary navigation lights all round when visibility falls below two miles
- fog signals cannot be given
- any element of the aids to navigation is operating on stand-by because of a failure of primary systems.

If a total failure of the navigational aids occurs, the installation must be temporarily marked by one of two methods:

a) A temporary battery powered system of lights and fog signals must be installed that meets the requirements for light and fog secondary systems

or

b) two cardinal buoys are to be located on either side of the installation at one cable length from the installation.

The buoys must be provided with a light with an intensity of not less than 70 candelas and a fog horn consisting of a wave activated whistle or bell.

3.4.4 Helideck Illumination

The requirements for the illumination of helidecks are outlined in the *International Standards and Recommended Practices for Aerodromes, Volume II Heliports* which is issued by the *International Civil Aviation Authority*. These requirements are also detailed in the document CAP 437, *Offshore Helicopter Landing Areas – Guidance on Standards* issued by the *Civil Aviation Authority* in the UK.

Figure 3.47(a) – Yellow perimeter lights

Figure 3.47(b) – Green perimeter light

Figure 3.47(c) – Wave-off light

Perimeter lighting – The safe landing area must be delineated by green lights mounted around the perimeter of the helideck at intervals of no more than 3 metres. A minimum of four lights on each side of the helideck are to be provided and the lights are to have an intensity of at least 30 candelas. The lights are to be supplied from an uninterrupted power supply (UPS) to ensure that they are available even under emergency conditions. Older installations may have been fitted with orange or yellow lights but it is recommended that these are replaced.

Wave-off lights – Wave-off lights or system status lights are to be provided when a situation may arise that could jeopardise the safety of the helicopter and its occupants. The wave-off lights are normally provided to give warning of a gas release and they are to be activated automatically and supplied from an uninterrupted source of power so that they will function under emergency conditions. The wave-off lights are to be fitted in pairs, red in colour, have an intensity of 700 candelas and flash with a frequency of 120 flashes a minute. A facility should be provided so that the lights can be dimmed to 60 candelas whilst the helicopter is on deck.

Floodlights – Floodlighting is to be provided so that the whole of the landing area can be illuminated after dark. The lights must be carefully positioned so that they provide maximum illumination whilst not blinding the helicopter pilots and they should be subject to the control of the helicopter landing officer.

Obstacle lighting – All fixed obstacles that could present a hazard to helicopter operations such as crane booms and flare stacks are to be illuminated and painted if required so that they can be clearly seen from the air. If painted, an alternating black and white, black and yellow or red and white colour scheme should be selected, the coloured bands not being less than 0.5 metres and not more than 6 metres in width. The lights are to be omni-directional, red in colour and have an intensity of at least 10 candelas. Additionally, any object that exceeds the height of the helideck by more than 15 metres must be provided with red lights at 10 metre intervals. Furthermore, an omni-directional red light of an intensity of 25 to 200 candelas is to be fitted to the highest point on the installation. Where this is not practicable (eg on top of the flare) the light should be fitted as near to the extremity as possible. Like the helideck lights, the obstacle lights are to be powered via the UPS system.

3.4.5 Identification Panels

Identification panels should be fitted that display a registered name or unique number by which the installation is registered. Figures should be 1 metre (40 inches) high on a yellow background and be arranged so that one panel is visible from any direction.

The panels should be illuminated or constructed from retro-reflective materials.

The helideck must also be clearly identified with the installation name and number.

Figure 3.48 – *'Windermere'* gas platform

3.5 Hazardous Areas

3.5.1 Hazardous Areas and their Determination

The classification of hazardous areas was originally developed to provide guidelines for selecting electrical equipment in onshore oil refineries. It has since been adapted for similar application on offshore installations for both electrical and mechanical equipment. The hazardous and non-hazardous areas will be clearly marked on general arrangement drawings of the installation.

The definition of various zones is provided by the present IEC (International Electrotechnical Commission) Standard IEC 60079-10.

Non-hazardous or safe areas are defined as locations where explosive hydrocarbon mixtures are not expected to occur in quantities sufficiently large to present a hazard.

> It should be noted that it is necessary for all accommodation spaces to be located in a safe area.

3.5.1.1 Hazardous areas

A hazardous area is a location where flammable gas and air mixtures may occur in large enough quantities so that special precautions have to be taken in the selection, installation and use of electrical and mechanical equipment. Hazardous areas are generally subdivided into three categories.

A. Zone 0 is where an explosive atmosphere consisting of a mixture of air and flammable substances in the form of gas, vapour or mist is present continuously or for long periods.

   It is unlikely that a Zone 0 area would be encountered on an offshore platform. On an FPSO, the free space above the crude oil in the cargo tanks and slop tanks is normally classed as a Zone 0 area. All electrical equipment should be excluded, or at least be of an intrinsically safe design.

B. Zone 1 is where an explosive atmosphere consisting of a mixture of air and flammable substances in the form of gas, vapour or mist is likely to occur during normal operating conditions.

   These regions occur around flare booms and vent stacks. If a horizontal circle is drawn with its centre at the gas source and with a radius of 15 metres, all regions in this circle will typically be classed as Zone 1. The zone extends vertically upwards 7.5 metres and vertically downwards 9 metres, or to the nearest solid deck for gases that are heavier than air. While the primary constituent of natural gas is methane, which is lighter than air, it often contains heavier gas fractions such as butane and heptane. When large volumes

Figure 3.49 – Accommodation block and helipad on an oil platform

   of gas are vented (not flared), the extent of the hazardous area is established from gas dispersal calculations. When large volumes of gas are flared, the hazardous area boundaries in the flame out condition should be considered.

C. Zone 2 is where an explosive atmosphere, consisting of a mixture of air and flammable substances in the form of gas, vapour or mist, is unlikely to occur during normal operations. If it does, it will persist for only a short period of time.

   Hydrocarbon processing areas and the main deck above the cargo oil tanks on an FPSO up to 3.0 metres are normally classed as Zone 2 areas. While hydrocarbon releases during normal operating conditions are not expected, it is always possible for a flammable substance to be released from a pump shaft seal, a leaking valve stem or a flanged joint.

There are differences in the interpretation of area classification from country to country and company to company. For example, America still divides hazardous areas into two divisions. Division 1 is the more hazardous of the two divisions and includes Zone 1 and Zone 2 areas. Zone 2 and Division 2 areas are broadly similar. In Europe, Part 10 of the *'IEC 60079 Standard, Electrical apparatus for explosive gas atmospheres – Classification of hazardous areas'*, provides guidance on classifying hazardous areas, as does the Institute of Petroleum (IP) *'Model Code of Safe Practice, Part 15 - Area Classification Code for Installations Handling Flammable Fluids'*.

3.5.1.2 Determination of hazardous areas

Plant design

The design of the plant should keep Zone 0 and Zone 1 hazardous areas to a minimum. To reduce the extent of the hazardous areas, the design and operation of the plant should keep the amount of

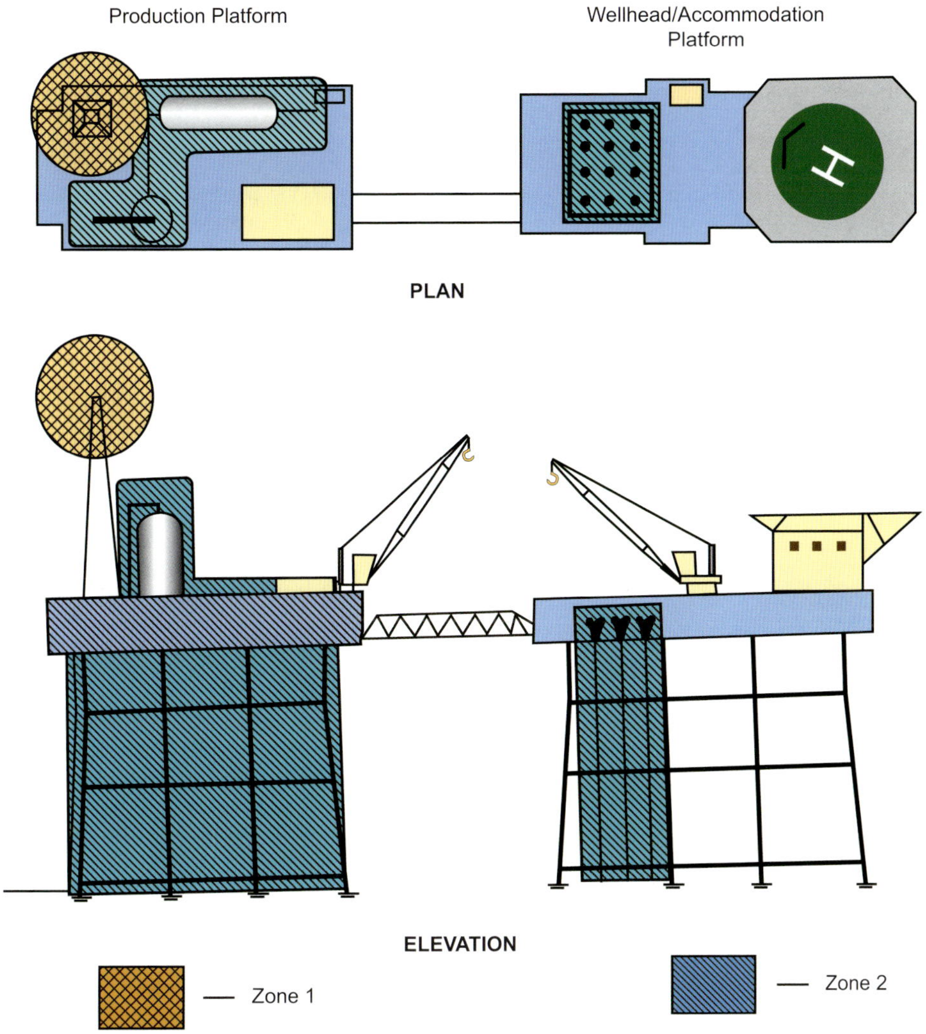

Figure 3.50 – Hazardous areas

flammable substances released into the atmosphere to a minimum even when operating abnormally. Ideally, the plant should comprise mainly Zone 2 and non-hazardous areas.

It is rarely possible to decide by simple examination which hazardous area designation should be allocated to a particular part of the plant. A detailed approach is normally required and should be carried out by a team of engineers who are familiar with the properties of the flammable materials concerned, the process and the equipment. The basic steps that should be carried out when establishing the type and extent of the various hazardous areas are:

- Identify the sources of release
- determine the grade of release
- calculate the extent of the release.

Sources of release
Each item of process equipment containing flammable materials must be considered as a potential source of release and, whether it is a tank, pump or pipeline, must be identified on a general arrangement drawing of the plant. Some items, such as the welded pipe spools, are not considered to be a potential source of release because, while they contain flammable materials, there are no joints that could leak. For items that may release flammable materials, the first task is to determine the grade of release.

Grades of release
There are three basic grades of release and it is the grade of release that provides an indication of the likely frequency and duration of the release.

1. **Continuous grade of release**
   This release is continuous or is expected to occur frequently or for long periods. The surface of a flammable liquid in a closed tank and a vent that is used continuously or frequently are two examples of a continuous grade of release. These examples are normally classed as Zone 0 hazardous areas.

2. **Primary grade of release**
   This can be expected to occur periodically during normal operations. Sources that could cause a primary grade of release are sample points, pig launcher doors, outlets from relief valves, vents and pumps, and compressors with standard seals. These locations would normally be classed as Zone 1 hazardous areas.

3. **Secondary grade of release**
   This is not expected to occur in normal operations and, if it does, is likely to be infrequent and for short periods. Typical examples are pumps and compressors with high integrity seals, valves, flanges and fittings containing flammable substances. These locations are normally classed as Zone 2 hazardous areas.

Once the source and grade of release are established, it is necessary to determine the release rate and consider any other factors that could influence the extent of the hazardous area zone.

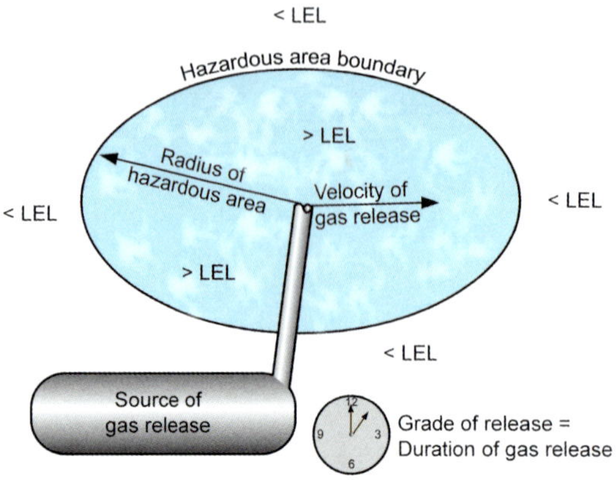

Figure 3.51 – Factors affecting determination of hazardous area boundaries

Release rate

The greater the release rate, the greater the extent of the zone will be. The release rate depends on a number of parameters (Figure 3.51):

- Geometry or source of release, related to the physical characteristics. For example, the release of flammable vapours from the surface of an open tank will be different to the release from a leaking pipe flange that is under pressure
- release velocity. The size of a cloud of flammable gas or vapour is determined by the rate of release and dispersion. Gas and vapour that flow from a leak at high velocity will develop a cone-shaped jet that will entrain air and be self-diluting. The extent of the explosive gas atmosphere will almost be independent of wind velocity. However, if the material is released at a low velocity or if its velocity is reduced by impinging on a solid object, it will be carried by the wind and its dilution and extent will then depend on the wind velocity.

Extent of the zone

The extent of the zone depends on the estimated distance that an explosive atmosphere covers before it disperses into the air at a concentration below its lower explosive limit (LEL). The extent of the zone is affected by the chemical and physical parameters of the release. Some are intrinsic properties of the flammable material, such as density of the gas, and others are specific to the process, eg the process pressure.

Gases that are lighter than air, such as methane, are those that have a relative density of less than 0.8 of the ambient air density, and they will disperse fairly rapidly upwards. Heavier than air gases, such as hydrogen sulphide (H_2S), have a relative density greater than 1.2 of the ambient air density and they may travel a considerable distance over solid decks before being safely dispersed by atmospheric conditions.

Effective ventilation reduces the extent of the zone while obstacles that hinder ventilation, particularly where ventilation is natural, will cause the boundaries of the hazardous area to increase. Natural ventilation in some areas of the world, such as the North Sea, can be considerable but in some tropical environments the ventilation may be virtually non-existent.

Documentation

When the type and extent of hazardous areas on a plant or offshore installation have been identified, they should be represented on hazardous area drawings. Both plan and elevation drawings of the installation will be required, drawn in sufficiently large scale to show all the main items of equipment and all the compartments. The positions of all openings, such as doors, windows and ventilation inlets and outlets, should be clearly shown and the various steps that were taken in arriving at the layout of the drawings should be documented. This documentation should include:

- Recommendations from relevant codes and standards
- gas and vapour dispersion calculations
- a study of the ventilation characteristics.

The documentation should provide the ignition temperatures of the fluids and gases that are likely to be encountered and their temperature class and gas group so that the electrical equipment can be correctly specified.

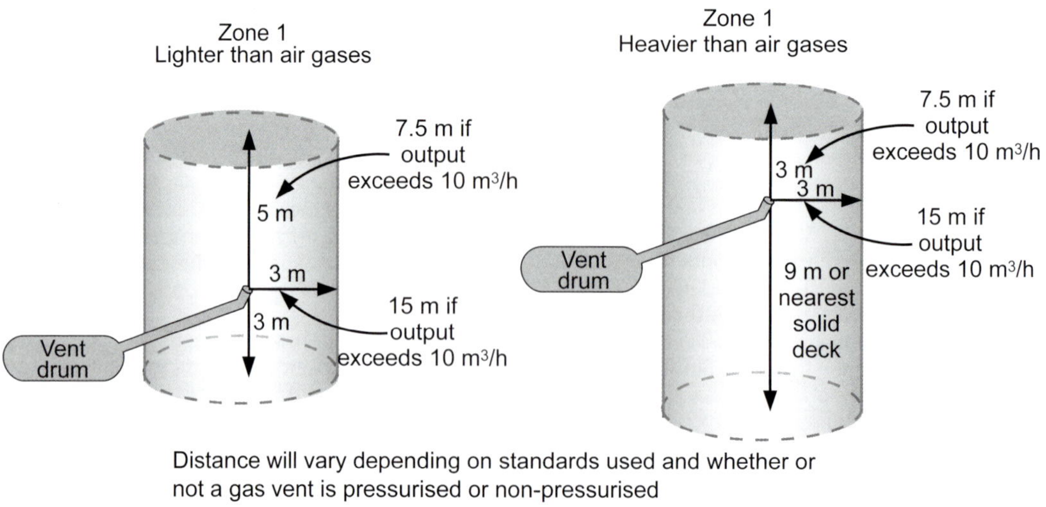

Figure 3.52 – Typical hazardous area boundaries for occasional gas venting

Examples

While the development of hazardous area drawings for a large plant is a complex process, there are some basic guidelines in the codes of practice that can be used to determine boundaries for the more basic items of process equipment and these should be taken into consideration.

Machinery

Items of equipment such as pumps and compressors fitted with high integrity seals, classed as sources of secondary grades of release, would result in a Zone 2 hazardous area. The hazardous area boundary covers an area of 3 metres horizontally and up to 5 metres vertically.

Figure 3.53 – Crude oil distribution pump

Vents

A non-pressurised vent would be classed as a Zone 0 hazardous area extending 3 metres, while a pressurised vent discharging gas in excess of 10 m^3/ hour would extend 15 metres.

Pig launcher

A pig launcher door, considered to be a primary source of release, would result in a Zone 1 hazardous area extending a distance of 3 metres, and occasionally it would be classed a Zone 2 area extending a further 2 metres beyond the Zone 1 area.

Figure 3.54 – Pig receiver door

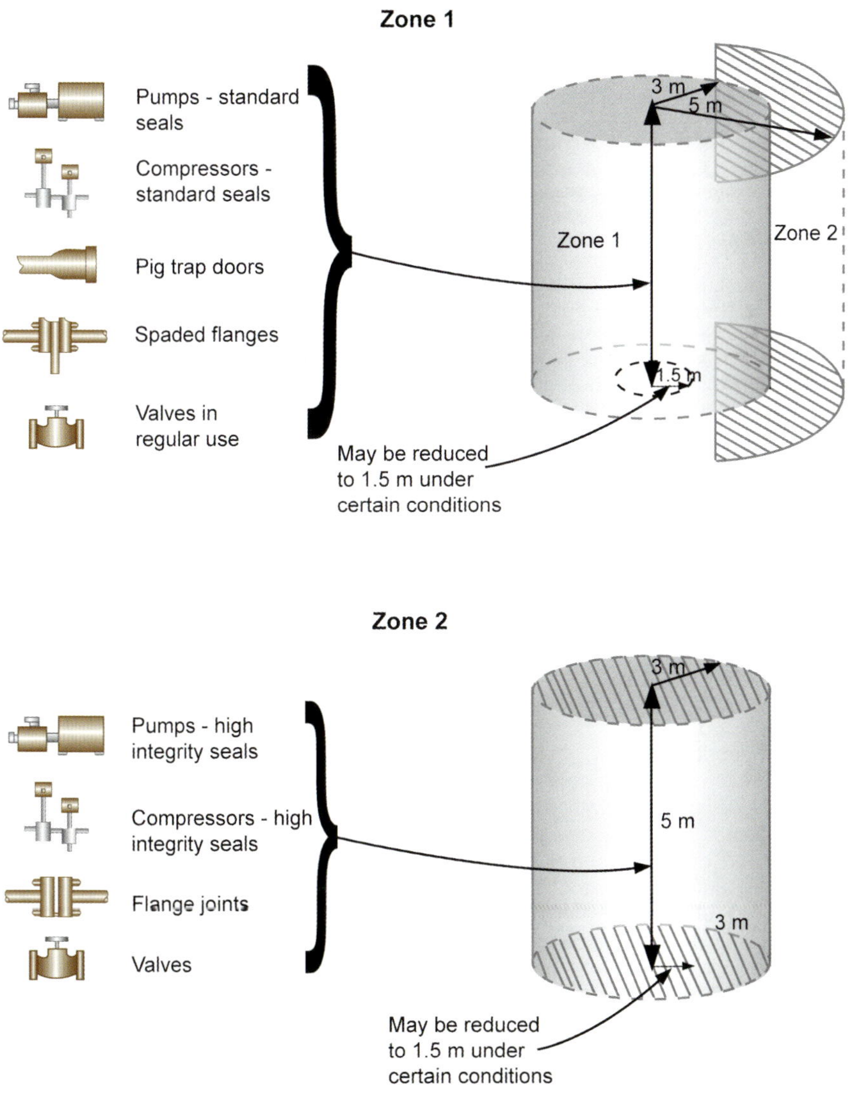

Figure 3.55 – Classification of equipment in Zone 1 and Zone 2 areas

Valves
Well-maintained and infrequently used valves are unlikely to leak, so they can be considered a secondary source of release. This would result in a Zone 2 hazardous area boundary set at 3 metres. Frequently used valves are more likely to leak so, as a primary source of release, a Zone 1 rating may be more appropriate.

Flanges
The majority of flanged joints are rarely broken and should be considered as secondary grades of release, ie Zone 2 with a radius of 3 metres. Flanges that are broken during normal operations, such as removing or replacing spades, should be regarded as primary grades of release, Zone 1 with a radius of 3 metres.

The radii quoted here may, in certain circumstances, be reduced to 1.5 metres. In the case of individual valves and flanges, the likelihood of release is so small that their location may not warrant being classified at all.

All distances quoted for hazardous area boundaries are based on open deck conditions that are considered to be adequately ventilated. This is when natural ventilation is sufficient to prevent significant quantities of flammable gas mixtures building up. Enclosed areas must be considered separately due to the greater possibility of a gas build-up occurring. The figures quoted are taken from, or correspond to, those in various national industrial codes and have been provided as a guide only. In all cases, the relevant code for the task being carried out should be referred to.

3.5.1.3 Enclosed and semi-enclosed spaces

This section is based on guidance from the Institute of Petroleum (IP) *'Model Code of Safe Practice Part 15 – Area Classification for Installations Handling Flammable Fluids'*. The IP15 code is a document that has been in use for over 30 years and has often been used as the basis for codes of practice recently issued.

Open space
This space is open on at least three sides, of which two are opposite sides and can be adequately ventilated by a natural means.

Semi-enclosed space
This is a space that, while open on two opposite sides, cannot be considered to be adequately ventilated due to deck obstructions that may slow down the dispersion of a gas release.

An enclosed space
This is any compartment or area that is enclosed on all or five sides. With no artificial ventilation, the air movement will be limited and a flammable atmosphere, if one occurs, will not be dispersed naturally. When there are potential sources that may release flammable vapours or gases into the space and natural ventilation is inadequate, mechanical ventilation must be provided.

An enclosed space or a semi-enclosed space can be considered to be an open space when it is mechanically ventilated, ie with a minimum of 12 air changes an hour, and when it is confirmed that there are no stagnant pockets of air in the space.

Any spaces containing a flammable source of release should be ventilated with at least 12 air changes an hour, while 6 air changes an hour will suffice for most other locations.

Adequate ventilation
This is either natural or mechanical ventilation that prevents the accumulation of significant quantities of gas and air mixtures, ie in concentrations above 20% of their lower explosive limit (LEL). This typically equates to at least 12 air changes an hour for 95% of the time. Mechanical ventilation systems should be designed with reliability in mind, for example including an automatic start-up of a standby ventilation fan if the primary fan fails. Power for the main and standby fans should not come from a common supply.

Ventilation sources
In principle, air supplies must be taken from and expelled to a safe area. They can, however, be expelled to a hazardous area, Zone 1 or Zone 2, so long as the ventilation dampers are of a gas-tight design and a spark arrestor is provided. All dampers must be provided with a facility for both local and remote closure.

Enclosed spaces
Two situations must be considered when discussing enclosed spaces, those that contain either a primary or secondary source of flammable release and are located in a safe area, and those that do not contain any source of flammable release and are located in a hazardous area. The former are referred to as enclosed hazardous spaces, either Zone 1 or Zone 2, and the latter as non-hazardous enclosed spaces.

Non-hazardous enclosed spaces
Non-hazardous enclosed spaces are spaces that do not contain a potential source of flammable release. On an FPSO, there may be a number of these compartments in hazardous areas. Many contain electrical control panels, switchboards, etc. Electrical equipment of this type cannot be constructed to meet the requirements of the hazardous area equipment codes, so precautions need to be set in place so they can function safely in what may become a hazardous environment.

Pressurisation
Pressurisation allows an enclosed space that does not contain any flammable sources of release to be located within a Zone 1 or 2 hazardous area and still be classed as a non-hazardous location, even if it contains uncertified electrical equipment. In all cases, the enclosed space must be maintained at a positive pressure of 25-50 Pa (5-10 mm water gauge) in respect of its surroundings. The ventilation system should also be designed with reliability in mind.

Figure 3.56 – UPS batteries

Note: Battery rooms are normally provided with 12 air changes per hour.

Figure 3.57 – Pressurisation gauge and pressure sensor

A major discussion point that relates to the design of pressurised compartments on offshore installations is the action to be taken when pressurisation is lost, ie the requirement to initiate loss of pressurisation alarms and isolate electrical equipment.

Loss of pressurisation – Zone 2 areas
Loss of pressurisation of a non-hazardous compartment in a Zone 2 hazardous area does not constitute a hazard. A hazard is realised when there is a flammable gas leak in an adjacent area, with the potential for gas to enter the compartment, and where there is a potential ignition source from uncertified electrical equipment. In view of this, it is not normally necessary to isolate uncertified electrical equipment on loss of pressurisation. An audio visual alarm should be given, indicating a loss of pressurisation at a manned control station, and a written procedure provided detailing the action to be taken to isolate uncertified electrical equipment if the situation persists. A master switch is normally available to provide electrical isolation when required.

When all the electrical equipment has been certified for hazardous area usage, there is no need to isolate it when pressurisation of the compartment is lost, or even to provide a loss of pressurisation alarm, although usually one is fitted.

Loss of pressurisation – Zone 1 areas
When pressurisation is lost in a non-hazardous compartment in a Zone 1 area, the loss of pressurisation alarm should initiate and all uncertified electrical equipment should be isolated immediately.

Compartments in a Hazardous Area

Scenario 1 – Pressurisation Considerations
Scenario 2 – Electrical switch room located in Zone 2 hazardous area contains uncertified items of electrical equipment. No sources of flammable release.

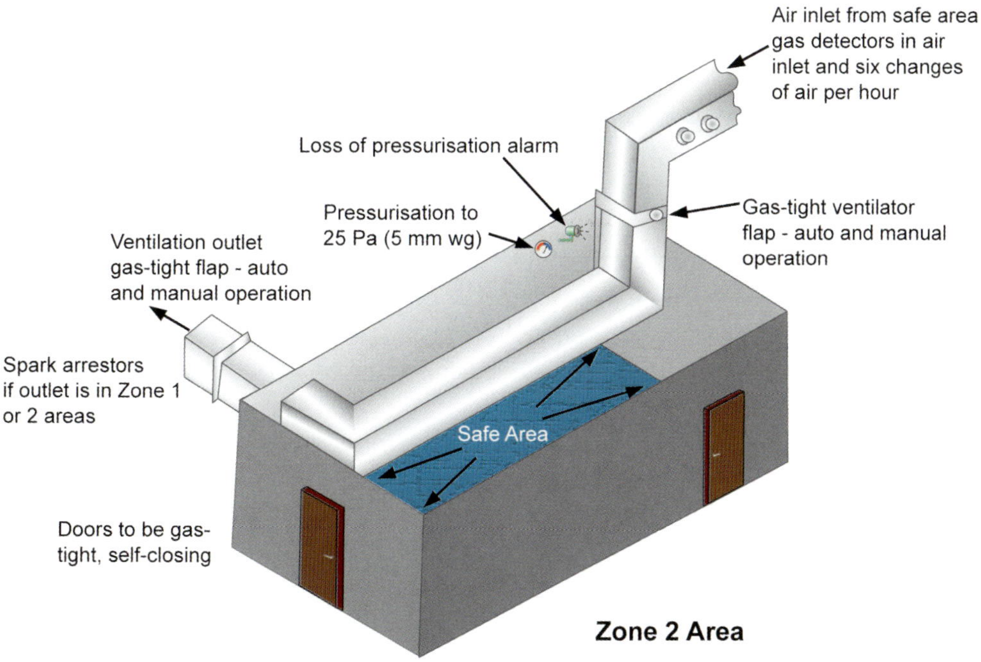

Hazard 1 – Loss of pressurisation
Action – Alarm only + procedure for dealing with this event + provision for manual/electrical isolation
Hazard 2 – Confirmed gas in air inlet ducting
Action 1 – Shut ventilation dampers and stop fans or put the system on recirculation
Action 2 – Isolate uncertified electrical equipment

Note: It would be most unusual to site such a compartment in a Zone 1 area but it is quite acceptable to do this once the following modifications have been carried out:

1. Doors – air locks to be provided
2. Pressurisation – automatic isolation of non-certified electrical equipment.

Figure 3.58 – Compartments in a hazardous area

This is primarily due to a Zone 1 area being defined as a location where an explosive gas and air mixture is likely to occur during normal operating conditions and, as a result, there is risk of gas entering the compartment and being ignited by uncertified electrical equipment. If all the electrical equipment has been certified for hazardous area usage, it is not necessary to isolate it on loss of pressurisation, but a loss of pressurisation alarm should still be provided. Pressurisation should be restored as soon as possible to prevent a build-up of gas within the compartment which effectively has now become an unventilated enclosed space.

Gas detection

Gas detection must be provided in the ventilation air intakes of all non-hazardous spaces in hazardous areas. On sensing gas, these should initiate a shut down of the ventilation system, fans and dampers and isolate all electrical equipment not certified for a Zone 1 application (on loss of ventilation, an enclosed area located in a Zone 2 area is classed as a Zone 1 area).

(a) **Gas detection, air inlet**

(b) **Gas detector close up**

Figure 3.59 – Gas detection at an air inlet

3.5.1.4 Hazardous enclosed spaces located in safe areas

Occasionally, it will be necessary to locate a hazardous enclosed space, such as a gas turbine powered generator, in a safe area. This is perfectly acceptable so long as adequate ventilation is provided.

Adequate ventilation ensures that the enclosure can be properly classified as a Zone 2 hazardous area. A mechanical ventilation system must be provided to maintain the space at a negative pressure of 50 N/m^2 (5 mm WG) in respect of its surroundings to prevent a gas leak from the compartment reaching the surroundings. In addition to the above design features, for a reliable ventilation system and to ensure that an under-pressure is maintained at all times, the supply fans should be interlocked so they cannot run unless the exhaust fan is running. One of the standby exhaust fans should be fed from the emergency supply. The ventilation outlets should lead to a location where a gas leak can be vented safely and gas detection should be provided in the outlet's ducts.

Figure 3.60 – Gas detector, turbine enclosure

All electrical equipment in the compartment should be suitably certified for use in a Zone 2 hazardous area as a minimum. It should be noted that, if ventilation is lost, the enclosure will be classed as a Zone 1 hazardous area. Most electrical equipment on an offshore installation is rated as either 'Ex d' or 'Ex e', both of which are suitable for a Zone 1 location. Either an audible or visual alarm system should be provided to warn when the ventilation system is not functioning correctly and pressurisation has been lost.

3.5.1.5 Hazardous enclosed spaces located in hazardous areas

An enclosed space containing a potential source of a flammable leak would normally be categorised as a Zone 2 hazardous area. If the enclosed space is located in a hazardous Zone 2 area, it must be classed as a Zone 1 area. Additionally, the surrounding area for 7.5 metres will also be considered to be a Zone 1 area.

Unless suitable ventilation and pressurisation systems are provided, an enclosed area in a Zone 2 or 1 area must be classed as a Zone 1 area, even if it does not contain a source of flammable release. This is due to the possible build-up of gas that could occur in the compartment in the event of a localised gas leak.

3.5.1.6 Access to hazardous areas from enclosed and semi-enclosed non-hazardous areas

It is not advisable to have access doors between the non-hazardous and hazardous areas and between Zone 1 and Zone 2 areas. Where there is a door, the spaces either side of the door are considered as Zone 1 and the most onerous area classification is to be adopted. By taking certain precautions, doors may be provided, enabling the original hazardous area ratings to be maintained. A description of these precautions follows.

Safe area to Zone 2 area
Enclosed non-hazardous spaces providing access to a Zone 2 area may retain their non-hazardous area status as long as:

- The access doors are a self-closing gas-tight design with no hold back arrangements
- the non-hazardous space is maintained at a positive pressure with respect to the hazardous area
- a loss of pressurisation alarm is provided, repeating in a manned control station.

Safe area to Zone 1 area
Enclosed non-hazardous spaces providing access to a Zone 1 area may retain their non-hazardous area status provided:

- The access arrangements are through a mechanically ventilated airlock fitted with two self-closing gas-tight doors with no hold back arrangements. (If one of the doors needs to be weathertight and the self-closing arrangement is considered to be impractical due to this, the self-closing device may be dispensed with provided that an appropriate notice is posted on the door)
- the non-hazardous space is maintained at a positive pressure with respect to the hazardous area
- a loss of pressurisation alarm is provided, repeating in a manned control station.

If an airlock is not provided but all the remaining safeguards listed above are met, the enclosed space is regarded as a Zone 2 area rather than a non-hazardous area.

Note: All doors leading from a safe area to a hazardous area must bear a sign that reads 'HAZARDOUS AREA' in letters approximately 50 mm (2 inches) in height.

3.5.2 Ventilation Systems and Machinery in Hazardous Areas

3.5.2.1 Ventilation systems

The ventilation inlets and exhaust must be suitably positioned to prevent the ingress of rain, snow and seawater. Fans with non-overloading, non-stall characteristics should be fitted.

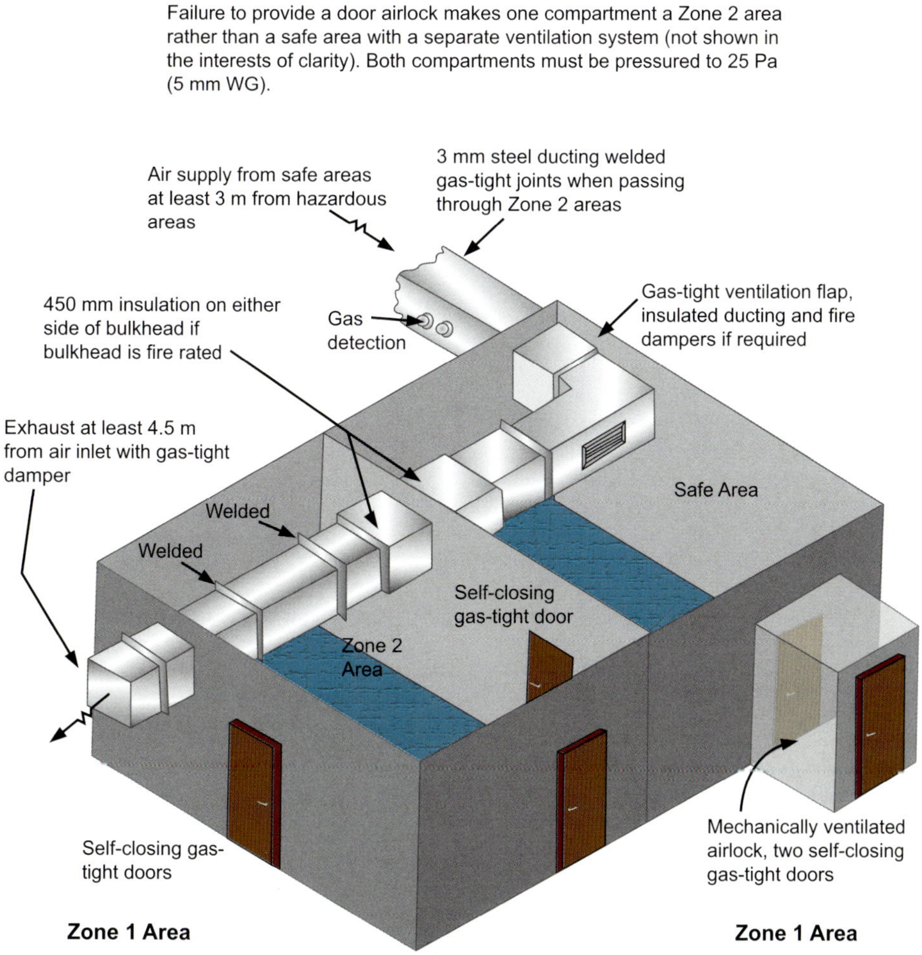

Figure 3.61 – Access to hazardous areas from enclosed and semi-enclosed non-hazardous areas

The ventilation ducting requirements are straightforward and some of the key points follow:

- Air inlets must be located in non-hazardous areas and at least three metres from hazardous area boundaries. It is recommended that there is a distance of at least 4.5 metres between the inlet and exhaust vents minimising the risk of cross-contamination
- ventilation systems and ducting for hazardous area spaces and non-hazardous areas must be completely separate. Ducting for non-hazardous areas should not be routed through hazardous areas. If this is unavoidable, the trunking should be a gas-tight welded design and be maintained at a pressure in excess of the hazardous area compartment it passes through
- if electric heating elements are required, they must be fitted with an automatic temperature control device with an independent sensor, cut-out and manual reset. The maximum surface temperature of the heating element should be less than the temperature of any flammable gas that may be encountered and less than 200°C
- gas turbine exhausts are to be routed well clear of the structure and positioned so that helicopter operations are not affected
- when air is supplied to a compartment for combustion purposes or for cooling the engines, it must be via a system separate from the compartment ventilation system.

Ventilation systems fitted on FPSOs must take into account the fact that they may weathervane.

3.5.2.2 Ventilation of gas turbine powered enclosures

The turbine hall must be designed as a non-hazardous space, mechanically ventilated with at least 12 changes of air per hour. An over-pressure of at least 50 Pa should be maintained in relation to the turbine hood and any adjacent area or space with openings to the turbine room.

The turbine hood must be mechanically ventilated with at least 12 changes of air per hour and should be maintained at an under-pressure of 50 Pa in relation to the turbine hall. A standby extraction fan must be provided, that starts automatically on failure of the duty fan and initiates an alarm signal in the main control room. Hood ventilation fans must be interlocked with the turbine starting arrangements so at least five air changes per hour are carried out to purge the turbine hood before any electrical equipment is energised. When the turbine is shut down, the duty ventilation fan should continue to run so the hood is purged until the turbine comes to rest. At least one ventilation fan must be supplied from an emergency power source.

When the turbine hood is ventilated at 12 air changes per hour, it will be regarded as a Zone 2 area while the turbine is running. This is conditional on no surface temperature in the hood exceeding, 200°C or 80% of the auto-ignition temperature of the fuel gas or diesel fuel. If these temperatures are exceeded, the air flow through the hood should be increased to 90 air changes an hour. Under these conditions, the hood does not need to be classified as a Zone 2 area. If any equipment in the turbine hood remains energised after the turbine shutdown or if the hood ventilation fails, it is to be rated for a Zone 1 hazardous area.

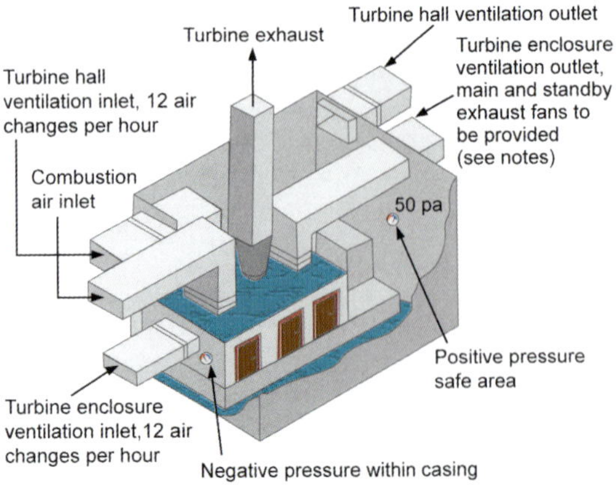

(a) Turbine hall and turbine enclosure ventilation arrangement

(b) Gas turbine (fuel rail clearly visible)

Figure 3.62 – Turbine hall ventilation

3.5.2.3 Machinery in hazardous areas

Machinery should not be installed in hazardous areas but, if it is, it should be constructed to reduce the risk of sparking due to friction between moving parts, static electricity forming or the generation of hot spots such as exhausts that may become a source of ignition. Air compressors should not be installed in hazardous areas but, if they are, the air inlets should be located in a safe area and fitted with a gas detection arrangement.

Diesel engines in hazardous areas

Diesel engines must not be installed in Zone 0 or Zone 1 hazardous areas. While they may be installed in Zone 2 areas, they must comply with additional safety regulations set out in the National Standards, Government Regulations or Codes of Practice relevant to the position of the installation. Additional safety fittings are likely to include, as a minimum:

1. A shut-off in the air intake system that can be operated both manually and automatically from an engine over-speed signal. It is also recommended to activate the shut-off valve when gas is detected in the compartment.
2. An approved corrosion resistant flame arrestor in both the air intake and exhaust system.

3. An approved spark arrestor in the exhaust system downstream of the flame arrester.
4. For engines over 370 kW, prototype explosion tests to a recognised standard must be carried out on the air intake and exhaust systems. A hydraulic proof pressure test should be carried out on all production systems, at a pressure of 5 bar or 1.5 times the maximum pressure obtained in the explosion test.
5. The engine breather must be led to a safe area and should not be routed into the engine's air intake.
6. The engine exhaust should not have a temperature higher than 200°C when on full load. A water cooled exhaust manifold may have to be installed to achieve this temperature.
7. Ventilation and radiator fan blades and drive belts must be an anti-static type and the combination of fan blade and protective housing or guard must be non-sparking.
8. The engine starting system should not be able to introduce a source of ignition externally to the engine. Air or hydraulic starting systems are preferable to electric systems.
9. The engine should not be able to run in reverse.
10. A remote emergency stop button should be provided.
11. An engine fuel shut-off valve should be fitted. It should be activated if any of the following occurs:
    - Engine over-speed
    - high exhaust temperature
    - high engine cooling water temperature
    - low lubricating oil pressure.

The alternative and more effective arrangement to a fuel shut-off valve is the activation and closure of the air intake device.

The above guidance applies to engines located in open deck, Zone 2 hazardous areas. When they are installed in an enclosed hazardous area, the compartment they are housed in should have at least 20 air changes an hour when the engine is running and 12 air changes an hour when it is not. Fire and gas detection sensors should also be provided.

3.5.3 Hazardous Area Equipment

3.5.3.1 Selection of equipment

The equipment should comply with the local government Administration or Flag State regulations.

The selection of equipment used in hazardous areas is governed by the hazardous area zone where it will be located and the gases that it may come into contact with.

Apparatus that is to be used in hazardous areas should be selected according to the following criteria:

- Classification of the area
- temperature classification
- apparatus group
- environmental conditions.

A typical equipment certificate number would be Ex e IIB T4(IP 66). The build-up of this code can be explained as follows:

- Ex e refers to the type of protection selected
- IIB to the suitability for particular gas groups
- T4 refers to the surface temperature rating of the equipment
- IP 66 to the protection provided by the enclosure against the ingress of solids and water.

3.5.3.2 Equipment protection

The generic term for all methods of hazardous area protection used in Europe is 'flameproof', while American practice is to use the term 'explosion-proof'. They are basically the same thing.

Selection of Apparatus and Systems According to Hazardous Area Zone		
Zone	Type of Protection	Marking
0	Intrinsically safe apparatus or system	Ex ia
	Special protection (specifically certified for use in Zone)	Ex s
1	Flammable enclosure	Ex d
	Intrinsically safe apparatus or system	Ex ib
	Pressurisation, continuous dilution and pressurised room	Ex p
	Increased safety	Ex e
	Special protection	Ex s
	Any equipment suitable for Zone 0	
2	Non-incendive	Ex n
	Oil immersion	Ex o
	Sand filling	Ex q
	Any equipment suitable for Zone 0 and Zone 1	

Table 3.1 – Selection of apparatus and systems according to hazardous area zone

Compliance with one or more of the CENELEC (European Committee for Electrotechnical Standardization) standard types of protection is represented by the symbol EEx. In both cases, the symbol is accompanied by one of nine lower case letters indicating the type of protection given, the definitions of which are provided in BS EN 60079-14 – *'Explosive atmospheres. Electrical installations design, selection and erection'*.

Of the nine protection categories, the most frequently encountered are Ex d (flameproof), Ex e (increased safety) and Ex i (intrinsically safe).

Flameproof
The principle under which flameproof Ex d equipment operates is to contain the explosion, if one occurs within the equipment, preventing the fire from spreading elsewhere. This involves constructing extremely robust equipment that tends to be heavy and expensive. There is a limit to the size of what can be produced to contain an explosion and the limit for Ex d motors is 500 kW. This is the favoured method of protection in the UK, while in Europe and the USA Ex e equipment is preferred.

(a) Ex d circuit breaker (note robust explosion-proof construction)

(b) Explosion-proof junction box

(c) Ex d II B T5 explosion-proof junction box

Figure 3.63 (a-c) – Explosion-proof junction boxes

Increased safety

The increased safety Ex e equipment is designed so that the tendency for arcing and sparking is reduced. Electric machines that use brushes are deemed as sparking devices and are excluded from the e type concept. Preventing arcing and sparking removes the potential ignition source and hazard. This is achieved by improved design measures, such as using a higher specification of insulation and improved security of the electrical terminals.

Intrinsically safe

Intrinsically safe Ex i equipment is designed so that, should a fault condition arise, an incendive spark will not occur. They are essentially lower power devices.

It should be noted that no Ex rated equipment is gas-tight or relies on being gas-tight as a way of providing protection. The only equipment that relies on preventing gas entering is the pressurised equipment Ex p. Gas is prevented from entering by keeping the component or compartment at a positive pressure.

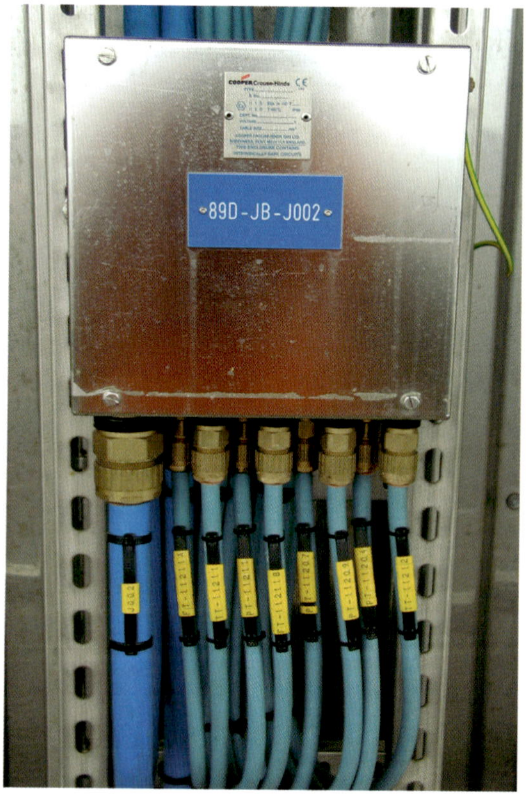

Figure 3.64 – Ex e – Increased safety junction box

Compound	Gas Group	State at N.T.P	Density (Air = 1)	Flammable Limits Lower %	Upper %	Auto Ign °C	Temp Class
Methane	IIA	Gas	0.55	5.0	150	535	T1
Ethane	IIA	Gas	1.04	3.0	125	515	T1
Propane	IIA	Gas	1.56	2.1	9.5	466	T1
Butane	IIA	Gas	2.05	1.5	8.5	265	T2
Pentane	IIA	Liquid	2.48	1.4	8.0	285	T3
Hexane	IIA	Liquid	2.79	1.1	7.5	233	T3
Heptane	IIA	Liquid	3.46	1.1	6.7	215	T3
Kerosene	IIA	Liquid	4.50	0.7	5.0	210	T3
Hydrogen	IIC	Gas	0.07	4.7	74.2	560	T1
Ethylene	IIB	Gas	1.00	2.7	34.0	425	T2
Hydrogen sulphide	IIB	Gas	1.10	4.3	45.5	260	T3

Table 3.2 – Data on typical flammable products

Air is taken from a safe area and, as a result, is protected. Uncertified electrical equipment can be utilised. Pressurised control rooms are used largely with suitable alarms, providing a warning and electrical power isolation if pressurisation is lost.

Apparatus groups
The appropriate equipment group must be determined for all flammable liquids, gases and vapours when selecting equipment for hazardous area usage.

Gas groups
There are two basic apparatus groups, the Roman numeral I being allocated to mining activity, where the biggest risk is from methane and coal dust, usually called firedamp.

All surface industry equipment, including equipment used offshore, is marked with the Roman numeral II and sub-divided into gas groups IIA (propane), IIB (ethylene) and IIC (hydrogen). The gases shown in brackets are a representation of typical gases. An Ex e II rating without reference to the letters A, B or C means that the equipment is suitable for all gas groups.

The gas groupings consider only the spark ignition and flame propagation aspect of the explosion protection, ie the amount of energy or the size of spark needed to ignite a gas and for the flames to spread. It should be noted that there is no correlation between the classification of gases for ignition energy and ignition temperature. For example, hydrogen, which is classified as Group IIC (ignition energy 19 µJ), has an ignition temperature of 560°C while acetaldehyde is classified as Group IIA (ignition energy approximately 150 µJ) and has an ignition temperature of 140°C. Most petroleum products are in Group IIA, but where rechargeable batteries are stored, if hydrogen gas can be discharged, it should be considered whether to specify the equipment as Group IIC.

BS EN 60079-20 provides the temperature classification of the more common industrial gases and is the most recent authoritative list.

Combustible liquid groups
Combustible liquids must be assessed in a similar way to combustible gases and categorised accordingly. Combustible liquids found on an installation or FPSO in any significant volumes will be, in the main, crude oil, diesel oil, kerosene and methanol. Electrical equipment located in the immediate vicinity of these liquids should be certified in accordance with the requirements of gas group IIA and temperature rating T3. Some of these liquids, in particular crude oil, will generate hazardous areas, others may not. The American Petroleum Institute (API) and the Institute of Petroleum (IP) in the UK both carry out tests on liquids to determine whether they are able to generate a hazardous area and categorise them accordingly.

The API RP 505 'Recommended Practice for Classification of Locations for Electrical Installations at Petroleum Facilities Classified as Class 1, Zone 0, Zone 1, and Zone 2', provides definitions for the categorisation of combustible liquids that have been tested in accordance with test procedures outlined in NFPA 30. A similar exercise was carried out by the IP and is outlined in their publication, *'Area Classification Code for Installations Handling Flammable Fluids'*. The coding system is generally similar and gives an indication of how the various hydrocarbon liquids should be considered in respect of hazardous area requirements.

Class I Liquids
There are liquids that have a closed-cup flashpoint below 37.8°C (100°F) and a vapour pressure not exceeding 276 kPa (40 psig) at 37.8°C (100°F). Class I liquids are usually handled at temperatures above their flashpoint, ie they will always produce a vapour in the flammable range in air even at temperatures far below the ambient. As a result, areas handling these liquids must be classed as hazardous. Gasoline, crude oil and methanol are examples of typical Class I liquids.

Class II Liquids
These have flashpoints at or above 37.8°C (100°F) and below 60°C (140°F). Kerosene and diesel oil are typical Class II liquids.

Class IIIA Liquids
These have flashpoints at or above 60°C (140°F) and below 93°C (200°F).

Class IIIB Liquids
These have flashpoints at or above 93°C (200°F).

Establishing a specific classification for crude oils in respect of liquid classes is difficult as they are made up of a mixture of hydrocarbons of widely varying composition. For instance, some crude oils may include volatiles such as butane and propane. Crude oil is usually classified as a Class I flammable liquid, its flashpoint generally being between –6.7 to 32.2°C (20 to 90°F).

One point that must be considered when determining hazardous areas in respect of flammable liquids is what happens to them when they are pumped under pressure. When flammable liquids are released in a spray form or mist, which may occur if there is a leak in a piping system, they will often ignite at temperatures far below their flashpoint. When a Class II or III liquid cannot be raised to its flashpoint, or be released in a flammable mist or spray, then it does not need to be considered as causing a hazardous area as the probability of a flammable vapour occurring in the ambient temperature range is unlikely. However, if it is possible for a flammable mist or spray to form, then for the purposes of determining hazardous areas, the liquid concerned should be considered as if it were in the next highest category, which may mean that it will be considered as a hazardous area. In these cases, it must be confirmed that the surroundings cannot raise the liquid's temperature if a leak occurs, eg if the liquid comes into contact with a hot surface or another source of ignition.

The ambient conditions must always be considered. For example, in mainland UK, a maximum ambient temperature of 30°C can be assumed whereas a temperature of 24°C is typically used for platforms in the North Sea. Elsewhere in the world, the temperatures are likely to be much higher. Kerosene used to re-fuel helicopters has a flashpoint of 38°C. This is not a problem in Europe but care should be taken in tropical locations.

Temperature classification

T Class	Temp °C
T-1	450
T-2	300
T-3	200
T-4	135
T-5	100
T-6	85

Table 3.3 – Relationship between T Class and maximum surface temperature

Gas and air mixtures can be ignited by coming into contact with hot surfaces. This must be taken into consideration when selecting equipment. The maximum surface temperature that can be generated by the apparatus must not exceed the ignition temperature of the gases or vapours that may come into contact with it. As a result, all electrical equipment used in hazardous atmospheres is classified according to the maximum surface temperature reached during prototype tests. The tests are carried out in an ambient temperature of 40°C. If the equipment is to be used at higher temperatures than this, the temperature classification should be re-assessed.

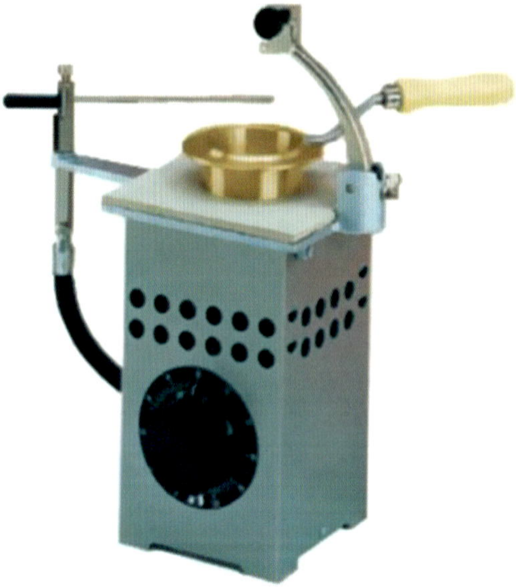

Figure 3.65 – Lighted taper used to determine flashpoint

It is important to note that flashpoint and ignition temperature are not the same thing. Flashpoint is the temperature at which a liquid gives off sufficient vapour for ignition to occur when it comes into contact with a naked flame. It is significantly lower than the ignition temperature (often referred to as the auto-ignition temperature). This is the temperature where ignition will occur without an ignition source being present. For example, kerosene has a flashpoint of 38°C and an ignition temperature of 210°C, while methane gas has an ignition temperature of 538°C.

It is also worth noting that some highly flammable gases have relatively high ignition temperatures. Methane (the main constituent of natural gas) is 538°C and hydrogen is 560°C. As a result, T-1 rated equipment with a permissible surface temperature of up to 450°C may be used with these gases. Hydrogen sulphide (a constituent of sour gas) has an ignition temperature of 260°C, so T-3 certified equipment must be used. This has a maximum permissible surface temperature of 200°C.

The temperature classification awarded to simple electrical apparatus and components will usually be T-4. This means the maximum surface temperature will not exceed 135°C. Junction boxes and switches may be certified with a T-6 designation as they do not contain heat dissipating components, will not get hot and their maximum surface temperature will not exceed 85°C. Because of this, they are suitable for most gases.

Ingress protection (IP)

0	No protection	0	No protection
1	Objects > 50 mm	1	Vertically dripping water
2	Objects > 12 mm	2	75-90° angled dripping water
3	Objects > 2.5 mm	3	Sprayed water
4	Objects > 1.0 mm	4	Splashed water
5	Dust protected	5	Water jets
6	Dust light	6	Heavy seas
		7	Effect of immersion
		8	Indefinite immersion

Table 3.4 – Ingress Protection (IP) against contact and ingress of foreign bodies and water

Electrical apparatus should be constructed to provide a degree of protection against both electrical and mechanical failure. Protection should be provided against the effects of the elements, heat from the adjacent plant and solvents.

A coded classification scheme has been developed indicating the degree of protection given by an electrical enclosure against liquids and solid materials entering. The classification is supplementary, not an alternative, to the types of protection that are necessary to ensure protection in explosive gas and air mixtures, and should not be confused with explosion protection designations.

A two number coding system is used. The first number designates the degree of protection given to stop solid materials entering the system, and the second number designates the degree of protection provided to stop liquids entering, as shown in Table 3.4.

The IP rating will not appear on all equipment as it is not specifically relevant to hazardous area equipment. It appears on some equipment and only if it exceeds the minimum requirement of IP 56. It is found extensively on junction boxes.

Example of Apparatus Marking

Typically, electrical equipment is rated as follows:

eg – Ex d II B T-3

Ex d Flameproof equipment (suitable for use in Zones 1 and 2)

IIB Surface/offshore equipment complying with ethylene gas groups

T-3 Has a maximum surface temperature of 200°C (suitable for all gases that have an ignition temperature of more than 200°C).

3.5.3.3 Certification and marking

Marking
All electrical equipment installed in hazardous areas must be certified as suitable for the location. The equipment must be clearly marked so the user can assess the suitability of the equipment for its intended location. There have been many changes to the marking requirements of the various types of explosion protected apparatus, but in Europe the markings are now standardised and can be found in BS EN 60079-0.

All equipment must have a plate that carries the following information:

- Manufacturer's name or trademark
- equipment type
- identification of the type of protection, eg 'e', 'd', 'p', etc
- name or symbol of the test house and the certificate reference. This includes the EEx mark. When either a Certificate of Conformity to one of the harmonised standards listed in the EEC Directive 79/196/EEC has been provided, or an Inspection Certificate has been issued by an approved body recognised by the EEC, the Distinctive Community Mark may also be affixed
- apparatus group
- temperature T class or maximum surface temperature
- any other relevant information.

Certification
All hazardous area equipment must be accompanied by a manufacturer's certificate that refers to the type of approval tests of the component concerned and the prototype test certificate number, which will have been issued by an approved test house following the type approval tests.

A certificate reference number and the manufacturer's name must be on the equipment, tying it back to the approval certificate.

When conformity with EEC Directives is required, certification must be carried out by an approved body listed by the Commission.

3.5.3.4 Commissioning, inspection, maintenance, testing and repair of electrical equipment

General
Electrical installations in hazardous areas have features that allow them to operate safely in these atmospheres. It is essential that, during installation and throughout the life of the installation, the integrity of the special features is preserved.

The Installation, inspection and maintenance of electrical hazardous area equipment should only be carried out by experienced personnel who are trained in the various types of protection and installation practices and who are familiar with the rules, regulations and general principles of hazardous area classification. Evidence of the experience and training of these individuals should be available.

Pre-Commissioning
BS EN 60079-17 requires that all electrical apparatus, systems and installations located in hazardous areas should be inspected before commissioning to confirm that the selection and installation of the equipment is appropriate for the location and is in accordance with the specifications. The standard provides useful checklists covering the scope of these inspections and checklists should be maintained as a permanent record of the inspections.

Testing
All electrical installations should be tested before commissioning and these tests should include the following:

- Insulation resistance measurement
- earth electrode resistance measurement
- earth loop impedance measurement
- setting and, where appropriate, operation of protective devices.

The test results should be recorded and included in the record of inspections.

Maintenance and inspection
To maintain the hazardous area protection provided by a piece of equipment, it must be effectively inspected, maintained and tested. Correct functional operation does prove that the equipment is in good protective order. Inspections should also be carried out once any repairs or modifications have been completed.

The scheduling of routine inspections depends on the type of equipment, the factors governing its deterioration and the results of previous inspections. The IEC Code recommends that inspections are carried out at intervals not exceeding three years without expert advice. While a sampling inspection process may be used for basic items of equipment, such as light fittings and junction boxes, as a minimum it is recommended that all components are visually examined. It may be possible to extend the inspection interval beyond three years when sufficiently comprehensive maintenance records can be provided to justify an extended time interval. The records should contain details of any defects found and any remedial action taken.

The standard BS EN 60079-17 provides an alternative to regular periodic inspections, in some cases using the more frequent presence of skilled personnel to ensure the ongoing integrity of the apparatus. Continuous supervision is the alternative and is aimed at detecting faults early. This process must be overseen by a technical executive. It is recommended that the standard is referred to if a fuller appreciation of this operation is required.

All inspection activities should be carried out by appropriately trained personnel who are independent of the demands of maintenance activities so that their ability to report the findings of their inspections reliably is not prejudiced.

3.5.3.5 Records

The standards require comprehensive records to be maintained for all items of hazardous area equipment. This includes the location where the equipment is installed. These records are referred to as the hazardous area register. The hazardous area register contains the full hazardous area specification of each component and its geographic location on the plant. From the hazardous area register, it is possible to cross-reference each component with its original certificate and inspection certificates, both initial and periodic, and with the manufacturer's instructions.

Equipment testing authorities

France	CERCHAR, LCIE
Italy	CESI
Belgium	INIEX
Germany	PTB, BVS
Denmark	DEMKO
USA	Underwriters Laboratories (UM), Factory Mutual (FM)
Canada	CSA

3.5.3.6 Guidance notes and reference standards

1. Department of Energy Offshore Installations: Guidance on Design Construction and Certification (discontinued in 1997 but still used as a valuable reference document).
2. Institute of Petroleum: Code of Safe Practice Part 1, 1965.
3. Institute of Petroleum: Area Classification Code for Installations Handling Flammable Fluids.
4. BS 5345 - Code of practice for selection, installation and maintenance of electrical apparatus for use in potentially explosive atmospheres (now obsolete and superseded by BS EN 60079 but still an excellent reference document).
5. BS EN 60079-0 - Electrical apparatus for explosive gas atmospheres – General requirements.
6. BS EN 60079-10 - Explosive atmospheres. Classification of areas. Explosive gas atmospheres.
7. BS EN 60079-13 - Electrical apparatus for explosive gas atmospheres – Construction and use of rooms or buildings protected by pressurisation.
8. BS EN 60079-14 - Explosive atmospheres. Electrical installations, design, selection and erection.
9. BS EN 60079-17 - Electrical apparatus for explosive gas atmospheres. Inspection and maintenance of electrical installations in hazardous areas (other than mines).
10. BS EN 60079-19 - Explosive atmospheres. Equipment repair, overhaul and reclamation.
11. BS EN 60079-20 - Explosive atmosphere. Material characteristics for gas and vapour classification. Test methods and data.
12. NFPA 70 - National Electrical Code (USA).
13. NFPA 496 - Standard for Purged and Pressurized Enclosures for Electrical Equipment.
14. CP 1003 - Electrical apparatus and associated equipment for use in explosive atmospheres of gas or vapour other than mining applications.
15. API RP 505 - Recommended Practice for Classification of Locations for Electrical Installations at Petroleum Facilities classified as Class1, Zone 0, Zone 1, and Zone 2.
16. The IEC Rules for Offshore Structures provide practical guidance and recommendations on electrical equipment and installation practices for offshore work.
17. EEMUA No. 107 - Recommendations for the Protection of Diesel Engines in Zone 2 Hazardous Areas.

3.6 Emergency Systems

3.6.1 Emergency Shutdown System (ESD)

The ESD system is the means by which the oil and gas processing system is made safe in the event of an emergency. It involves isolating the oil supply and venting the gas to minimise the hydrocarbon inventory, reducing the hazard and resulting in the following:

- Protection of personnel
- protection of the environment
- protection of the operating plant and equipment
- maximisation of oil and gas production.

The ESD system operates selectively in stages to bring about the following final events:

- Shutdown of all production and utility systems
- closure of all wellhead valves
- depressurisation of hydrocarbon process equipment (blow-down)
- closure of surface-controlled subsurface safety valves (SCSSVs).

To achieve these objectives, the ESD system usually provides up to four levels of shutdown, although these may vary from operator to operator and, to a certain extent, are dependent on the complexity of the installation. Each level of shutdown deals with a particular level of risk to avoid wasted time and lost production that could result if the ESD system initiated an unnecessary complete shutdown. It can take days to bring an installation back on line from a complete shutdown due to the time it takes to pressure up the hydraulic valve control lines on long outreach wells and due to the extreme care that must be exercised when recommencing the flow of oil if hydrate formations are to be avoided.

Unit Shutdown
This is not strictly part of the ESD system. This occurs when there is a malfunction of a particular part of the process plant or utility system, or a single fire or gas detector goes into alarm condition (a confirmed fire or gas signal requires two coincident detectors to alarm) in the immediate vicinity of the unit. The affected unit will shut down but other similar units will continue to function normally. This limited shutdown action provides the operator with an opportunity to take corrective action while minimising lost production.

ESD 1 – Surface process Shutdown
Typically, a level 1 ESD is initiated by a disturbance in the hydrocarbon process system, such as an excessively high (or low) pressure or liquid level in one of the process pressure vessels, or the crude

oil distribution pumps failing. The oil and gas processing systems are put on hold with selected process valves closing to isolate the hydrocarbon inventory. The riser boarding valves are closed to prevent oil being brought onto the vessel and all process pumps are stopped. When the reason for the process disturbance has been identified and corrected, the oil and gas processing can resume.

ESD 2

A level 2 ESD is typically initiated by a confirmed fire or gas signal being activated. This is potentially a serious situation and requires not only the shutdown of the process equipment, ie a level 1 ESD, but also the removal of the hydrocarbon inventory by blowing down to flare the gases from the process pressure vessels, pipework, pumps and compressors. To isolate the oil supply further, the valves on the subsea christmas trees will close, as well as the riser boarding valves on FPSOs.

ESD 3 – Total Platform Shutdown (TPS)

A level 3 ESD is usually initiated manually prior to abandoning the installation. As well as the actions taken under ESD 1 and 2, the surface-controlled sub-surface safety valves (SCSSV) in the wells will close and all non-essential platform utility systems will shut down, including the main power generator. The emergency generator and fire pumps will continue to run if needed and all systems provided with power from a UPS system (batteries) such as the navigation aids.

Blow-down

The blow-down of the process system via the flare or vent stack that takes place under a level 2 ESD shutdown is how the gaseous hydrocarbon inventory is removed from the installation. The rapid depressurisation of the topside process pressure vessels and piping headers is aided by dedicated vent valves or blow-down valves. In an emergency situation, such as a confirmed fire or gas release in a process area, the vent valves open and release the contents of the vessels to the atmosphere via the flare/vent stack. It is a requirement of API RP 521- *'Guide for Pressure-Relieving and Depressuring Systems'*, that the pressures within the hydrocarbon process systems are reduced to 100 psig (6.9 barg) within 15 minutes. A means to manually initiate depressurisation of the process system must also be provided. This is normally via the operation of a manual push button located in the main control room.

Figure 3.66 – Vent stack

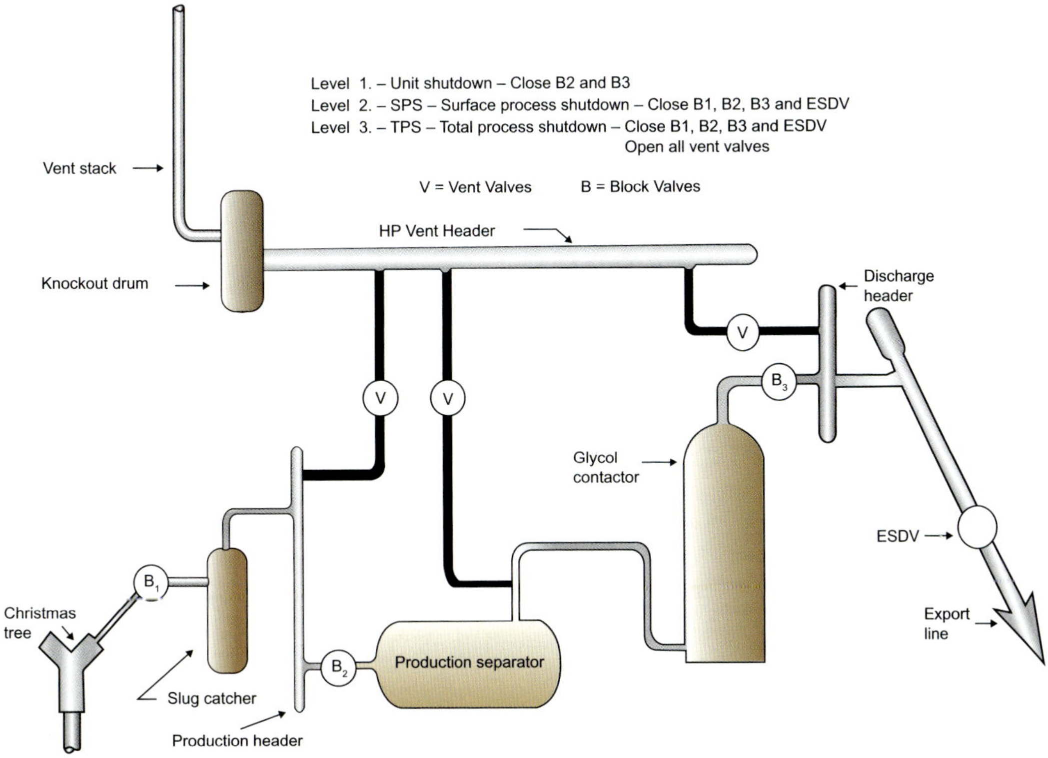

Figure 3.67 – Schematic – emergency shutdown system (shows only one train)

It should be noted in respect of FPSOs that the subsea flowlines and risers are not normally depressurised as part of the ESD process. Due to the large volume and high pressure of hydrocarbons contained in the flowlines, pipelines and risers, depressurising them could take a long time and require a large flare system. Closure of the riser boarding valves will normally provide adequate isolation.

Figure 3.68 – Flare stack

3.6.2 Process ESD Valves

Generally, all valves that form part of the ESD and blow-down system must be certified as fire safe and be designed to be fail-safe. Each valve is initially held closed by a spring loaded hydraulic actuator. To open the valve, hydraulic oil is supplied under pressure to push the piston open in the actuator. The pressure must be maintained the whole time the valve is open. As soon as the oil pressure is released, spring pressure will close the valve. The valve control system on the subsea christmas trees is a little more complex.

There are a number of valves on a subsea christmas tree, some that can be operated from the installation and some that can only be operated by an ROV. The upper master valve, choke and SCSSV need to be operated remotely and they are fitted with hydraulic actuators to facilitate this. Hydraulic fluid is supplied to the actuators via the control umbilical. Accumulator bottles are located on the subsea manifolds to minimise the time delay when opening the valves by providing a ready supply of stored energy, otherwise the time to pressurise the actuators would be prohibitive. Electrically operated solenoid control valves on the manifold can be operated from the installation and it is through these valves that hydraulic fluid is supplied to the various valve actuators. The hydraulic fluid is a water based composition, so when the valves need to close, the fluid in the actuators can be vented to the sea rather than bled back. This improves the response time.

3.6.3 Gas Detection

The object of the gas detection system is to identify a gas leak and take appropriate action before the gas reaches the point when an explosion can take place, ie before the concentration of gas in the air reaches the lowest concentration that an explosion can occur, referred to as the lower explosive limit (LEL). Consequently, the gas detectors are usually set to raise an alarm when a gas release is recorded at 20% of the LEL. At this point, an alarm will be initiated at the manned control station and an indication of the location of the gas leak given. If the gas release continues, a second alarm will be initiated when the gas concentration reaches 50% of the LEL. Automatic operation of the ESD will commence in order to safeguard the installation.

Figure 3.69 – Gas line of sight detector

To minimise any unnecessary shutdowns from false alarms, the gas detection system operates on a voting system and normally requires two out of three separate gas detectors to be activated at a particular location before a gas leak is considered to be confirmed and an ESD initiated.

The siting of the gas detectors in open deck areas is critical if they are to be effective. They should be located as near as is practical to potential leak sources to detect significant gas accumulations, and far enough away from known leak sources such as pump glands, valve stems, etc, to avoid false alarms and unnecessary shutdowns of equipment. The gas detector's position must also take into account whether the leaks are likely to be gases that are lighter or heavier than air.

Siting gas detectors in enclosed spaces also needs careful thought and the location of the space and whether the space contains a flammable source of release must be considered. Where enclosed safe areas are located in hazardous areas and do not contain a flammable source of release, gas detection is normally placed in the ventilation inlet ducting. If gas is detected, the appropriate action must be taken, normally the initiation of an audible and visual alarm, shutdown of the ventilation fans, closure of ventilation dampers and isolation of uncertified electrical equipment. The principle is different for enclosed spaces that contain flammable sources such as turbine hoods. Here, gas detection is normally sited in the ventilation outlet ducting. The action, once gas is detected, would be to initiate an alarm, shut down the unit and isolate the gas supplies and any uncertified electrical equipment. The ventilation system is designed to continue running to prevent build-up of a significant volume of gas.

There are a number of different types of gas detector in use and infra-red detectors are a favourite due to their reliability, stable operation and fast response times. The detectors rely on the comparison of an infra-red beam being absorbed by the gas sampled with that of a similar beam or the same beam in a prescribed reference to the gas. The difference in intensity of the two beams measures the gas concentration.

More recently, acoustic gas detectors have been successfully introduced. These operate by sensing the distinct high frequency acoustic ultrasound that is transmitted by all high-pressure gas leaks. Based on sophisticated microphone gas sensing technology, the detectors need only to detect the sound of a gas leak to operate and usually work regardless of ambient conditions, changing wind direction or fast dilution of the gas cloud.

Figure 3.70 – 'Gassonic' acoustic gas detector

3.6.4 Fire Detection

The overall objective of the fire detection system is to provide continuous automatic monitoring functions so that personnel are alerted to the presence of a fire, allowing control actions to be started manually or automatically and minimising the likelihood of escalation.

On an offshore installation, fires can be detected by using devices that monitor for flames, smoke and heat. Smoke and heat detectors are used primarily in compartments while flame detectors are used to provide cover in open deck areas. Flame detectors are, however, widely used in machinery spaces containing diesel engine or gas turbine powered machinery.

Figure 3.71 – Fire detector

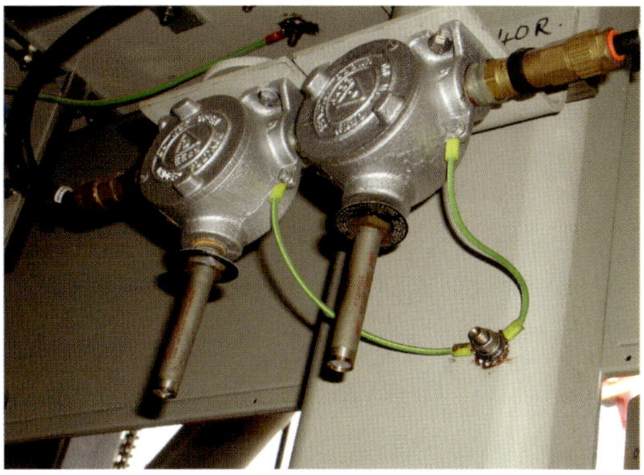

(a) Heat detector

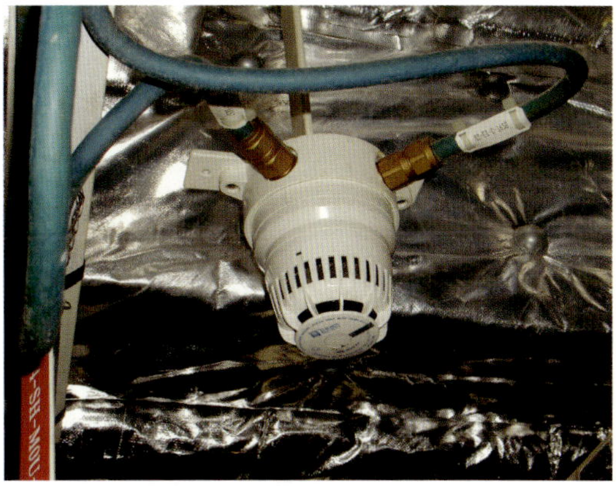

(b) Smoke detector

Flame detectors

There are several types of flame detector, with one of the most popular types known as an infra-red detector. Rather than generating an infra-red beam and measuring its absorption, infra-red flame detectors are set-up to look for the infra-red radiation that is produced by combustion. Fires involving carbonaceous materials, ie materials containing carbon-like hydrocarbons, produce large quantities of carbon dioxide (CO_2) gas. The CO_2 then produces very high levels of infra-red radiation in a narrow wavelength region centred around 4.3 µm and it is by monitoring for this wavelength of infra-red radiation that the flame detectors work.

(c) Flame detector (infra-red)

Figure 3.72 (a, b, c) – Fire detection

Flammable mist detectors
These are used in enclosed spaces containing machinery where there is the potential for a release of high pressure flammable liquids.

3.6.5 Containment Systems

These are designed to collect escaped liquid hydrocarbons and initiate platform shutdown systems. This is carried out through the provision of bunds around storage tanks and machinery where hydrocarbons could escape.

Figure 3.73 – Oil mist detector

SOLAS requires a coaming (known as a fish plate) to be provided that encloses the cargo deck on tankers. It also requires provision of a fixed foam fire-fighting system capable of depositing a layer of foam over any oil spillages in the area contained by the coaming. The foam systems must be designed to achieve a minimum foam density of 10.2 litres/min/m^2 to protect the cargo deck from pool fires.

3.6.6 Fire Loop

A pneumatic fire loop is often included as part of the ESD system and as a further safety precaution. It provides a very basic and foolproof heat activated way of initiating the ESD system, consisting of either a stainless steel instrument piping loop fitted with fusible plugs or a small bore plastic pipe loop.

The loop extends through all the process areas and is pressured to 3.5-7.0 bar (50-100 psi). If a fire occurs, the fusible plugs or plastic pipe will melt and release the air pressure. This activates the ESD system via an electrical, pneumatic or hydraulic solenoid. Fusible plugs are also often fitted to key items of equipment, such as the wellhead control panel. Failure of the plugs releases the hydraulic pressure that holds the christmas tree valves and sub-surface safety valves open.

Sub-surface safety valve (SSSV)
The SSSV, also known as the SCSSV, is located in the well bore approximately 20-30 metres below the seabed level. It is installed in the production tubing and comprises a ball or flapper valve that is held open against a spring by hydraulic pressure, making it fail-safe. It is regarded as the last safety barrier and will protect the environment if the christmas tree becomes dislodged by a ship's anchor or fishing boat's trawl board. To ensure reliable operation, it is recommended that the valve is cycled at least once a month and tested for leaks on a regular basis, perhaps every three months.

3.6.7 Emergency Shutdown Valves – Pipelines

The Emergency Pipeline Valve Regulations were introduced as many offshore installations share common subsea pipelines for transferring produced oil and gas to a reception facility on the mainland. The ESDV is designed to isolate an installation if an emergency situation such as a fire occurs, and to prevent a recurrence of a *'Piper Alpha'* type disaster, where the initial fire was fuelled by oil and gas pumped in by adjoining platforms.

On fixed platforms, the ESDV should be located below the lowest deck of the installation and as far down the pipeline riser as is practical, while still above the highest anticipated wave crest and available to be maintained. On an FPSO, the ESDVs are often referred to as 'boarding valves' and are located on the side of the hull above the waterline or on the seabed if the water depth is not too great. The ESDV must be of a rapid acting, fail-safe design that operates automatically on signals generated by the ESD system, and which can be operated at a control point adjacent to the valve.

The valve and its actuating mechanism must be protected from damage caused by fire, explosion or the impact of dropped objects. The aim is to ensure that the valve closes under all foreseeable conditions. To define the type and extent of fire protection needed, the operator has to consider the type, severity and duration of anticipated fires as well as the minimum duration that the integrity and operability of equipment to be protected must be maintained (see Figure 3.74).

3.6.8 Guidance Notes and Reference Standards

1. API RP 14C -'Recommended Practice for Analysis, Design, Installation and Testing of Basic Surface Safety Systems for Offshore Production Platforms'.
2. API RP 14G – 'Recommended Practice for Fire Prevention and Control on Fixed Open Type Offshore Production Platforms'.
3. API RP 521- 'Guide for Pressure-Relieving and Depressuring Systems'.
4. Pipelines Safety Regulations (PSR) - SI 1996/825.
5. Guidance on Regulations, Schedule 3 - The Pipelines Safety Regulations (SI 1996 No. 825).
6. BS EN ISO 13072:1999 – 'Petroleum and natural gas industries – Control and mitigation of fires and explosions on offshore production installations – Requirements and guidelines'.
7. Department of Energy Offshore Installations: Guidance on Design, Construction and Certification.

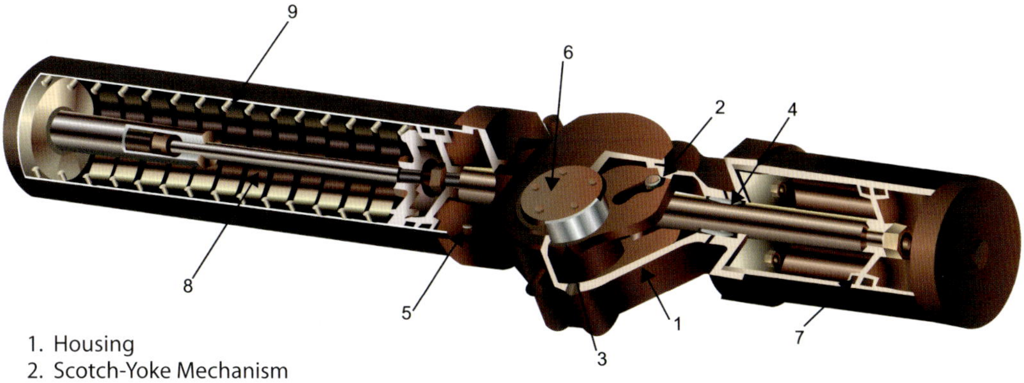

1. Housing
2. Scotch-Yoke Mechanism
3. Journal Bearings
4. Dual Rod Bushings
5. Bi-directional Travel Stops
6. Vented Fill Plug
7. Piston Seal(s)
8. Xylan Cylinder Coating
9. Spring Cartridge

Figure 3.74 – Fail safe valve actuator

3.7 Deepwater Horizon – Impact for the Industry

3.7.1 Background

On the 20th April 2010, a major explosion of escaping methane gas occurred on the *'Deepwater Horizon'* drilling rig operated by BP in the *Macondo* field in the Gulf of Mexico, killing 11 out of the 126 crew on board, making world headlines and rocking the oil industry back on its heels.

Figure 3.75 – The *'Deepwater Horizon'* drilling rig

The *'Deepwater Horizon'* was built in 2001, a fifth generation, dynamically-positioned semi-submersible drilling unit with a fully automated drill floor and a BOP system rated to 15,000 psi. In the nine years preceding the *Macondo* blowout, it had successfully operated in water depths up to 9,000 feet and had drilled wells up to 35,055 feet in depth. The *Macondo* well was located in *Mississippi Canyon Block 252*. The water depth was 4,992 feet and the well was drilled to a depth of 18,350 feet, parameters that were within the capabilities of the *'Deepwater Horizon'*.

In addition to the loss of life, the blowout created what was described as the worst oil spill in US history. Before the well was finally resealed four months later, the US government estimated that approximately 5 million barrels of oil had spilled into the Gulf, impacting wildlife, coastal environments and the livelihoods of those in the local fishing, tourism and related sectors.

3.7.2 The National Commission on the BP Deepwater Horizon Oil Spill and Offshore Drilling

Figure 3.76 – President Barack Obama addresses the media at Coast Guard Station Venice, 2nd May 2010

On 22nd May 2010, President Obama established the National Commission on the BP Deepwater Horizon Oil Spill and Offshore Drilling to review the *'Deepwater Horizon'* event and its consequences. The Commission reported in January 2011, proposing significant changes to government and oil industry attitudes to safety and highlighting the need for more effective post-event response technologies.

The report blamed systemic failures in risk management and the general safety culture of the industry. They believed that both oil companies and the government agencies involved with the oil sector had become complacent about safety, while the greater part of their attention had been paid to increasing revenue generation. In their recommendations, they want to see industry and government focussing on improving risk assessment, risk management and safety procedures. They also stated that government and industry need to demonstrate a better understanding of environments around deepwater oil producing areas and to increase their knowledge and technical capability to respond to oil spills. The *'Deepwater Horizon'* spill response clearly demonstrated that they were relying on the same technologies as 20 years ago – booms, dispersants and skimmers.

The Commission emphasised that future change should address the ability of companies to respond to any incidents quickly and efficiently, not solely prevention improvements, and that the industry needs to build containment and response approaches into their safety policy.

Four major oil companies now have invested in the Marine Well Containment Company, a non-profit organisation created in July 2010 which will provide a central source of spill containment equipment. Other major companies are considering joining the venture. A similar smaller scale service is also being offered by Helix Energy Solutions Group.

The Commission also referred in its investigation to safety experience and practices from other deep sea environments and compared them with the US approach. Using examples from the North Sea (*'Piper Alpha'* platform explosion, which killed 167 people in 1988, and the capsize of the *'Alexander Kjelland'* flotel with the loss of 123 lives in 1980) and from Canada (the *'Ocean Ranger'* drilling rig which sank with all 84 crew members off Newfoundland in 1982), they highlighted key differences in the overall approach to offshore safety in these regions.

These countries established a radical change of emphasis in responsibility for safety following the disasters described above. Operating companies are required to demonstrate full consideration of all risks associated with offshore exploration, development and production activities. The regulator will only provide operating consent when they are satisfied that the operating company has demonstrated that they have critical safety and risk management systems in place - known as the safety case.

In the US, there has been more of a standard setting approach, with the government's Mineral Management Service (MMS) taking responsibility for overseeing all drilling and operating activities in the US. Various attempts had been made over the years to increase the safety and risk management element of its regulation, without much real change being introduced. The Commission felt that the MMS had generally fallen behind the industry in its understanding of the fast-developing technologies involved in deepwater oil exploration and production and that the oil companies operating in US waters needed to dramatically improve the safety aspects of their operations, including self-policing mechanisms.

Within a month of the *'Deepwater Horizon'* blowout, the MMS organisation was replaced by the Bureau of Ocean Energy Management, Regulation and Enforcement (BOE), which has three separate agencies:

- Bureau of Ocean Energy Management – responsible for leasing areas of the Outer Continental Shelf for conventional and renewable energy resources
- Bureau of Safety and Environmental Enforcement – responsible for ensuring oversight, safety and environmental protection in all offshore energy activities
- Office of Natural Resources Revenue – responsible for royalty and revenue management, including collection and distribution of revenue, auditing and compliance, and asset management.

3.7.3 The Events Leading up to the Incident

By all accounts, the *Macondo* well had been successfully drilled and indications were that it would be a very productive well. As was normal practice, the well was to be temporarily abandoned pending the preparation of field development plans. Temporary abandonment involves replacing the drilling mud with conditioned seawater and the installation of a cement plug in the production tubing to provide an additional safety barrier should there be any hydrocarbon leakage into the well.

Prior to plugging the well, it had to be subjected to what is referred to as a 'negative pressure test' to confirm the structural integrity of the well. This involves reducing the hydrostatic pressure in the well to below that of the reservoir pressure in order to test the effectiveness of the cement seal at the base of the well and the mechanical casing seals. It was while carrying out this test that the disastrous chain of events commenced.

As the hydrostatic pressure in the well was reduced by slowly replacing some of the heavy drilling mud with seawater, the pressure in the well was monitored from pressure sensors in the drill string and at the kill line returns from below the BOP. While the kill line returns did show an absence of pressure, the pressure measured within the drill string was in excess of 1,400 psi. Despite this, the toolpusher and driller mistakenly deemed the test to be a success, inexplicably attributing the pressure in the drill string to a phenomenon called 'bladder effect', a phenomenon they seem to have invented.

On completion of the negative pressure test, the annular BOP was re-opened placing the well once again in the overbalanced condition and the process of circulating out the drilling mud with seawater commenced. When the hydrostatic head had fallen to reservoir pressure, oil started entering the well. It is believed that this was due to the failure of the cement shoe at the base of the well. The first indication that hydrocarbons were entering the well was the increased flow of drilling mud back to the mud tanks. Unfortunately, it was some time before this was noticed because the mud loggers who should have been monitoring this had apparently ceased work some 7½ hours previously. The escalation of the pressure in the drilling riser and drill pipe went unnoticed for a period of half an hour during which time corrective action could have been taken. Eventually drilling mud started to spill from the drilling riser onto the drill floor and it was then that corrective action was attempted by closing the annular BOP in order to stop any more oil entering the riser. However, the BOP failed to seal

Figure 3.77 – The *Macondo* well had already proved a difficult one, like many deepwater wells. By 20th April, the operation was six weeks behind schedule and $58 million over budget.

and the flow of oil, gas and drilling mud continued. The drilling mud returns were then diverted into the mud returns separator when they should have been directed overboard, a critical error. The mud system was overwhelmed by the sheer volume of mud, oil and gas which was then directed up the vent pipe inside the drilling derrick from where it subsequently cascaded onto the deck of the rig. The oil and gas flooded the work area and entered the machinery space causing a diesel generator to over-speed, self-destruct and in so doing ignite the main body of gas causing a massive explosion and fireball. At this point the fail-safe BOP shutdown system on the seabed should have operated and sealed the well completely, but it proved to be faulty and ineffective. The BOP was eventually partially closed 33 hours later by intervention using an ROV.

3.7.4 Analysis of the Incident

The reason why the *Macondo* well suffered a blowout has been analysed by a number of different groups, including BP itself, its key partners in the well, Transocean (who owned the rig) and Halliburton (who provided major offshore services including the well cement), the National Academy of Engineering and the National Commission on the BP Deepwater Horizon Oil Spill and Offshore Drilling. The investigation team looked at a number of issues that could have contributed to the blowout of the *Macondo* well such as liner design and the number of casing centralisers. However, the conclusion was that it was the design of the cement mix used to seal the completion liner that was the root cause of the well failure and this was confirmed by the laboratory that tested the proposed mixture prior to its application. Had the negative pressure test been conducted correctly, the failure of the cement seal would have been apparent and corrective action could have been taken. Even after the blowout had commenced, a disaster could have been avoided if the well fluids had been directed overboard rather than into the mud separator. The BOP which is designed as a final safety barrier to deal with these very situations could have saved the day had it been properly maintained and worked as intended.

Figure 3.78 – The '*Deepwater Horizon*' fire

While there were many aspects of the *Macondo* well where better technical or operational solutions may have been more effective and could have reduced the risks involved, the investigators put the majority of the blame on human error. Investigators assessed that management failed to:

- Set clear processes and procedures
- operate an effective risk management policy and procedures

- ensure effective communications (particularly between partners in the operation)
- adopt a consistent and comprehensive approach to change procedures, and
- ensure the skills levels of all individuals working on the facility.

It was also felt that decision making around cost and time-saving opportunities did not take full account of the risks associated with these decisions, particularly where these decisions were made on site. The blame fell not only on the operating company and their technical partners, but on commercial partners and government agencies.

It was evident that MMS regulations were inadequate to address the risks of deepwater drilling. Key areas of drilling operations were left to the industry to decide without agency review and agency staff did not have the full expertise and training needed to enforce those regulations effectively.

3.7.5 The First Changes After the Incident

On 12th October 2010, seven weeks ahead of the original plan, the Department of the Interior lifted the US moratorium on deepwater drilling that was put into place on 12th July 2010. However, they did not reach this decision without introducing some significant changes to their offshore regulatory structure.

All deepwater rigs wishing to operate now have to undergo a Bureau of Ocean Energy Management, Regulation and Enforcement (BOE) inspection before being awarded a permit. The two most recent rules introduced were announced on 30th September 2010:

- The drilling safety rule sets out a full set of standards for well cementing, casing and the use of drilling fluids to secure well bore integrity. It also stipulates further safeguards for shutting the flow of oil and gas with a blowout preventer and its components, including subsea remotely operated vehicles (ROVs), pipe rams and shear rams. Operators will have to convince independent inspectors of the effectiveness of their well designs and flow intervention mechanisms before being allowed to proceed. A loophole which exempted many Gulf of Mexico offshore operators from presenting plans for worst-case discharge scenarios has been removed
- the workplace safety rule requires operators to present and maintain a safety and environmental management system (SEMS) to reduce human errors and ensure that all staff are well-drilled on procedure in case of an emergency. The SEMS is similar to the safety cases that are already required for companies operating in UK and Norwegian waters.

Major changes are also likely in the area of liability with a vote by the Senate Environment and Public Works Committee to eliminate caps on liability that oil companies would face for damages stemming from offshore spills. The change, if approved and made into law, would apply retroactively to BP. Drillers are currently protected by a $75 million cap for compensating communities for economic losses and the impact of offshore spills on natural resources. The Senate Energy and Natural Resources Committee is expected to vote on several bills relating to the spill, including measures that would increase financial requirements for drillers and raise civil and criminal penalties. The Committee is also considering measures that would require the Interior Department to ensure that any companies bidding on new leases have met all safety requirements for their existing operations and have complied with any obligations for damages in prior accidents. This

Figure 3.79 – Drill ship and other offshore marine units on site at the scene of the '*Deepwater Horizon*' explosion

measure, in particular, could affect those companies involved in the *'Deepwater Horizon'* rig spill.

In Europe, government is also reviewing industry practice and regulation in light of the *'Deepwater Horizon'* disaster. The European Commission has reviewed its international regulations for offshore drilling. The Commission had also pushed for a similar deepwater moratorium to the US, but the proposal lost momentum. Although Norway is currently the only country supervising drilling at depths matching the *'Deepwater Horizon'*, the UK is planning wells off the Shetland Islands and Faroe Islands at depths of 1,600 m and 1,100 m respectively.

The European Energy Commission proposed that best practices already existing in Europe will become the standard throughout the EU and there is likely to be tighter regulation from the EU as a consequence. Companies operating in EU waters have been asked to provide assurances that they are working as hard as possible to ensure safe practice and that they are able to take on full responsibilities for environmental and other damage if an incident were to occur. New proposals from the Commission will require companies to guarantee that they have the funds – or insurance – to cover the costs of a spill clean-up operation if the worst should happen. In addition, the Commission is pushing for the distance from the shore at which companies can be made liable for spill costs to be increased from 12 nm to 200 nm. Draft legislation on the new rules is expected to be presented in 2011.

In the UK, the Energy Secretary has initiated a doubling of the number of inspections of drilling rigs and has asked the newly-formed Oil Spill Prevention and Response Advisory Group to report on its findings on the UK's ability to prevent and respond to oil spills. He also commissioned a review of the indemnity and insurance requirements for operating in the UK Continental Shelf.

At present, the investigation of the *'Deepwater Horizon'* disaster is still ongoing. While some changes in offshore oil regulations have already been made, more change is likely to come, with pressure from all quarters – government, industry, environmental agencies and the public – to prevent similar accidents. This will inevitably result in further new regulations, new 'best practice' and new technologies, and is likely to include:

- New mandatory procedures, testing and checks for all elements of well development and completion
- increased safety margins on all well completion processes
- increased attention to spill risk assessment and disaster planning
- more frequent training for offshore workers
- new approval processes with external agencies and within internal company decision-making
- investment in assessing and understanding the environment around field operations and the potential impact of oil spills
- companies working more closely with 'neighbouring' communities.

All of the above will have a significant impact on the time to first commercial oil and on operating costs. However, deepwater oil is crucial to satisfying the world's energy demands and both governments and industry are keen to progress deepwater development. In 2011, some 35 new deepwater drilling rigs are expected to be put into action.

4 Piping Systems and Process Pressure Vessels

4 Piping Systems and Process Pressure Vessels

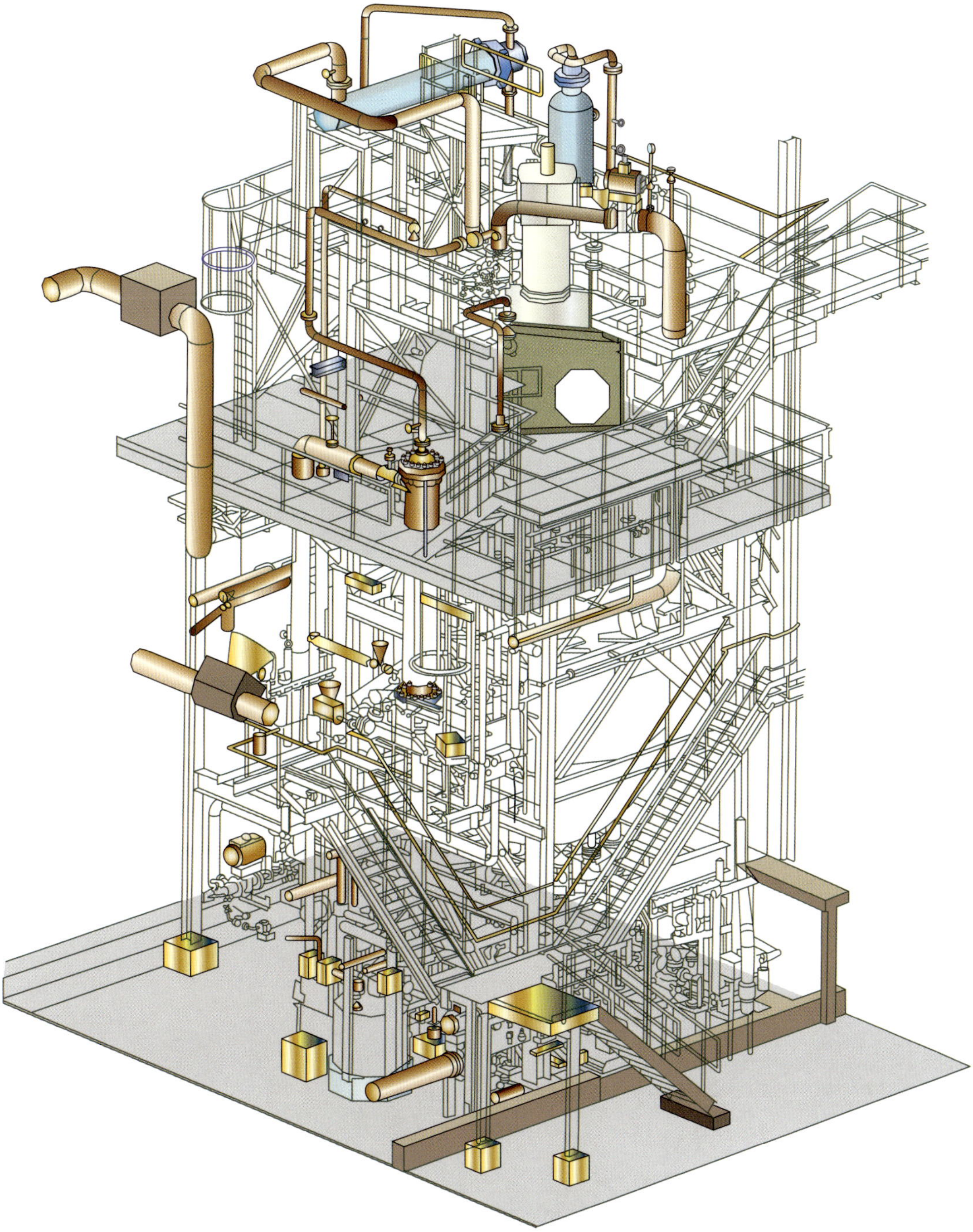

Figure 4.1 – Computer generated drawing of process pressure vessel and pipework

4.1 Piping Systems

The piping systems on an offshore installation may be divided into two basic categories:

1. Hydrocarbon process.
2. Utility.

All pipework should be colour coded to assist in system identification and the coding system particulars and P&IDs (piping and instrumentation diagrams) should be available for review as supporting documents.

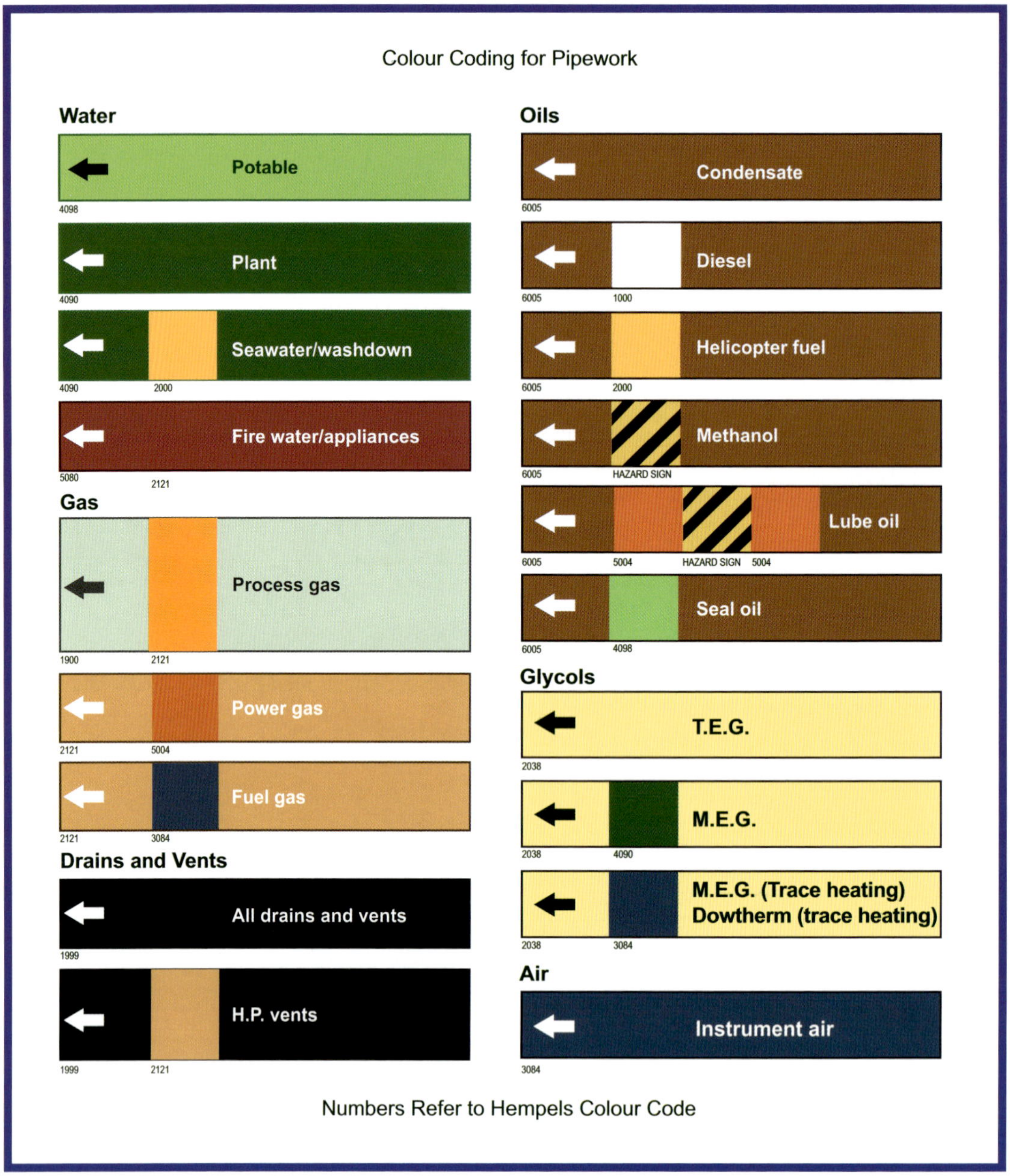

Figure 4.2 – Piping systems, typical colour coding plan

4.1.1 Hydrocarbon Process

Gas/oil process

The main objective of an offshore installation is to separate hydrocarbon products into liquid or gaseous forms and to remove any impurities that will inhibit transportation. A more detailed account of the gas and oil production processes can be found in Chapter 5.

Condensate

Condensate is a clear, highly volatile liquid produced as a by-product of the gas production process. The condensate system is primarily concerned with the removal of water prior to the injection of the condensate into the gas subsea pipeline for transportation to the onshore reception facility.

Fuel gas

Electrical power on a large offshore installation is generated by either gas turbine or gas engine driven alternators. Both systems run on process gas, as do the glycol reboilers. The fuel gas is taken either from the main outlet header, or produced as a by-product

Figure 4.3 – Gas turbines

of the various refinement processes and is subject to further separation, drying and filtration before entering the fuel gas main.

Power gas

Older installations frequently utilise process gas to provide the motive power for the actuation of valves. New installations tend to favour hydraulics or compressed air systems, which create fewer restrictions on hazardous area boundaries.

Vent system

Figure 4.4 – HP and LP vents

Vent systems are generally associated with gas production installations and both high (HP) and low (LP) pressure systems are employed.

(a) The HP vent system is used primarily to depressurise process pressure vessels during an emergency shutdown (ESD) situation, or to relieve excess vessel pressure via pressure safety valves (PSV).

The system consists of a large diameter ring main, a knock out (KO) drum or separator vessel designed to remove entrained liquids, and a vertical pipe or vent stack that will release the gas to the atmosphere at a safe location above the installation. The vent system is continuously purged with gas to ensure a rich and non-explosive atmosphere prevails. It is protected against fire by a dry powder, carbon dioxide or halon equipped fixed fire-fighting system.

(b) The LP vent system is virtually a duplicate of the HP vent system, designed to operate at lower pressures. The system provides a route for the release of gases from PSVs and atmospheric vents.

Flare system

Figure 4.5 – Flare system

The flare system fulfils a similar function to the vent system, the obvious difference being that the gas is ignited as it leaves the end of the flare boom.

The flare system will only be encountered on oil production installations where the quantities of gas requiring disposal are often considerable and would constitute a hazard if simply released into the atmosphere through an open vent.

Corrosion inhibitor

The use of a corrosion inhibitor provides some protection to the internal surfaces of carbon steel pipework. The inhibitor, frequently Cronox 638S, is mixed with methanol or MEG (20%/80%) and may be injected into wells, process plant and subsea pipelines.

Hydrate inhibitor

Both methanol and monoethylene glycol (MEG) are used extensively by the offshore industry to combat hydrate formation. They may be injected into the process plant on an occasional basis during periods of peak gas production, or continuously metered into the subsea pipeline, a practice frequently encountered on small satellite platforms without water separation equipment. The inhibitors are eventually recovered from the wet gas at the onshore reception facility or the mother platform.

When employed on an occasional basis, stocks can be replenished (bunkered) by supply boat but the quantity consumed by continuous pipeline metering necessitates the installation of a small diameter (3 inch/75 mm) subsea pipeline that is attached (piggy-backed) to the main gas export line so that the inhibitor can be supplied direct from the onshore reception facility or the mother platform. Prior to use, the inhibitors are stored in atmospheric tanks, the methanol tanks frequently employing a blanket of fuel gas to reduce the air space and thus the flammability.

Methanol

Flashpoint 16°C (60°F). Methanol is the most effective of the hydrate inhibitors and it is the only inhibitor capable of melting a hydrate, should one occur. For this reason, reserves of methanol may be encountered on installations that rely primarily on MEG for routine hydrate inhibition. The effectiveness of methanol as an inhibitor is due primarily to its willingness to vaporise. To be effective, an inhibitor must be delivered to the location where condensation of water is likely to occur. Unfortunately, this characteristic hinders the separation of the methanol at the onshore reception facility and up to 50% of the methanol is unrecoverable. Heavy methanol consumption can prove both expensive and upsetting to the effectiveness of subsequent items of process equipment.

Monoethylene glycol (MEG)

Flashpoint 116°C (266°F), boiling point 197°C (388°F), freezing point minus 13°C (9°F). Monoethylene glycol is hygroscopic, ie it attracts water from the atmosphere. Up to 90% of the MEG injected into the gas stream can be recovered at the onshore reception facility and it makes an efficient and economic hydrate inhibitor.

Triethylene glycol (TEG)
Flashpoint 160°C (320°F), boiling point 288°C (550°F), freezing point minus 7°C (20°F). Triethylene glycol is not generally employed as a hydrate inhibitor, being more suited as a drying agent in a dehydration process where it is chosen in preference to other glycols because of its ability to withstand the elevated temperatures associated with the regeneration process.

The subjects of hydrates and hydrate formation are discussed more fully in Chapter 5.

4.1.2 Utility Systems

Firewater
The firewater system can be extremely complex, particularly on large installations, and the subject is discussed in greater detail in Chapter 3.

Seawater service
In its simplest form, the seawater service system consists of a small diameter (4 inch/100 mm) ring main that is maintained at a constant pressure by an electrically powered deepwell submergible pump. The water is used primarily for wash down purposes and for the flushing of domestic toilets, but it may also be used to maintain the firewater ring main at a constant pressure of approximately 100 psi (7 bar) to ensure an instantaneous supply of water prior to the activation of a fire pump.

The general service seawater system associated with a large oil production installation is considerably more complex than that previously described. It is often the largest single piping system on the platform, supplying water to the utility and production process coolers and also to the enhanced oil recovery plant.

Sodium hypochlorite
Sodium hypochlorite solution is frequently injected into seawater pump suctions to combat the growth of marine organisms and the algae that contribute towards the fouling of filters and pipelines.

The sodium hypochlorite is produced by the electrolysis of seawater within a hypochlorite generation cell, the seawater being forced to flow between two concentric titanium tubes that are connected to a DC (direct current) power supply. The resulting sodium hypochlorite solution is extremely corrosive and must be stored in stainless steel or GRP (glass reinforced plastic) tanks prior to use. The electrolysis of seawater also produces hydrogen gas, but the quantities generated are insufficient to constitute a hazard, provided that the storage tanks are located in a well ventilated open deck area.

Potable water
Potable water for domestic consumption is stored in freshwater tanks and supplied to the accommodation via a pressurised surge vessel. Freshwater may be bunkered from supply boats, or produced from seawater by desalination, or a reverse osmosis plant.

Compressed air
Compressed air is normally supplied from large air receivers that are maintained at a constant pressure of approximately 100 psi (7 bar) by multi-stage reciprocating air compressors. The discharge manifold connecting the air receivers divides into two separate systems.

Figure 4.6 – Compressed air tank

Instrument air
It is most important that the vast array of delicate instruments employed on a modern offshore installation are provided with a supply of clean, dry air if they are to function reliably. Consequently, air drawn from the main air receivers passes through an electrically heated desiccant drying medium and fine filtration package prior to entry into the instrument air main.

Utility air
Utility air is provided for various operations such as the starting of diesel engines and pressurising of water distribution vessels. Hand tools, pumps and other items of portable equipment can be operated from hose stations located throughout the installations.

Diesel fuel
Cranes, lifeboats, standby and emergency generators are all powered by diesel engines and diesel consumption can be considerable, particularly during process shutdown periods when fuel gas is unavailable. The diesel oil is bunkered from supply boats and may be stored in purpose-built tanks or in structural void spaces such as crane pedestals and jacket legs.

Figure 4.7 – Diesel engine powered standby generator

Helifuel

A number of offshore installations provide helicopter refuelling facilities. The aviation spirit is delivered by supply vessel and stored in purpose-built fuel tanks.

Drains

The design of a drainage system on an offshore installation is another area that tends to reflect individual company philosophy, and the lack of standardisation often leads to confusion. However, the basic function of the system is to ensure that pollutants are removed from waste waters prior to their discharge into the sea.

The drainage system may be divided into three categories. Open drains, closed drains and the sewage disposal system.

Open drains – Non-hazardous

A less confusing title would be 'safe area open drains' because the system simply provides for the removal of surface deck water from locations designated as safe areas. The waters are relatively free of contaminants and are piped directly into the overboard drains disposal caisson, a discharge pipe extending into the sea a distance of approximately 25 metres (90 feet).

Open drains – Hazardous

In practice, these drains rarely constitute a hazard but they are named as such because they originate from within locations designated as hazardous areas. They are used primarily for the disposal of water wash, rain and deluge waters and one or other of two systems may be used, depending on the age of the installation.

Generally speaking, the hazardous areas open drains on installations constructed prior to 1985 are discharged directly into the overboard drains disposal caisson alongside the non-hazardous open drains. To reduce the possibility of contaminating safe areas with potentially hazardous substances, the hazardous area drains discharge pipe is terminated above the non-hazardous area drain discharge pipe, both being submerged in seawater. On newer installations, the hazardous open deck drains are collected in a dedicated hazardous area open deck drains sump tank, entirely eliminating the possibility of contaminating safe and hazardous area drains. The hazardous area open drains sump tank acts purely as a holding vessel, the contents being transferred to the oily water separator (OWS), where the differences in the specific gravity of the oil and water are used to effect separation. The water gravitates into the open drains caisson while the oil is pumped into the gas export line or returned to the process plant.

The open drains disposal caisson provides a final opportunity for any remaining condensate/oil to separate from the water. The oil floats to the top of the caisson where it can be transferred into the slop oil tank or the oily water separator by a pump that operates automatically, activated by an oil/water interface sensor. An oil content of less than 15 ppm is required before the water can be discharged into the sea.

Closed drains

The closed drains system provides the means by which a process vessel may be manually emptied and should not be confused with the produced water system, which provides for the automatic removal of liquids separated from the gas/oil production process. The vessel drains are used on an occasional basis while the produced water system is in constant use and may be used to dispose of quantities of water varying from several tonnes to several thousand tonnes a day.

The closed drains system consists of a number of separate collection headers (pipelines) that are graded in accordance with the pressure ratings of the pressure vessels that they serve, the headers being protected against over-pressure by restrictive orifice plates located in the pressure vessel outlets. The headers terminate within the slop oil tank, or a dedicated closed drain drum, depending on the design of the system.

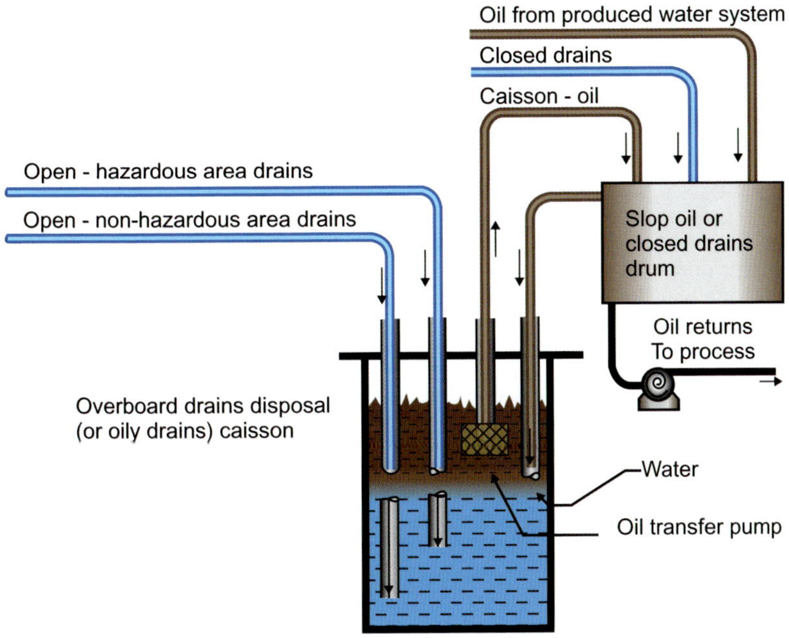

Figure 4.8 – Drains systems – older installations

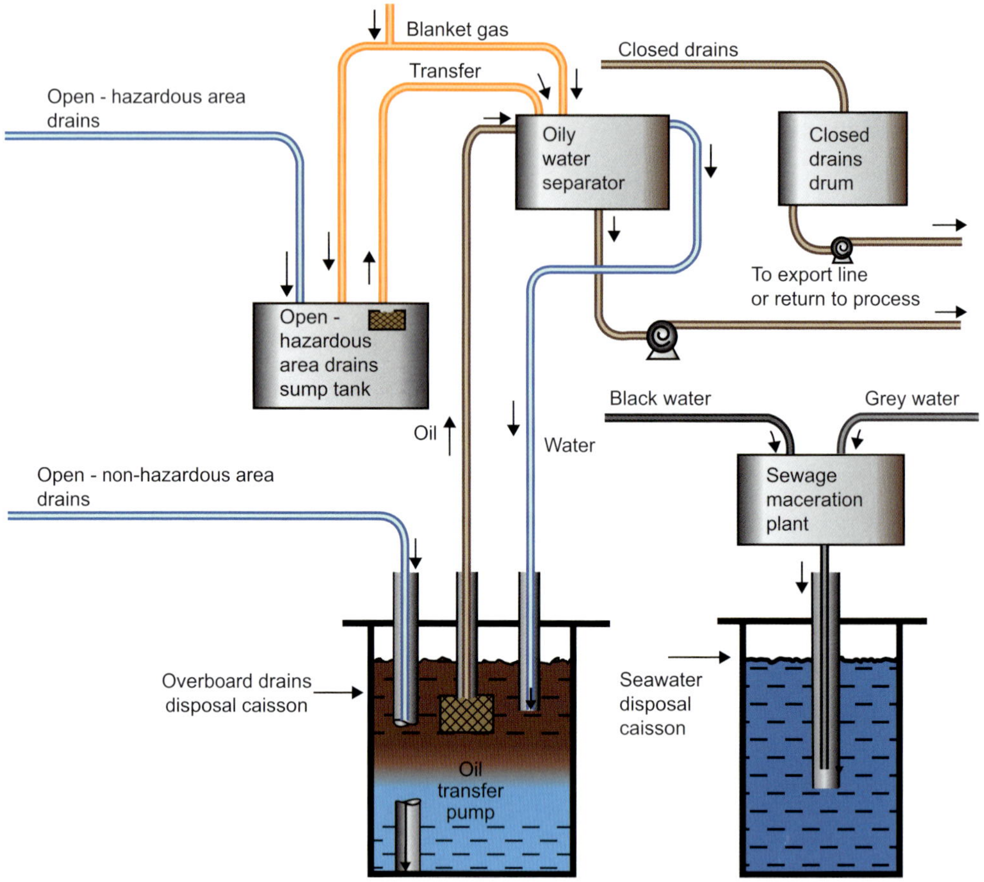

Figure 4.9 – Drains systems – newer installations

The slop oil tank featured on older installations frequently permits a degree of water/oil separation with the water gravitating to the overboard drain disposal caisson while oil products are disposed of in the same manner as the contents of the closed drains drum, ie pumped into the export line or back into the process system for re-circulation.

Sewage system

The sewage system can be divided into grey water, which consists of drainage water from showers and sinks, and black water or raw sewage. Both systems combine prior to their entry into the sewage maceration plant and subsequent discharge into the seawater disposal caisson.

4.2 Process Pressure Vessels

The vast majority of process vessels employed on an offshore installation are remarkably simple in design and rely on very basic principles of fluid dynamics to effect the separation of oil (or condensate), gas, water and sand.

Generally speaking, the vessels are common to both oil and gas production installations and, with the exception of the specialist dehydration, NGL removal, stabilisation and sweetening system vessels, they fall into four main categories:

1. Gas/oil (condensate)/water separators
   i) Production separator
   ii) Test separator
   iii) Condensate separator
   iv) Flash/degassing/vacuum separations
2. Gas/liquid separators
   i) Slug catcher
   ii) Scrubber/suction drum
   iii) Knock out drum
   iv) Suction boot
3. Oil/water separators
   i) Oily water separator
   ii) Flotation unit
   iii) Skimmer
   iv) Coalescer
   v) Hydrocyclones
4. Holding tanks.

4.2.1 Gas/Oil (Condensate)/Water Separators

The production separators lie at the very heart of the oil and gas production processes and are virtually identical to each other in design and to the other separators included in this category.

4.2.1.1 Production separator-gas process

The production separator (gas) is primarily concerned with the removal of liquids (water and condensate) from the gas stream and their subsequent separation prior to discharge from the vessel. The process gas entering the separator is forced to undergo a sharp change in direction and the momentum possessed by the entrained liquids ensures that they are thrown out of the gas stream and into the base of the vessel. Given time, the separation of the two liquids will proceed naturally due to the variation in their specific gravities. The

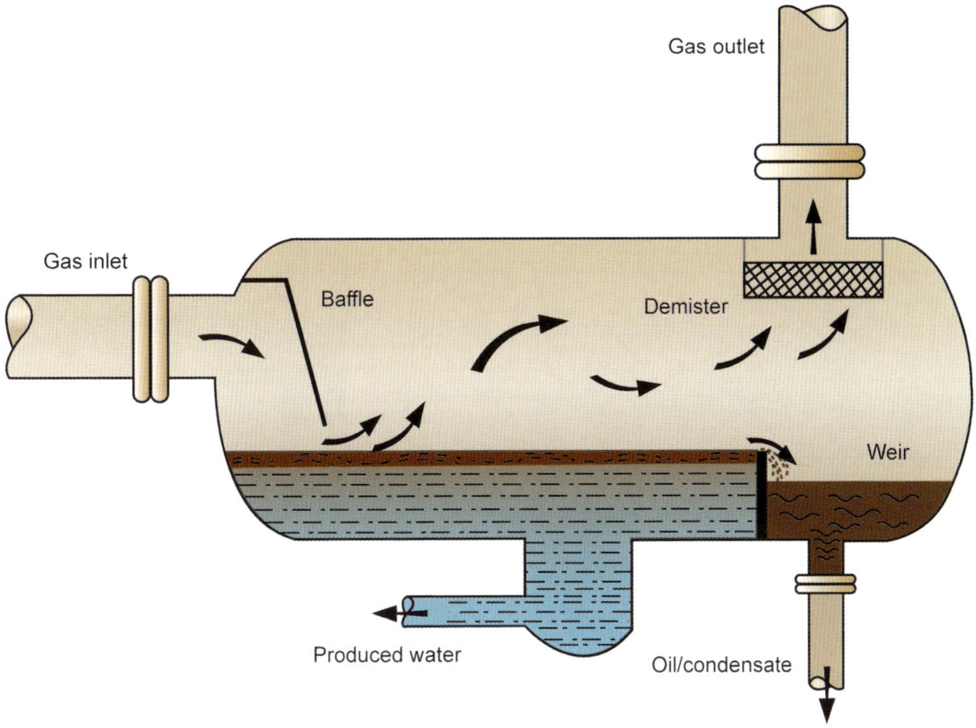

Figure 4.10 – Production separator

liquid levels are monitored by an oil/water interface sensor to ensure that the water is maintained at a level that permits the flotation of oil over a weir located in the end of the vessel.

4.2.1.2 Production separator-oil process

The production separator (oil) is primarily concerned with the separation of gas and water from the crude oil. As the crude oil enters the separator, the reduction in pressure facilitates the release of dissolved gases, which are removed from the top of the vessel for further processing. The separation of oil from water relies on the variation in specific gravities of the two liquids in a process identical to that employed by the gas process production separator.

4.2.1.3 Test separator

The test separators are identical to the production separators in all but size, and are designed to process the contents of only one well at a time. A manifold arrangement provides the way that each individual well can be routed into the separator while metering devices located in the vessel discharge nozzles permit the measurement of oil/condensate, gas and water. The results obtained provide an indication of the performance of a particular well, and to the condition of the reservoir.

4.2.1.4 Condensate separator

The condensate separator is identical in operation to the gas process production separator and facilitates the removal of natural gas liquids (NGL) from the process gas stream.

4.2.1.5 Flash/degassing/vacuum separators

These vessels are all smaller versions of the oil process production separator that operate at, near or below atmospheric pressure.

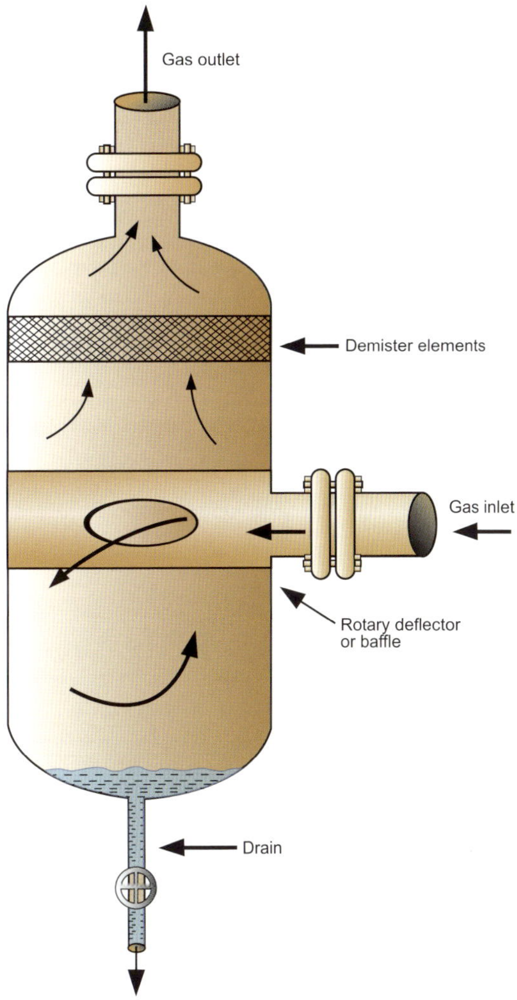

Figure 4.11 – Gas scrubber or suction drum

4.2.2 Gas/Liquid Separators

The vessels in this category are used to remove liquids from a gas stream and employ a primitive baffle or diffuser to divert the gas and eject the heavy liquid phases. No attempt is made to separate the liquids removed, they are simply discharged into the produced water system or routed back for further processing.

4.2.2.1 Slug catcher

The slug catcher will be found on the end of subsea pipelines and on gas production installations where the reservoir generates large quantities of water and/or sand. It prevents slugs (surges) of water or sand flooding the gas process plant and upsetting the more delicate items of equipment.

4.2.2.2 Scrubbers/suction drums

Identical to one another, the scrubber is located after a heat exchanger while a suction drum precedes a compressor. They are a more refined version of the slug catcher, operating on identical principles, but employing demister pads (metal mesh screens) more suited to the removal of condensed liquid droplets rather than large liquid slugs.

4.2.2.3 Knock out drums (or pot)

The KO drum is identical to the slug catcher but is employed at the end of vent headers (manifolds) to remove hydrocarbon liquids that could otherwise damage the vent stack, or constitute a hazard if released into the atmosphere.

4.2.2.4 Suction boot

The suction boots are normally associated with gas compressor suction pipelines and are best explained with a sketch. The vertical suction pipe is extended downwards so that moisture deposited during the ascent of the gas into the compressor can be safely removed from the system.

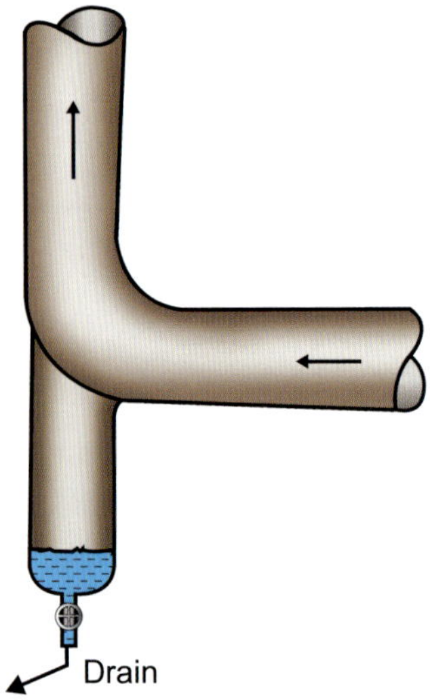

Figure 4.12 – Suction boot

4.2.3 Oil/Water Separators

The vessels in this category are used primarily for the separation of oil or condensate from water and, with the exception of the coalescer, are associated with the produced water and drains systems. The vessels rely on the variation in the specific gravities of the two liquids to effect separation. With the exception of the coalescer (and the hydrocyclone), they operate at atmospheric or near atmospheric pressure.

4.2.3.1 Oily water separators

The sketch in Figure 4.13 is self-explanatory, the various baffles channel the separated water to the outlet connection while oil is skimmed from the surface and drained into a holding tank.

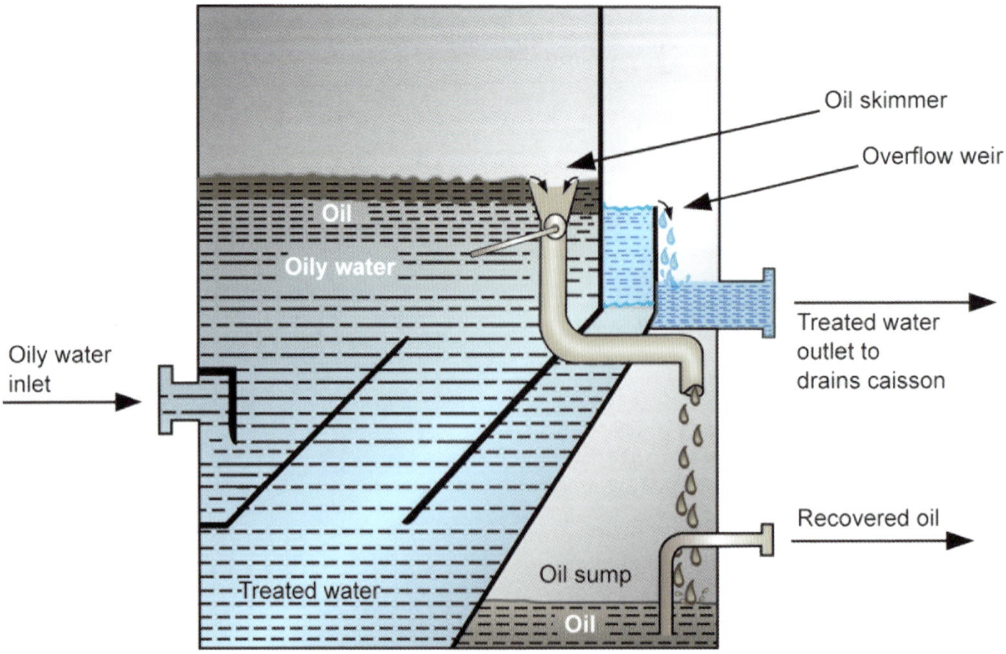

Figure 4.13 – Oily Water Separator (OWS)

4.2.3.2 Flotation unit

The operation of the flotation unit is identical to that of the OWS.

4.2.3.3 Skimmer

The flotation unit and skimmer are variations on the same theme. Some skimmers employ a separation weir while others are devoid of all internal equipment and rely on a level switch and an oil/water interface meter to effect separation of the two liquids via the control of drain valves.

4.2.3.4 Coalescer

Condensate received by the coalescer should be relatively water free. The coalescer performs a final polishing operation on the condensate prior to its injection into the subsea pipeline. The vessel operates at process pressure and contains metal screens, corrugated steel rolls or anthracite, which are all designed to increase the surface area of the vessel and encourage entrained moisture to coalesce and separate from the condensate.

4.2.3.5 Hydrocyclones

The operating principles of a hydrocyclone differ from those of a conventional separator, although they still rely on the variation in the specific gravities of the two liquids to effect separation. They can be configured to separate two liquids or to separate solids from a liquid.

Hydrocyclones operate under system pressure, and utilise pressure drop as the primary source of energy for separating oil from water. Within the vessel, a swirling motion is imparted to the water as it enters the tapered tubular liners through tangential ports, and the velocity of the water is further accelerated by the tapered shape of the liner. The water, being heavier than oil, is thrown to the walls of the liner by centrifugal force and exits the vessel through the discharge nozzle, while the oil forms a core at the axis of the column of water and is discharged from the inlet end of the vessel.

4.2.4 Holding Tanks

A number of holding tanks are incorporated in the design of the various piping systems and the bad oil, drains sump and drains drums are typical examples. Generally speaking, the tanks are atmospheric pressure vessels devoid of all internal equipment, which are used as temporary storage for oil, water and condensate products prior to subsequent processing. The vessel contents are maintained between pre-determined limits by a transfer pump activated by a level sensing device.

4.3 Piping and Pressure Vessel Design

There are a number of national standards organisations (NSO) and institutions that have developed specifications and codes of practice relating to the design, materials, fabrication, installation and testing of offshore piping systems, pipelines and pressure vessels:

1. BS — British Standards Institute
2. IP — Institute of Petroleum (UK)
3. API — American Petroleum Institute
4. ANSI — American National Standards Institute
5. ASME — American Society of Mechanical Engineers
6. NACE — National Association of Corrosion Engineers (USA)
7. ASTM — American Society for Testing and Materials

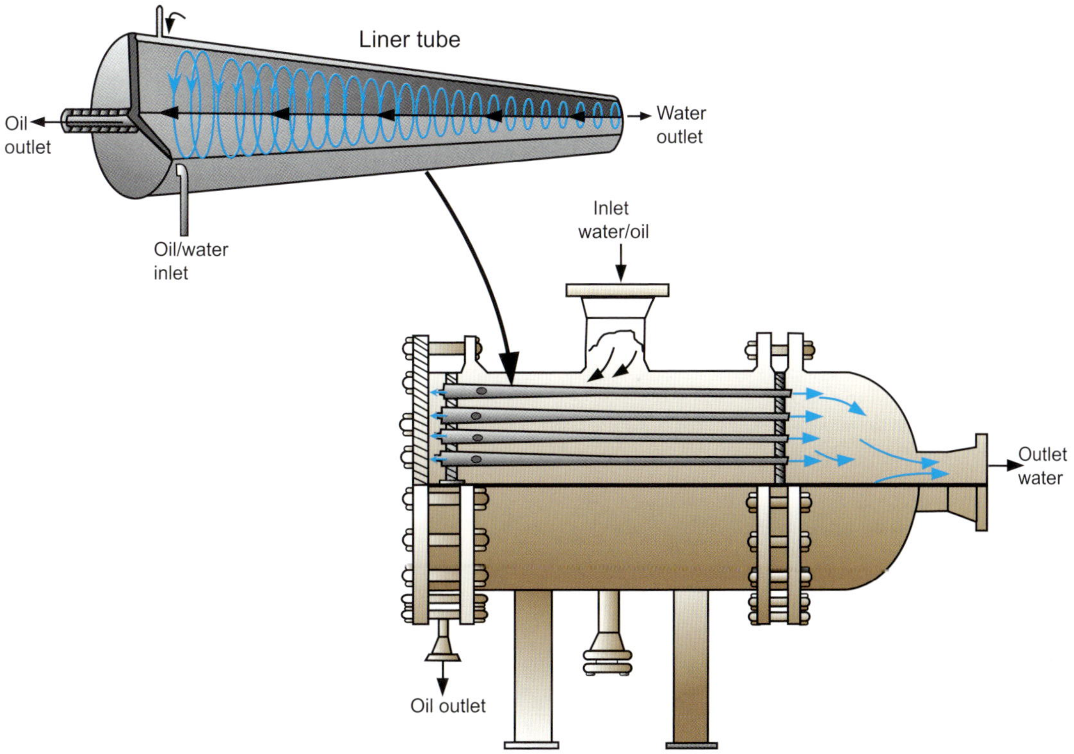

Figure 4.14 – Hydrocylone

8. MSS Manufacturers Standardization Society (USA)
9. AISI American Iron and Steel Institute.

Specifications may be loosely divided into three categories.

4.3.1 Piping Systems – Installation and Layout

The American Petroleum Institute (API) produce a number of Recommended Practices (RP), which provide guidance in the design and layout of offshore piping systems. They are informative, easy to read and used as the basis for industry standards worldwide. The more frequently used RPs are:

- **(i) API RP 2G** Recommended Practice for Production Facilities on Offshore Structures. (withdrawn)
- **(ii) API RP 14E** Recommended Practice for Design and Installation of Offshore Production Platform Piping Systems.
- **(iii) API RP 1111** Recommended Practice for the Design, Construction, Operation and Maintenance of Offshore Hydrocarbon Pipelines.
- **(iv) API RP 14C** Recommended Practice for Analysis, Design, Installation and Testing of Basic Surface Safety Systems on Offshore Production Platforms.
- **(v) API RP 520** Sizing, Selection and Installation of Pressure-Relieving Devices in Refineries, Parts I and II.
- **(vi) API RP 521** Guide for Pressure-relieving and Depressuring Systems.

4.3.2 Piping Systems – Design

In practice, we find that American standards predominate, regardless of the geographic location of the installations.

- **(i) ANSI/ASME B31.3.** Process Piping.
- **(ii) NACE MR-01-75** Sulphide Stress Cracking Resistant Material for Oilfield Equipment.
- **(iii) BS 3351** Specification for Piping Systems for Petroleum Refineries and Petrochemical Plants. (Withdrawn).

It would be true to say that virtually all hydrocarbon process pipework is manufactured in accordance with the requirements of ANSI/ASME B31.3. Reference to BS 3351 may occur on some older installations, and it provides some very valuable guidance, but this standard has been withdrawn.

When fabricating pipework for the conveyance of sour oil or gas, the requirements of the NACE MR–01–75 must be complied with in addition to ANSI/ASME B31.3. Sour reservoir products contain sulphur compounds, notably hydrogen sulphide (H_2S), which is a highly toxic, corrosive gas with the odour of rotten eggs. The UK fields produce predominantly sweet gas, but where H_2S does occur, the requirements of NACE must be strictly adhered to, particularly in the selection of materials and in the approval of welding procedures.

While it is permissible to use American Codes such as ANSI/ASME for process pipework on UK registered installations, the subsea pipelines come under the jurisdiction of the Pipeline Department of the *Health and Safety Executive* (HSE), and must be manufactured to the requirements specified in BS 4515 and PD 8010.

4.3.3 Pipeline System – Design

Pipeline standards, specifications and codes of practice
There are five documents that are likely to be referred to by the offshore industry for subsea pipeline systems:

(i) EN 14161:2003 – Petroleum and natural gas industries – Pipeline transportation systems.
(ii) BS EN 14161:2003 – Petroleum and natural gas industries – Pipeline transportation systems.
(iii) PD 8010-2:2004 – Code of Practice for Pipelines – Part 2: Subsea Pipelines.
(iv) API RP 1111 – Design, Construction, Operation, and Maintenance of Offshore Hydrocarbon Pipelines.
(v) The Pipeline Safety Regulations (PSR) – SI 825.

Details of each will be found in Appendix V.

EN 14161:2003 – Petroleum and natural gas industries – Pipeline transportation systems
When originally devised, the standard EN 14161 was intended to be the sole international standard in respect of pipelines, at least for the European Union. However, due to the major differences that exist between the requirements of member states, it was not possible to produce a single standard that suited all countries. Consequently, each member state was free to impose their own additional requirements on top of those specified in EN 14161.

While EN 14161 specifies requirements and gives recommendations for the design, materials, construction, testing and operation, maintenance and abandonment of pipeline systems, it should not be regarded as a design manual; rather it is intended to be used in conjunction with sound engineering practice and judgement. The standard is only applicable to rigid pipelines, not flexible ones.

BS EN 14161:2003 – Petroleum and natural gas industries – Pipeline transportation systems
The situation regarding the publication of BS EN 14161, the English language version of EN 14161, is somewhat strange. The UK voted against the adoption of the standard but was obliged to publish it in accordance with European Committee for Standardization (CEN) rules. It supersedes BS 8010-1:1989, BS 8010-2:1992 and BS 8010–3: 1993.

The British Standards Organisation recommend that BS EN 14161 is used in conjunction with their Published Documents (PD), these being;

- PD 8010-1:2004 – Code of practice for pipelines, Part 1 Steel Pipelines on land;
- PD 8010-2:2004 – Code of practice for pipelines, Part 2 Subsea pipelines.

PD 8010-2:2004 – Code of practice for pipelines, Part 2 Subsea pipelines
PD 8010 was prepared to cover a number of issues which are outwith the scope of BS EN 14161 and is intended to be used by designers, manufacturers and operators of pipelines. It gives recommendations and guidance on the design, use of materials, construction, installation, testing, commissioning and abandonment of carbon steel pipelines. Guidance on the use of flexible composite pipelines is also given.

A standard closely associated with PD 8010 is BS 4515-1:2004 - Specification for welding of steel pipelines. BS 4515 specifies the requirements for the welding of carbon, carbon manganese and low alloy steel pipelines with specified minimum yield strengths not exceeding 555 N/mm^2 and designed in accordance with PD 8010. It is applicable to both onshore and offshore pipelines and provides requirements for the testing of both welding procedures and welders.

API RP 1111 – Design, Construction, Operation, and Maintenance of Offshore Hydrocarbon Pipelines.
The American Petroleum Institute recommended practice, API RP 1111, has been in existence since 1976 and is now in its third edition and, as the title indicates, it covers the design, construction, operation and maintenance of subsea pipelines. A standard closely associated with API RP 1111 is API Std 1104 – Welding of pipelines and related facilities, and this is the standard to which all welding and inspection associated with the construction of the pipeline must be conducted. The API standards must be complied with when work is undertaken within the territorial waters of the USA but, like many of the British Standards, they are recognised for their excellence and are used all over the world.

While it is permissible to use American Codes such as ANSI/ASME for topside process pipework on UK registered installations, the subsea pipelines come under the jurisdiction of the Pipeline Department of the Health and Safety Executive (HSE), and must be manufactured to the requirements specified in British Standards. However, there are a number of API standards related to pipeline systems to which equipment is manufactured which can be used in UK waters. These relate primarily to flexible pipes used for risers and subsea connections, these being:

- API RP 17B – Recommended Practice for Flexible Pipe
- API Spec 17J – Specification for Unbonded Flexible Pipe
- API Spec 17K – Specification for Bonded Flexible Pipe
- API Spec 16C – Choke and Kill Systems.

4.3.4 Pressure Vessels – Design

ASME Boiler and Pressure Vessel Code
The ASME Code exists as a self-contained reference library which covers not only the design and manufacture of pressure vessels (Section VIII), but also the specifications for non-destructive examination (Section V), and approval of welding procedures and welders (Section IX).

The ASME Code consists of 11 volumes or sections which cover all aspects of fired and unfired pressure vessel construction. They are:

(i) Power Boilers
(ii) Material Specifications
  Part A – Ferrous Materials
  Part B – Non-ferrous Materials
  Part C – Welding Rods, Electrodes, and Filler Metals
  Part D – Properties
(iii) Subsection NCA – General Requirements for Division 1 and Division 2
  Division 1
  Subsection NB – Class 1 Components
  Subsection NC – Class 2 Components
  Subsection ND – Class 3 Components
  Subsection NE – Class MC Components
  Subsection NF – Component Supports
  Subsection NG – Core Support Structures
  Appendices
  Division 2 – Code for Concrete Reactor Vessels and Containments
(iv) Heating Boilers
(v) Non-destructive Examination
(vi) Recommended Rules for Care and Operation of Heating Boilers
(vii) Recommended Guidelines for the Care of Power Boilers
(viii) Pressure Vessels
  Division 1
  Division 2 – Alternative Rules
(ix) Welding and Brazing Qualifications
(x) Fibre-Reinforced Plastic Pressure Vessels
(xi) Rules for In-service Inspection of Nuclear Power Plant Components

The Sections covering NDE and welding qualification tests are referenced by numerous piping specifications, including ASME/ANSI B31.3.

PD 5500 Specification for Unfired Fusion Welded Pressure Vessels
The PD 5500 is not as self-contained as the ASME Code and must be used in conjunction with a number of British Standards to obtain information pertaining to NDE and welding requirements. The main supporting documents are:

(i)	**BS EN ISO 9606**	Approval testing of welders - Fusion welding. Previously BS EN 287
(ii)	**BS EN ISO 15607**	Specification and qualification of welding procedures for metallic materials - General rules. Previously BS EN 288
(iii)	**BS EN 970**	Non-destructive examination of fusion welds - Visual examination.
(iv)	**BS EN 1435**	Non-destructive examination of welds - Radiographic examination of welded joints.
(v)	**BS EN 1713**	Non-destructive testing of welds - Ultrasonic testing of welded joint.
(vi)	**BS EN 9934**	Non-destructive testing. Magnetic particle testing. General Principles.

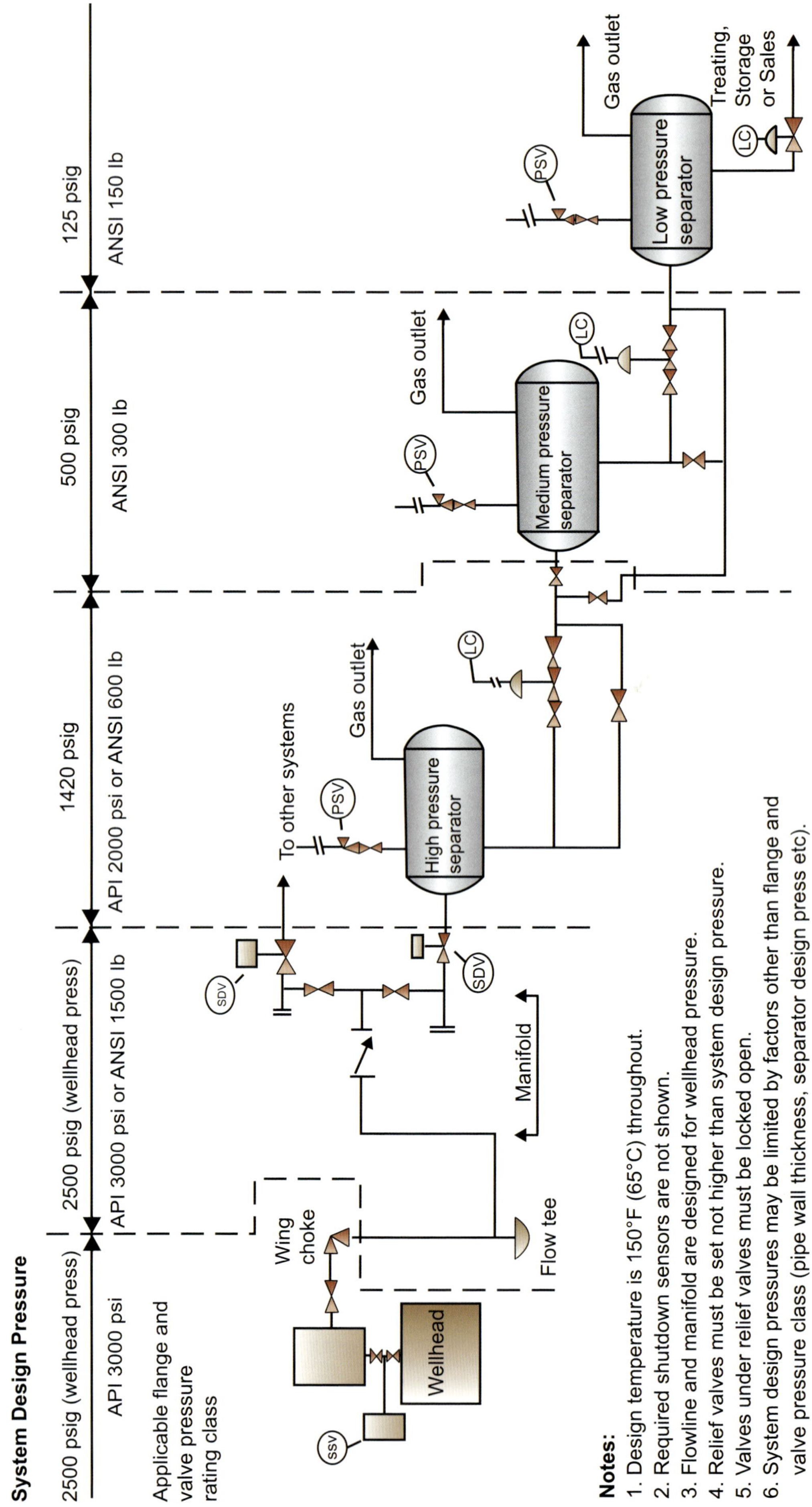

Notes:

1. Design temperature is 150°F (65°C) throughout.
2. Required shutdown sensors are not shown.
3. Flowline and manifold are designed for wellhead pressure.
4. Relief valves must be set not higher than system design pressure.
5. Valves under relief valves must be locked open.
6. System design pressures may be limited by factors other than flange and valve pressure class (pipe wall thickness, separator design press etc).

Figure 4.15 – From API RP 14e – Recommended Practice for the Design and Installation of Offshore Production Platform Piping Systems

4.4 Piping Systems – Construction

Figure 4.16 – Piping system

Offshore piping systems must be designed and constructed in accordance with a recognised fabrication specification. 'ANSI/ASME B31.3 Process Piping' tends to be used as the basis for the vast majority of process and utility piping systems. The specification stipulates parameters that cover the design, selection of materials, construction techniques, non-destructive testing (NDT) and hydrostatic pressure test requirements.

4.4.1 Piping Classes

A piping system must be designed to comply with a specific pressure rating or class, in accordance with the following table.

Class	Working Pressure		Test Pressure	
	psig	barg	psig	barg
150	285	20	450	30
300	740	51	1125	78
400	990	68	1500	104
600	1480	102	2225	154
900	2220	152	3350	230
1500	3705	255	5575	383
2500	6170	423	9275	639

Table 4.1 – Pressure rating

The figures quoted are for maximum non-shock working pressures at temperatures between minus 20°F (minus 29°C) and 250°F (121°C).

A piping system contains a selection of components categorised as either pipe, fittings or valves. These components are further categorised by the method of assembly used.

Figure 4.17 – Insulated pipework

Threaded
Threaded fittings are used extensively for low pressure utility systems and may be used for hydrocarbon service up to 2 inch (50 mm) nominal bore (NB).

Figure 4.18 – Butt weld and threaded flanges

Socket welded
Socket welded components use a fillet weld as the means of attachment, and may be used for utility pipework up to 3 inch (75 mm) NB and for hydrocarbon service up to 2 inch (50 mm) NB.

(a) Socket welded elbows

(b) Hub flanges

(c) Socket welded fittings and components

Figure 4.19 – Socket welded

Butt welded
Butt weld fittings are preferred for critical service applications such as high pressure hydrocarbon process systems. A butt weld offers maximum security and allows the application of more searching NDT techniques.

(a) Butt welded fittings

(b) Butt welded connection

Figure 4.20 – Butt welded fittings

4.4.2 Pipe Materials

Prior to selecting the pipe material specification, the service conditions of the system must be established. The service conditions relate primarily to the operating pressure, temperature and ability to resist corrosion. At one time, approximately 95% of offshore piping systems were manufactured from Grade B pipe complying with ASTM A106 or API 5L specifications, but today the use of ASTM A333 grade pipe has grown in popularity. These and other relevant specifications are outlined below.

ASTM A106 - 'Specification for Seamless Carbon Steel Pipe for High Temperature Service'
The vast majority of utility and process piping systems are made from pipe complying with this specification. Suitable for applications ranging in temperature from minus 29°C to 343°C (minus 20°F to 650°F), it is also used for systems operating in the normal operating temperature range of minus 29°C to 204°C (minus 20°F to 400 °F) and is available in sizes ranging from 2 inch (50 mm) NB to 24 inch (609 mm) NB.

API Spec 5L - 'Line Pipe Specification'
Line pipe to specification 5L Grade B is generally used for larger diameter pipeline systems operating in the ambient temperature range and is available in sizes up to 80 inch (2,032 mm) outside diameter. The specification includes a range of high strength steel pipes, such as API Spec 5L Grade X 52, a material that is frequently encountered in subsea

pipeline service. Due to the higher carbon content of these steels, a specialised welding procedure may be required and close supervision is necessary during production welding if sound welds are to be achieved.

Note: The numerical designation indicates the yield strength of the pipe, eg
X 52 = 52,000 psi (358 N/mm^2)
X 65 = 65,000 psi (448 N/mm^2).

ASTM A333 - Standard Specification for Seamless and Welded Steel Pipe for Low Temperature Service
There are a limited number of applications that require a pipe suitable for low temperature service and A333 is used extensively for high pressure vent systems due to the low temperatures generated by the gas as it expands during venting. The temperature range of the basic grade pipe is minus 46°C to 343°C (minus 50°F to 650°F) but grades are available for temperatures as low as minus 196°C (minus 320°F). As previously stated, the use of this material has grown in popularity in recent years and it has often been selected in place of Grade B materials for basic piping service. Many consider this to be an unnecessary expense as the Grade B pipe is usually quite adequate.

ASTM A312 - Standard Specification for Seamless, Welded and Heavily Cold Worked Austenitic Stainless Steel Pipe
Stainless steel pipes are used where resistance to corrosion or clinical cleanliness is required. The typical service applications are hydraulic control systems, chemical injection and helifuel distribution systems, where grade 316L is used extensively. It may also be specified for high temperature gas service lines, often associated with gas compression systems. When choosing 316L grades for service in a marine environment, the molybdenum content must be carefully monitored when purchasing the materials if surface corrosion is to be avoided. The molybdenum content can vary between 2-3% and materials with the higher molybdenum content should be specified.

ASTM A790 - Standard Specification for Seamless and Welded Ferritic/Austenitic Stainless Steel Pipe
Stainless steels in the A790 range are normally referred to as 'duplex' as they have significantly higher chromium content than the A322 stainless steels. The addition of nitrogen and molybdenum greatly enhances the mechanical properties of the steel, virtually doubling the yield strength and increasing the resistance to both corrosion and stress corrosion cracking. The price of materials increases in a similarly dramatic manner.

The combination of high strength and erosion and corrosion resistance has made duplex steels a popular choice for corrosive hydrocarbon service and they are being used increasingly for subsea pipelines and wellhead manifolds, particularly where sour reservoir products are encountered.

Cunifer
Cunifer is a 90/10 copper nickel alloy that is used extensively for constructing fire mains and general service seawater systems because of its ability to resist the corrosive effect of seawater.

Figure 4.21 – Cunifer firewater pipework

Glass Reinforced Plastic (GRP)
Glass reinforced plastics have slowly established a foothold on ships and offshore platforms in seawater service systems. The ballast lines that run through the ballast tanks are usually constructed of GRP, although steel transition pieces must be installed at the bulkheads. In the process areas, GRP is used on the low pressure side of the seawater injection system and for the overboard sections of the produced water lines.

Piping Schedule
Once the most suitable materials for a particular piping system and the diameter needed to provide a satisfactory flow rate is established, the wall thickness or pipe schedule required to withstand the design pressure and temperature can be determined from tables reproduced in the relevant specification. It is recommended that a minimum schedule 80 pipe thickness is used for hydrocarbon process systems of 3 inch (75 mm) NB or less to combat the effects of vibration and corrosion. The wall thickness quoted in the piping specification includes a corrosion allowance of 0.05 inch (0.2 mm).

Figure 4.22 – GRP pipe spools – seawater service

Material Identification

All pipes should be clearly marked with the material specification and grade, supported by suitable documentation, before being used for construction. On completion of fabrication, testing and coating, the piping systems should be colour coded. The particulars of the coding system and copies of the P&IDs (piping and instrumentation diagrams) should be available for review as supporting documents.

4.4.3 Fittings

The term 'fitting' includes items such as flanges, elbows, bends, reducers and branches and they should be manufactured in accordance with one of the following specifications.

(a)

(b)

Figure 4.23 – (a) Flanges – raised face and ring joint (b) Weld neck flanges

Flanges

ANSI B16.5 - Pipe Flanges and Flanged Fittings

There are seven flange ratings in the AISI/ASA series. These are used for virtually all utility and process services. These flanges are supplied with a raised face or profile (RP) sealing area up to and including AISI 400 rating while the higher pressure flanges are machined to accommodate ring joint (RJ) seals.

The seven available ratings are:

AISI 150, 300, 400, 600, 900, 1500, 2500.

The numerical part of the designation indicates the maximum non-shock loading working pressure, rating in psig at a temperature of 454°C (850°F).

The temperature range that most offshore piping systems are designed for is minus 29°C to 37°C (minus 20°F to 100°F). At these temperatures, the working pressures can be increased to the corresponding values shown in the table at the beginning of this chapter.

The unwritten rule for calculating the ambient working pressure of an AISI flange or valve is to multiply the designated class rating by 2.4:

eg AISI 300 × 2.4 = 720 psig

Also, the maximum test pressure can be approximated by multiplying by 3.5:

eg AISI 300 × 3.5 = 1050 psig

It should be noted that this rule does not apply to Class 150 components.

a. API Spec 6A - Specification for Wellhead and Christmas Tree Equipment
   The API range of flanges is used for the wellhead service and comes in five pressure ratings. The numerical designation indicates the maximum working pressure within the operating temperature range of minus 29°C to 121°C (minus 20°F to 250°F). The flanges are all the ring groove type and are available in the following pressure ratings:

   API 2000, 3000, 5000, 10000, 15000.

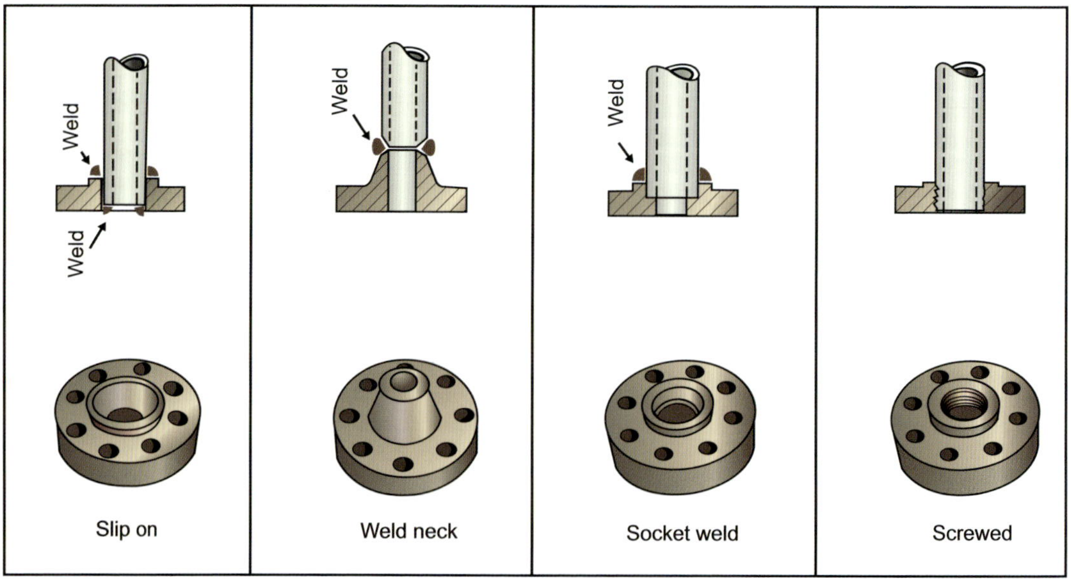

Figure 4.24 – Flange types

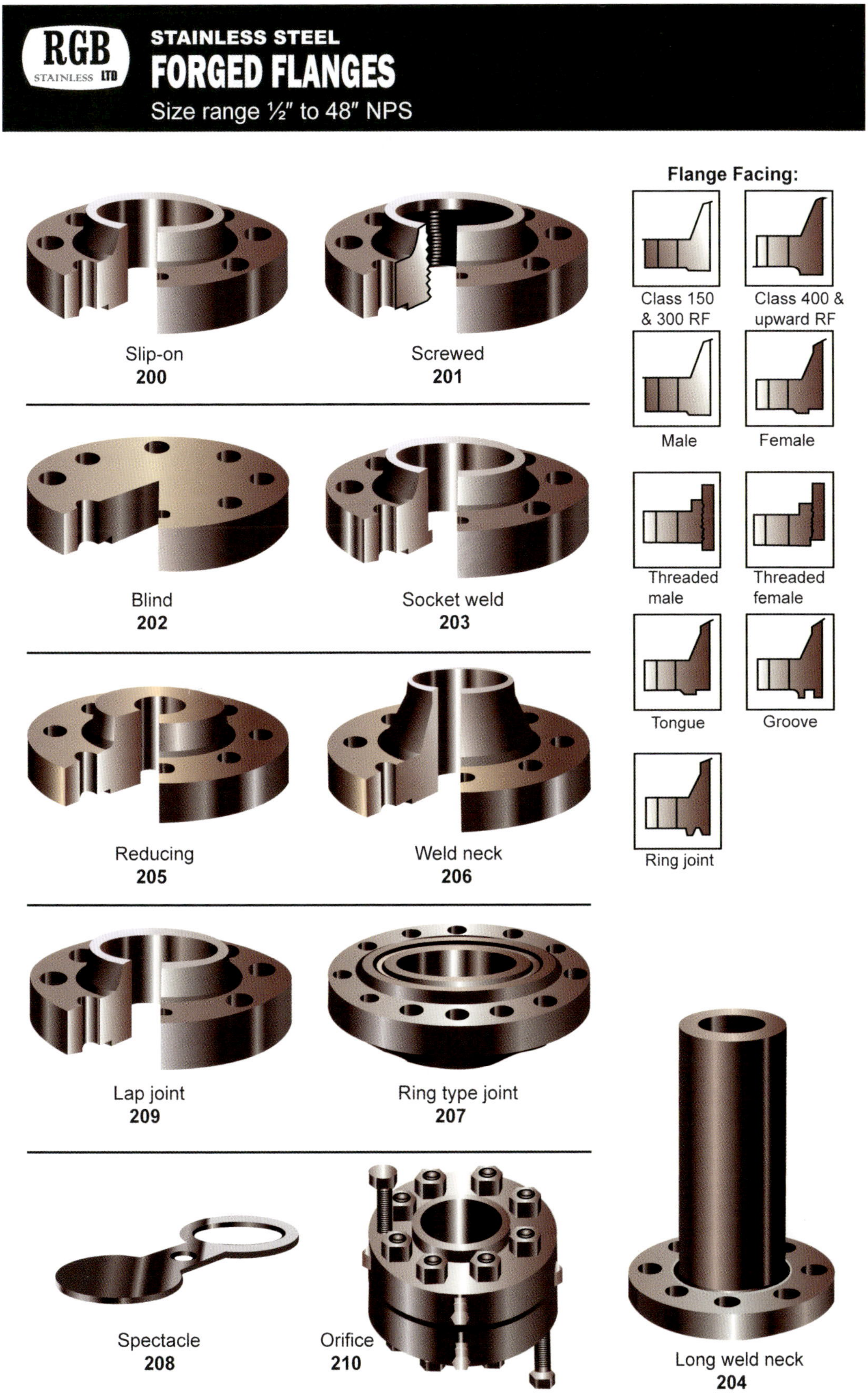

Figure 4.25 – Flange types

b. MSS SP-44 - Steel Pipeline Flanges
The Manufacturers Standardization Society (MSS) developed the Standard Practice (SP) 44 due to the demand for flanges larger than the 24 inch diameter covered by ANSI B16.5. The standard contains a selection of flanges varying in size from 12-38 inches and in pressure ratings corresponding to the customary ANSI classes, ie 150, 300, 400, 600 and 900.

c. Compact Flanges
In the last ten years, a number of specialised flange manufacturers have developed compact flanges. These do not comply with any of the recognised flange specifications, such as ANSI B16.5 or API 6A. They rely on a metal sleeve or bush seal rather than a ring joint or gasket on spherical faced nuts to ensure accurate alignment.

Compact flanges are physically smaller than the equivalent ANSI/API flanges and weigh up to 30% less. While they are more expensive to manufacture, the saving in material costs, particularly when duplex and titanium materials are used, tends to redress the balance. The reduction in the size of the flanges has been achieved by taking advantage of computer aided design techniques and improved manufacturing methods to optimise the design.

The main advantage of compact flanges is that they are lightweight and are favoured in subsea applications where the reduced weight simplifies installation. They are also used extensively as riser flanges attached to FPSOs, where the reduction in wave induced inertia forces on the flanges reduces the load on the risers.

d. Gaskets
Spiral wound asbestos gaskets are used as a sealing medium on AISI raised face (RP) flanges up to and including Class 400. The higher pressure rated flanges and the API flanges employ ring joint seals. Soft iron rings are used for 600 and 900 class flanges, and all other ratings use low carbon steel rings.

(a)

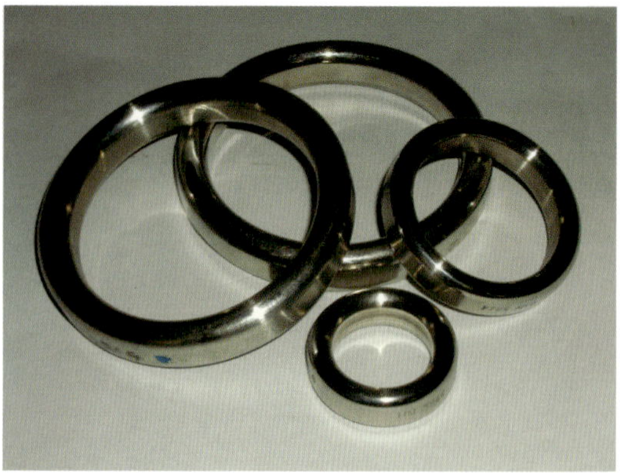

(b)

Figure 4.26 – (a) Spiral wound raised face gaskets (b) Ring type joints

Fittings other than flanges

The remaining fittings are covered by the following specifications, which are self-explanatory:

- ANSI B16.9 – Factory-made wrought steel buttwelding fittings
- ANSI B16.11 – Forged fittings, socket-welding and threaded

Fittings intended for use with Grade B pipe (A106 or 5L) should be manufactured from materials that comply with:

- ASTM A234 - Standard Specification for Piping Fittings of Wrought Carbon Steel and Alloy Steel for Moderate and High Temperature Service
- ASTM A105 - Standard Specification for Carbon Steel Forgings for Piping Applications.

4.4.4 Valves

Figure 4.27 – Valve arrangement. Note GRP grating.

The choice of valve for a particular piping application is dictated by the service conditions of the piping system and the operational mode required. A number of valve types are available.

(a) Ball valve with actuator

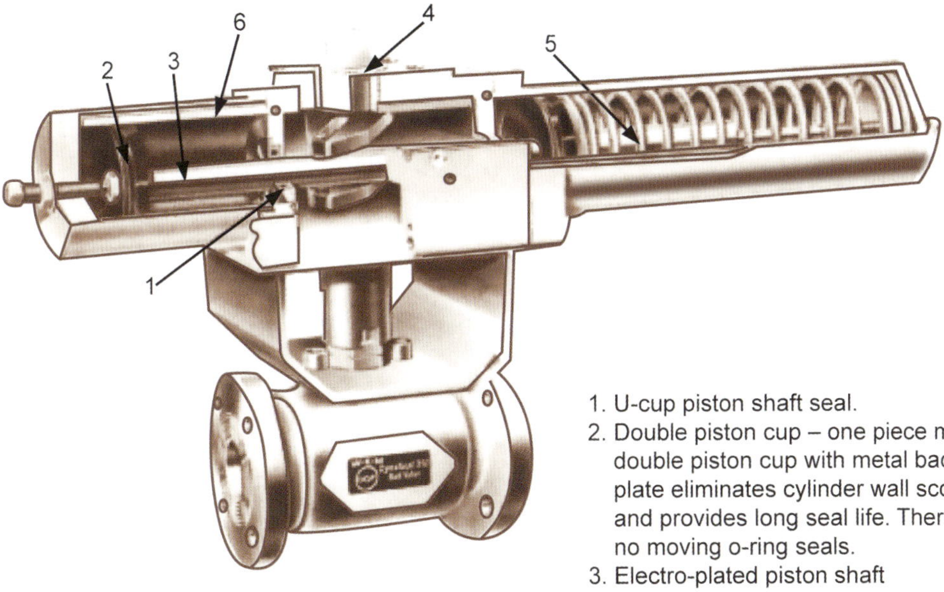

(b) Valve actuator

Figure 4.28 – Actuators

Ball valve

This is normally associated with hydrocarbon service as it provides an efficient on/off operation. The valve is ideally suited to automation due to the short distance that the actuator has to travel for full operation to take place.

(a) Ball valve

(b) Ball valve

(c) Globe control valve

(d) Globe control valve

(e) Ball valve

Figure 4.29 – Valves

Gate valve

The gate valve is used extensively in hydrocarbon service applications, and particularly for christmas tree valves. While the valve seats are prone to damage, they are relatively cheap and easy to replace.

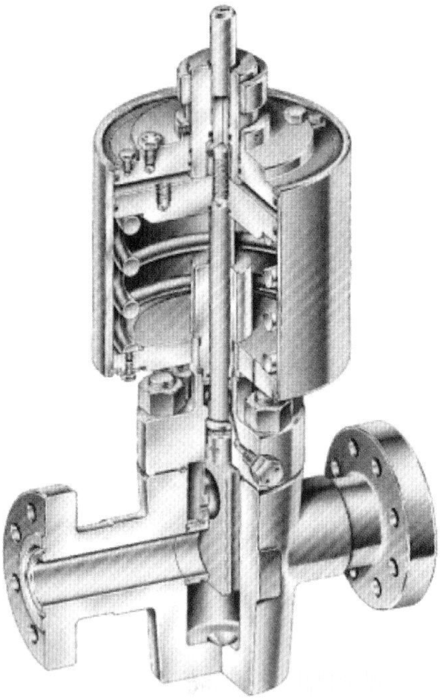

Figure 4.30 – Gate valve with PLS actuator

Plug valve

These have a similar action to a ball valve and are widely used for hydrocarbon systems of less than 3 inch (75 mm) NB. They are particularly suitable for drain lines due to their quick and efficient sealing action.

Choke valve

Manual and automatic chokes are used to regulate the flow of hydrocarbons in both process and drilling applications. On production installations, automatic chokes are often fitted to the christmas tree wing valves to reduce well pressures to the level required by the process plant.

Butterfly valve

Butterfly valves can be difficult to make leak tight, but using a neoprene sealing face enhances the sealing effectiveness. These are used predominantly as isolation valves on fire mains and seawater lines.

Figure 4.31 – Butterfly valve

Diaphragm valve

These tend to be restricted to low pressure seawater service, where they give an enduring and efficient seal.

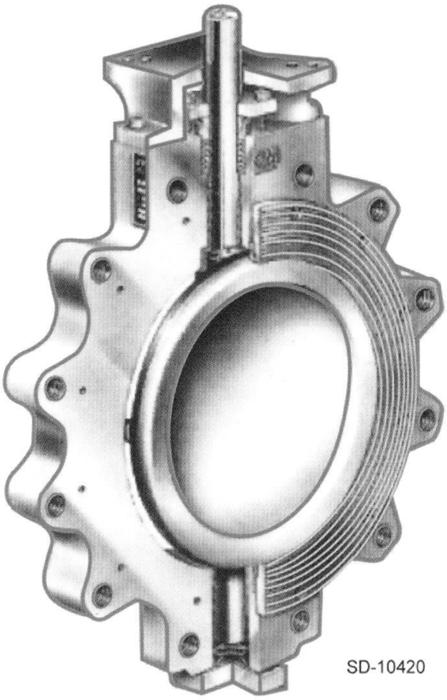

14"-24" class 150, 12"-24" class 300, 10"-12" class 600

Figure 4.32 – Diaphragm valve

Check valve

These can also be described as reflux, non-return, back pressure, retaining and check valves. The purpose of a check valve is to allow flow in one direction and to prevent it in the reverse direction. They are used extensively in both hydrocarbon and utility piping systems.

Valves intended for hydrocarbon process or fire main service must be approved in accordance with a fire test such as API RP 6FA or BS 6755.

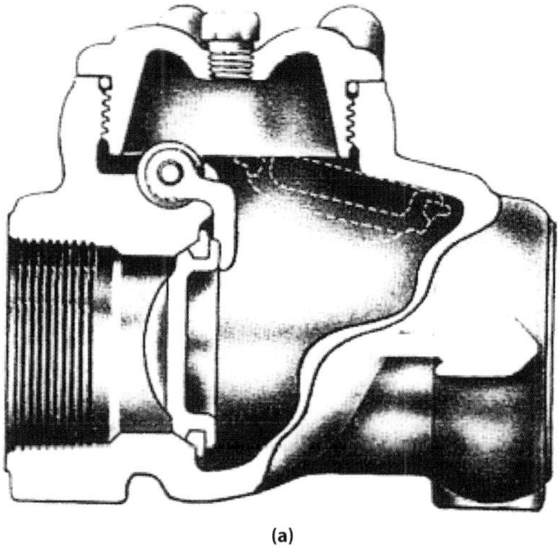

(a)

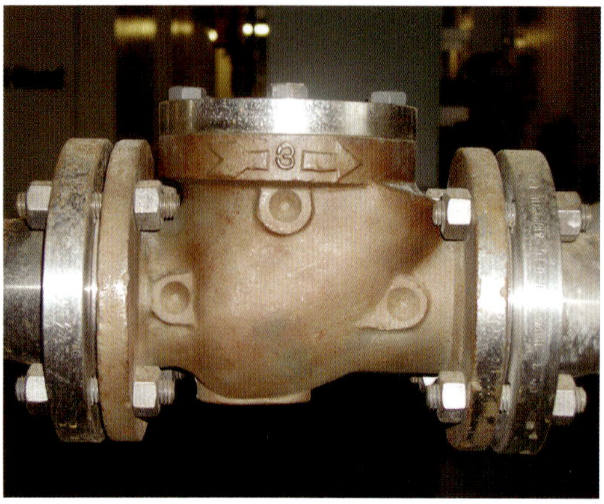

(b)

Figure 4.33 – Non-return valves

4.4.5 Bolting

Figure 4.34 – Bolting, cadmium plated and teflon coated

Stud Bolts

- ASTM A193 - Standard Specification for Alloy-Steel and Stainless Steel Bolting for High Temperature or High Pressure Service and other Special Purpose Applications
- ASTM A320 - Standard Specification for Alloy-Steel and Stainless Steel Bolting for Low-Temperature Service.

Nuts

- ASTM A194 - Standard Specification for Carbon and Alloy Steel Nuts for Bolts for High Pressure or High Temperature Service or both.

Nuts and bolts should be protected from corrosion by cadmium plating, hot dip galvanising or resin coatings. It is recommended that bolt tensioning equipment is used when tightening up large flange connections. A torque machine may also be used. Pounding with a hammer should only be undertaken as a last resort, when no suitable equipment is available.

Figure 4.35 – Flange tightening with hydraulic torque machine

4.4.6 Pressure Testing

The final test of the completed piping system is the hydrostatic pressure test, often referred to as a strength test or body test. This should be carried out at a pressure of 1.5 x design pressure and is intended to highlight any weakness in the structural integrity of the constituent components of the piping system. It should not be confused with a leak test, which may be carried out at 1.1 × design pressure after the strength test, the object of which is to ensure that there are no leaks from valves, gaskets, plugs or instrumentation that may have been installed after the strength test. A leak test using nitrogen gas with a trace of helium is often carried out before first oil or gas to ensure there will be no leaks on start-up. Each flange connection can be sniffed with a sensor to detect any helium traces as this gas will find a leak source that a liquid or methane gas test will not.

4.4.7 Pipeline Testing

On completion of installation, a pipeline must be hydrostatically strength tested to prove suitability for the intended service. The test or tests should be carried out against a written procedure that has been agreed by all the relevant parties and test records are to be retained by the operator for the life of the pipeline system.

Figure 4.36 – The Conoco-Phillips *'Mimas'* minimum facilities gas platform completing hydro-testing of the subsea pipeline. The test successfully completed, the water, which contains a green fluorescent dye, is pumped overboard.

The requirements for the testing of pipelines vary from code to code and a brief overview of the main requirements follows.

PD 8010-2: 2004 – Code of Practice for Pipelines
One of the most widely used standards for the pressure testing of subsea pipelines, particularly in the North Sea, is PD 8010 which requires pipelines to be subjected to a hydrostatic strength test at either 150% of the internal design pressure or the pressure that is expected to result in a hoop stress equal to 90% of the specified minimum yield stress, whichever is the lesser.

First, the pipeline is filled with inhibited seawater containing a fluorescent coloured dye that will give an immediate indication if there are any leaks present, the process being monitored by divers or ROVs. The pressure in the pipeline is then increased gradually and it is recommended that the rate of pressurisation should be no more than 1 bar/min up to 95% of the test pressure and 0.5 bar/min up to the test pressure. Hold periods for checking for leakage and for checking air content of at least 30 minutes duration are recommended at 60% and 90% of the test pressure. Throughout the test, the volumes of water added, the corresponding pressure rise and the time should be recorded, and the air content calculated. To ensure the detection of small leaks in large pipelines, and to cover the possibility of creep mechanisms causing failure after a number of hours, the test must be conducted for a period of at least 24 hours on completion of stabilisation.

EN14161: 2003 – Petroleum and Natural Gas Industries – Pipeline transportation systems
The pressure test requirements specified in EN 14161 are not as onerous as those specified by other standards. EN 14161 only requires pipelines to be strength tested to 1.25 × (maximum allowable operating pressure [MAOP] minus the external hydrostatic pressure) for a period of at least one hour after stabilisation. After a successful strength test, pipelines are to be leak tested at 1.1 × (MAOP minus the external hydrostatic pressure) for a period of at least eight hours. Alternatively, if the strength test is conducted for eight hours, the leak test can be waived. Furthermore, the time interval can be reduced to two hours if all parts of the pipeline or assembly being tested can be accessed for inspection.

Note: MAOP = Maximum allowable operating pressure.

API RP 1111 – Design, Construction, Operation and Maintenance of Offshore Hydrocarbon Pipelines
Except for fabricated items and components, all parts of an offshore pipeline should be subjected to a hydrostatic strength test of 1.24 maximum operating pressure (MOP) for a minimum period of 8 hours

4.4.8 Golden Welds

Occasions may arise when it is not feasible to hydrotest the closing weld or tie-in on a pipeline if, for instance, it is a new pipeline being joined to an extensive existing network. In these instances, what is referred to as a golden weld may be permitted. A tie-in weld that has not been subjected to a pressure test must be subjected to enhanced inspection arrangements, which include:

- Visual inspection of the preparation, the weld faces
- dye penetrant (DP) or magnetic particle inspection (MPI) of weld preparations to establish that there are no laminar or pipewall defects
- visual inspection of the total welding process
- 100% radiography and ultrasonic testing (UT), followed by MPI on completion of welding.

4.4.9 Pipeline Commissioning

On completion of pressure testing, the pipeline will normally be swept with a series of pigs and methanol/corrosion inhibitors. The pigs are pushed through by compressed nitrogen, which also dries the pipeline. Once the pigging operations have been completed and the pig launcher/receivers closed, all the relief valves and gauges that were removed for the pressure test can be reinstated.

5

Oil and Gas Production

5 Oil and Gas Production

The main function of the process plant on an offshore installation is to separate hydrocarbon products into liquid and gaseous phases and to remove any impurities that could inhibit transportation, such as water. The refinement process is loosely described as production.

The underground rock formations that create conditions conducive to the formation of an oil field are somewhat different to those that constitute a gas field and, for this reason, offshore installations are generally categorised as either oil or gas producing.

In the UK sector of the North Sea, all the gas fields are located in the southern sector where there is absolutely no oil. In the northern sector, the fields are predominantly oil producing but the oil is also associated with gas and water. Consequently, an oil production installation usually features a gas process plant in addition to the oil processing facility.

The production processes rely on very basic principles of fluid flow and thermo-dynamics and so you might expect to find a virtually standard design of offshore installation. In practice, this is certainly not the case, as while to a certain extent the process equipment must be tailored to suit individual reservoir conditions, it is often company policy that exerts the greatest influence on platform design. However, while equipment layout may vary from installation to installation, the general principles of operation remain the same.

To simplify the explanation of the oil and gas production processes, they have been dealt with separately.

5.1 Gas Production

The basic function of the process equipment on a gas producing offshore installation is to remove water prior to the transportation of the gas to the onshore reception facility.

Figure 5.1 – Gas installation

Figure 5.2 – Gas production rig

Figure 5.4 explains the gas production process. A more detailed description of the operating principles of the process vessels can be found in Chapter 4, Section 4.2.

5.1.1 Wellheads

The gas process commences at the christmas tree. This is the main isolation valve assembly through which wet gas at full reservoir pressure, 2,000–3,000 psi (140–200 bar), must pass before entering the production flowline. From the flowline,

Figure 5.3 – Christmas tree

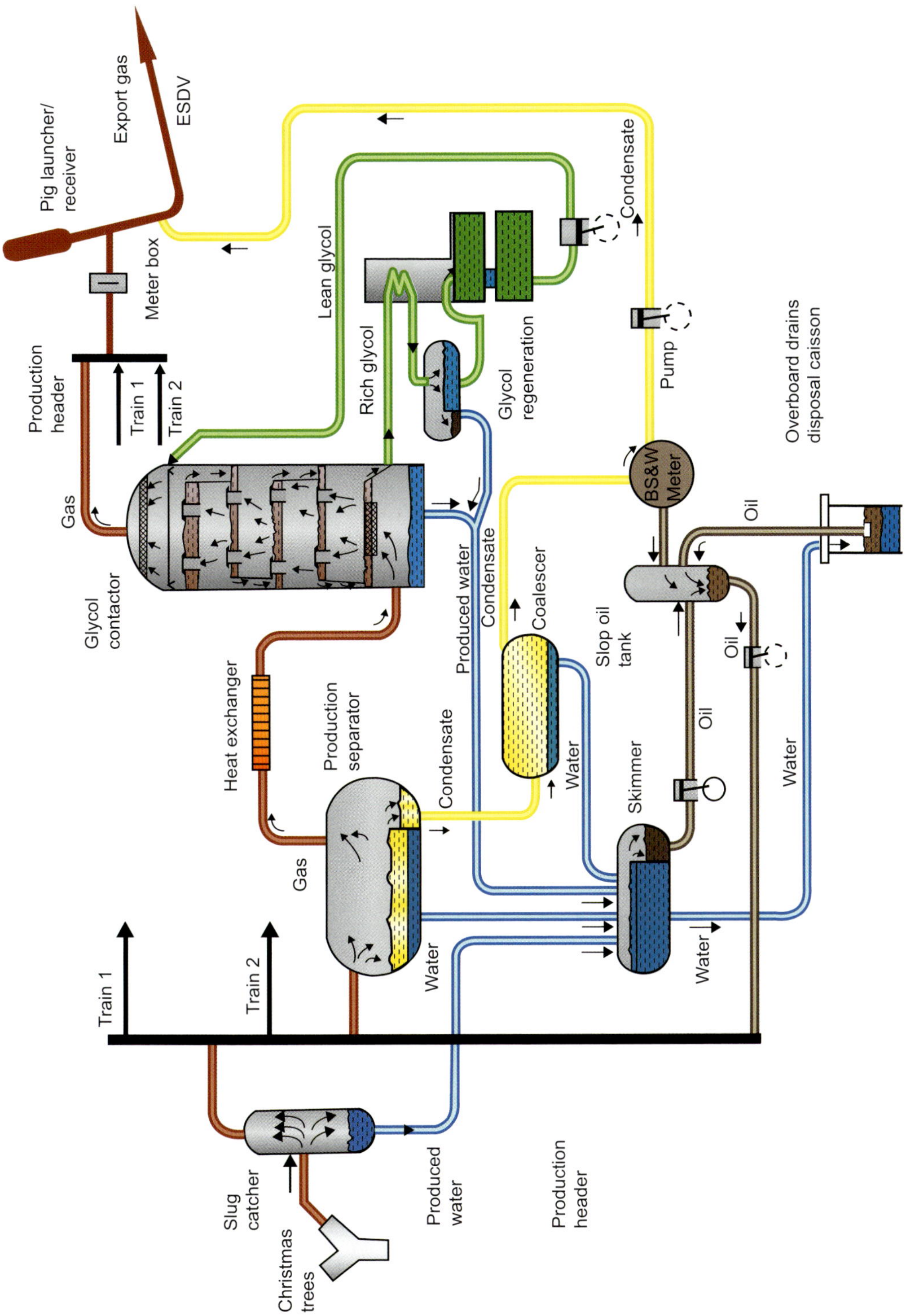

Figure 5.4 – Gas production process

the gas enters a slug catcher, where the first phase of liquid removal takes place.

5.1.2 Slug Catcher

The slug catcher removes large slugs of water and sand prior to the gas entering the production header.

In common with most gas process equipment, the slug catcher relies on very basic principles of fluid dynamics for its operation. The incoming gas is forced to make a sharp change of direction within the vessel, causing the heavy liquid phases to be thrown out into the base of the vessel. The liquids are disposed of by the produced water system.

5.1.3 Meter Run and Test Separator

Having removed any large slugs of liquid or sand that could affect the process equipment, the gas may flow through a metering orifice box prior to entry into the production header. Individual metering of each well permits output to be regulated and provides an indication of its performance. When more specific information is required, the well can be routed through a test separator, which can measure the exact quantities of gas, water and condensate produced. Regular condition monitoring of the reservoir provides an early indication of any problems likely to affect gas production and permits corrective action to be taken.

5.1.4 Production Header

The production header is a manifold that receives the gas from all the slug catchers (maybe 12) and redistributes it into a number of process trains (maybe 3). For reasons of clarity, only one process train is shown in Figure 5.4.

Some installations do not require slug catchers, so the production header receives gas direct from the christmas tree flowlines.

The next step in the liquid removal process takes place in the production separator. However, before the gas enters the production header it must be reduced in pressure to less than 1,440 psi (98 bar), the maximum operating pressure of the process train. The gas is reduced to this pressure due to the impracticality of constructing plant and equipment capable of operating at the full reservoir pressure.

Pressure reduction and the regulation of gas flow is through a control valve or choke, which is normally fitted between the slug catcher and production header. In the absence of slug catchers, the choke will be installed on the christmas tree outlet or wing valve.

5.1.5 Production Separator

The production separator separates gas and liquid in a similar manner to the slug catcher. However, the liquids thrown out as the gas changes direction settle and separate in the base of the vessel. The water enters the produced water system, while any condensate (condensed gas liquids) present is discharged into the coalescer.

The gas continues its journey by entering a heat exchanger.

5.1.6 Heat Exchanger

The heat exchanger produces gas at a temperature at which the water absorption process within the glycol contactor will be at its most efficient. The type of heat exchanger used will depend on reservoir conditions.

Some reservoirs produce gas at temperatures as high as 60°C (140°F) and, under these conditions, the gas will require cooling before it enters the contactor. This is normally carried out within a thermostatically controlled, fan assisted, fin-fan tube type cooler. Cooling the gas also assists in the removal of liquids that condense out as the temperature falls.

Once the gas is at the correct temperature it can be passed into the glycol contactor for final drying.

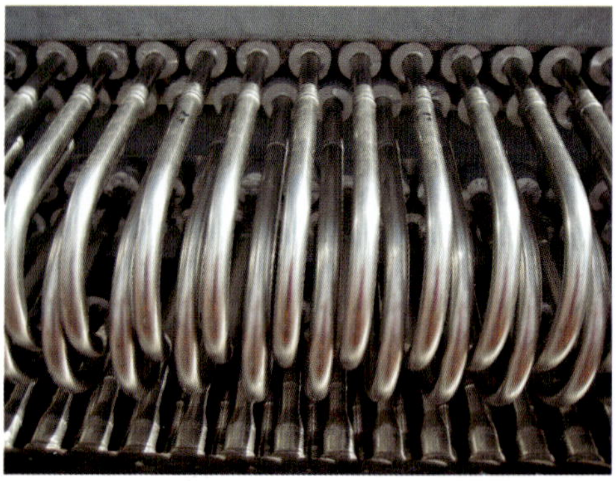

(a)

(b)

Figure 5.5 – Heat exchanger

5.1.7 Glycol Contactor (or Absorption Tower)

Figure 5.6 – Glycol contactor

This vessel may function as a combined scrubber and contactor, or the scrubber may be a separate vessel through which the gas passes prior to entering the contactor.

At this stage in the process the gas should be relatively dry. The scrubber will entrain any remaining water droplets within a series of fine metal mesh screens before the gas enters the contactor for the final mopping up operation.

The contactor consists of a vertical pressure vessel fitted with a number of horizontal trays over which triethylene glycol (TEG) cascades. As the gas flows upwards through the vessel it contacts the TEG, which acts like liquid blotting paper, absorbing any remaining moisture. The gas then passes through a mist extractor designed to remove any entrained glycol before entering the discharge header. The TEG leaves the base of the vessel for circulation through the glycol regeneration plant. (The dehydration process is described in greater detail under oil production.)

5.1.8 Discharge Header

The discharge header is a manifold that receives gas from the various process trains and redirects it into the export line.

5.1.9 Export Line

The export line usually contains a sphere launcher/receiver, a metering box, a condensate re-injection point and an emergency shutdown valve (ESDV).

5.1.9.1 Spheres (or pigs)

(a)

(b)

Figure 5.7 (a) – Pig spheres (b) Pig launcher/receiver

Over a period of time, liquids, scale and debris can accumulate in the horizontal section of the subsea pipeline, and eventually gas flow will be affected. To counter this problem large polypropylene spheres, or pigs, are periodically launched under gas pressure to sweep the line. The operation is referred to as 'pigging the line', the debris being pushed ahead of the pig into a slug catcher at the receiving installation or the reception facility on the beach.

Modern technology has spawned a new addition to the pig family, the 'intelligent pig'. These contain a magnetic eddy current measuring device energised by an intrinsically safe power source, and are employed to carry out an ultrasonic inspection of the subsea pipeline. The pig is propelled by gas (or oil) pressure and can quickly and efficiently detect any abnormalities in the condition of the pipeline.

Figure 5.8 – Condensate re-injection pumps

5.1.9.2 Condensate re-injection point

Condensate is pumped into the export gas for delivery to the beach, where it will again be separated for sale as a separate commodity to the gas.

5.1.9.3 Metering box

Gas leaving the installation enters a subsea pipeline, which is usually shared with other installations in the field. A metering box enables gas production to be calculated to ensure that an installation is properly credited.

Figure 5.9 – Metering box

5.1.9.4 ESDV

The ESDV is a fail-safe, fireproof valve designed to isolate the platform from the subsea distribution system. This ensures that, should an accident occur, the installation will not be fed with gas from other platforms. These valves were installed because this was the situation that arose on the *'Piper Alpha'* installation, a serious problem turning into a disaster as gas and oil from other platforms fuelled *'Piper Alpha's'* fires.

(a) ESDV fitted with H-120 fine protection jacket

(b) Cameron Iron Works (Cooper Oil Tools) emergency shutdown valve (ESDV) – installed on a subsea pipeline. The fail safe actuator can clearly be seen, as can the heat resistant cementations coating on the pipeline below the valve. Inset – a solid block Vetco Gray christmas tree. On this particular tree the lower master and swab valves are manually operated, while the upper master and wing valves are fitted with remote control actuators.

Figure 5.10 – Emergency shutdown valve

Figure 5.11 – Gas compression facility prior to installation offshore

5.1.9.5 Gas compression

Reservoir depletion eventually necessitates the installation of compressors to assist in the extraction of the gas (down to a pressure of less than 50 psi/3 bar) and to ensure that the gas is delivered to the national grid at a pressure of at least 1,000 psi (70 bar).

Frequently, one platform in the field will be provided with a compression facility and it will receive the gas from various satellite installations to compress for onward transmission.

Where space permits, a self-contained compressor package may be located within the confines of the existing installation. However, the larger developments may require an additional compression jacket to accommodate the compressor trains and associated process equipment. Compressor size and type will vary depending on field requirements, but the larger installations tend to use rotating compressors powered by marinised aero engines, manufactured by companies such as Rolls Royce. The gas engines run on fuel gas produced from process gas.

5.1.10 The Liquid Process

The basic function of the process equipment on an offshore gas installation is to remove water prior to transportation of the gas to the onshore reception facility.

Figure 5.12 – Gas compression station

The bulk of the water is deposited in the production separators in a relatively straightforward operation, but the process is complicated by the fact that the water is often accompanied by condensate, a valuable hydrocarbon by-product. In the production separators, the condensate and oil separate by gravitation prior to their discharge through separate outlets. The condensate is directed to the coalescer, where any remaining water is removed, while the water enters the produced water system where any remaining oil is removed.

Prevention of hydrates

Under certain conditions of temperature and pressure, water particles separate from the gas stream and freeze, trapping hydrocarbon molecules to form a solid ice-like substance known as a hydrate. Restrictions such as valve chests, orifice plates, pipeline reducers and bends exacerbate the problem as they create a throttling effect that further cools the gas and accelerates hydrate formation. If left unchecked, the hydrates will eventually form a blockage at these restrictions and, in extreme cases, complete sections of the subsea pipeline have been known to freeze.

Methanol and monoethylene glycol (MEG) are used extensively by the offshore industry to depress the dewpoint of the gas stream, reducing the possibility of hydrate formation. They may be injected into the process plant, subsea pipeline and the wells (methanol only) during periods of peak production when conditions of temperature and pressure approach values known to produce hydrates. The choice of inhibitor is largely dictated by individual company preference.

Should a hydrate blockage occur, it can be left to thaw naturally or melted by injecting methanol. Prior to thaw, the gas pressures either side of the blockage must be equalised if an explosive collapse of the hydrate is to be avoided. Propelled by gas pressure, a hydrate can cause a tremendous amount of damage, smashing valve chests, pipe bends and puncturing pressure vessels.

In time, the natural reservoir pressure will decline and remove the conditions associated with hydrate formation. This will permit the removal of certain items of process plant, such as the glycol regeneration system.

Prevention of corrosion

Because of the colossal costs associated with installing and replacing subsea flowlines and pipelines, every effort must be made to eliminate corrosion. While the major constituent of the gas dissolved in the crude oil is methane, normally around 92%, two of the least desirable constituents that may also be found are carbon dioxide (CO_2) and hydrogen sulphide (H_2S), the latter creating a condition referred to as sour gas.

CO_2 can account for up to 25% of the composition of the gas, and the carbonic acid produced in the presence of water is the primary cause of sweet corrosion in subsea pipelines. A water-soluble chemical is available that can be used to protect the subsea flowlines, pipelines and surface processing

Figure 5.13 – Hydrate formed in subsea flowline

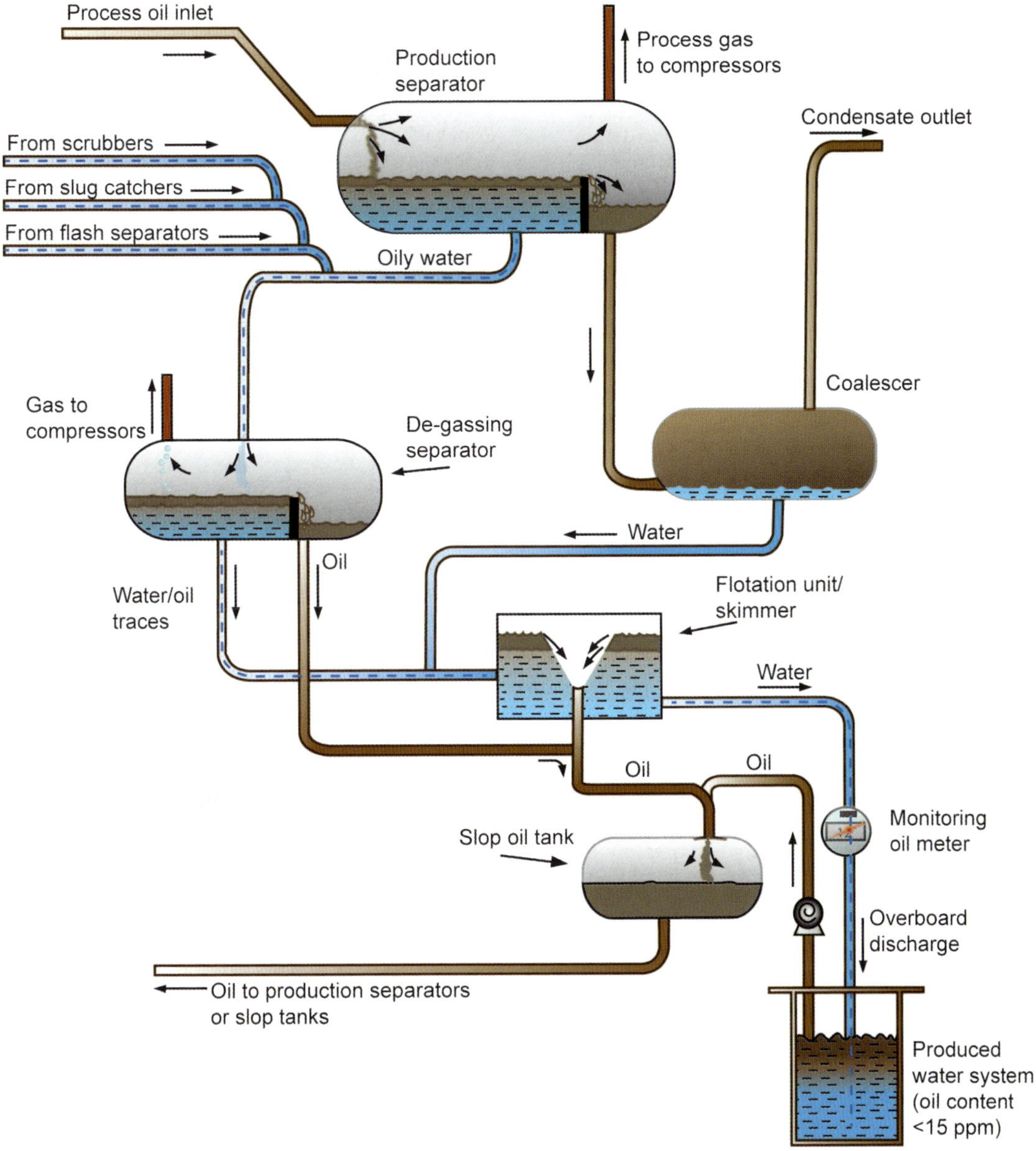

Figure 5.14 – Produced water and condensate system

equipment, the inhibitor remaining through to the water phase in the separators.

H_2S gas is extremely toxic and highly corrosive, so special precautions must be taken when fabricating pressure vessels and pipework that may be exposed to it. If present, H_2S must be removed if considerable damage to the process equipment is to be avoided and the welfare of platform personnel safeguarded. The removal process is referred to as gas 'sweetening'.

Corrosion inhibitors
The use of a corrosion inhibitor provides some protection to the internal surfaces of carbon steel pipework. The inhibitor (frequently Cronox 638S) is mixed with methanol or MEG (20%/80%), and may be injected into the wells, process plant and subsea pipelines.

5.1.10.1 Condensate system

Condensate is a clear, highly volatile liquid made up of the heavier hydrocarbon factions that condense out of the wet gas as it leaves the well. Liquid processing involves the removal of water from the condensate received from the production separators. The dry condensate can then be re-injected into the gas for delivery to the onshore reception facility ('the beach'), while the water is discharged into the sea via the produced water system.

Coalescer
The coalescer operates at process pressure and its function is to refine the condensate received from the production separator. The condensate passes through fine wafer pads, or demisters, onto which the water droplets coalesce before migrating into the base of the vessel.

The condensate passes through a BS&W (basic sediment and water) detection meter as it leaves the coalescer. If free of water, the condensate is pumped into the gas export line for transportation to the beach. If an unacceptable water content is recorded by the meter, the condensate will be diverted into the slop oil tank for further processing.

Slop oil tank
The slop oil tank is a holding vessel for water contaminated condensate, known as slop oil or bad oil. It is fed from the coalescer, skimmer and caisson drains and the contents will eventually be pumped back into the production header for recirculation through the process equipment.

5.1.10.2 Produced water system

Produced water consists of oil or condensate contaminated water removed from the production separators and other process vessels. The function of the produced water system is to clean the water prior to discharge into the sea.

Generally speaking, the production of water increases as the reservoir ages. A basic gas production installation could produce less than 100 tonnes of water a day while a large oil production installation may have to cope with the disposal of in excess of 20,000 tonnes of water a day. While the quantities of water may vary considerably, the basic equipment and operating principles remain the same for both oil and gas producing installations.

Pressure reduction
The first operation entails a reduction in the water pressure from that of the oil or gas process to atmospheric, or slightly above atmospheric, pressure. This may be achieved by the insertion of orifice plates into the process vessel produced water outlet connections and a degassing drum may also be included in the system.

Degassing drum
The degassing drum facilitates the separation of oil, water and gas. The reduction in pressure permits the release of dissolved gases, which are disposed of at the HP vent or flare stack, while the liquids are separated using an oil/water interface. The oil or condensate is transferred to the slop oil tank while the water is directed to a polishing unit for further refinement.

Polishing unit
The polishing unit, which may also be referred to as a produced water separator, flotation unit or skimmer utilises the different specific gravities of oil and water to achieve separation by gravitation. Any remaining oil is transferred to the slop oil tank, while the water is discharged into the sea via a dedicated produced water caisson or the overboard drains disposal water caisson.

Overboard drains disposal water caisson
The overboard disposal drains caisson provides the final opportunity for any remaining condensate or oil to separate from the water. The oil floats to the top of the caisson, where it is returned to the slop oil tank by a transfer pump that is activated automatically by an oil/water interface sensor. An oil content of less than 15 ppm (parts per million) is required before the water can be discharged into the sea.

Hydrocyclones
The produced water system described is typical of that found on the vast majority of offshore installations. However, in recent years hydrocyclones have been installed by some operators who are faced with the problem of disposing of very large volumes of produced water, and they have proven to be extremely effective and reliable in service, there being no moving parts. They are installed immediately after the production separator. Water exiting the vessel should be routed through a degasser/skimmer prior to discharge to remove any entrained gases.

5.1.10.3 Wet gas

Wet gas production is particularly common on small unmanned, minimal facility installations that produce back to a central processing facility (CPF). The platform typically accommodates from two to four wells and, following pressure reduction at the christmas tree, the gas enters the subsea pipeline with no further processing taking place. The pipeline is dosed with corrosion inhibitors and MEG to prevent hydrate formation, the liquids being removed on the CPF.

The operating life of these platforms varies from five to ten years, and adequate corrosion allowances are included in the design to support this mode of operation.

Figure 5.15 – Unmanned gas producing platform *'Cavendish'*, operated by RWE Dea in the UK southern gas basin

5.2 Oil Production

Figure 5.16 – Oil platform

Crude oil is found in an underground reservoir accompanied by large quantities of hydrocarbon gas and formation water. It is the gas, and to a lesser extent the water, that provides the driving force required to bring the oil to the surface. While the oil may be difficult to entice from the ground, the subsequent processing required prior to its export is minimal. However, the same cannot be said for the refinement of the gas associated with the oil and it becomes apparent why, in the formative years of the oil industry, the gas was simply burnt off at the flare stack (shown in Figure 5.17). Today, energy conservation policies and the realisation of the value of the gas, in terms of both operational and financial benefits, have reduced the routine flaring of gas.

Figure 5.17 – Flare stack

If the equipment plan of an offshore installation is studied, a complex picture emerges involving a number of process systems intermingled with one another. However, it should be remembered that the basic object of the exercise is to separate oil, gas and water and to refine them to an acceptable level of purity prior to their discharge from the installation.

5.2.1 The Oil Process

The refinement of crude oil to a quality suitable for transportation by subsea pipeline or oil tanker is a simple operation consisting primarily of the removal of associated gas and formation water within a series of production separators.

A sketch, in Figure 5.19, has been prepared to assist in the explanation of the hydrocarbon process equipment employed on an oil production installation. More detailed descriptions of the process vessels and principles of operation can be found in Chapter 4.

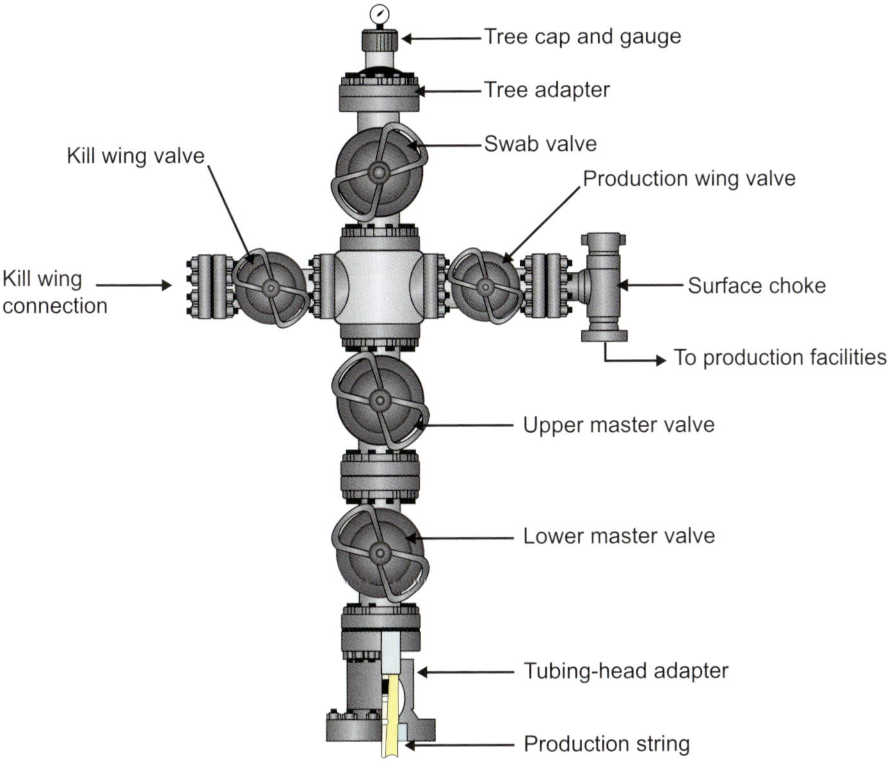

Figure 5.18 – Wellhead

Wellheads
The oil process commences at the christmas tree, where oil at full reservoir pressure is admitted to the installation and reduced in pressure to approximately 450 psig (30 bar) prior to entry into the production header or the well test separator.

Production header
The production header is a manifold that receives oil from all the wells on the installation (up to 60) and redistributes it to a number of process trains (up to 4). The process trains are identical and only one has been shown on the sketch in the interests of clarity.

From the production header the oil enters the first of a series of production separators.

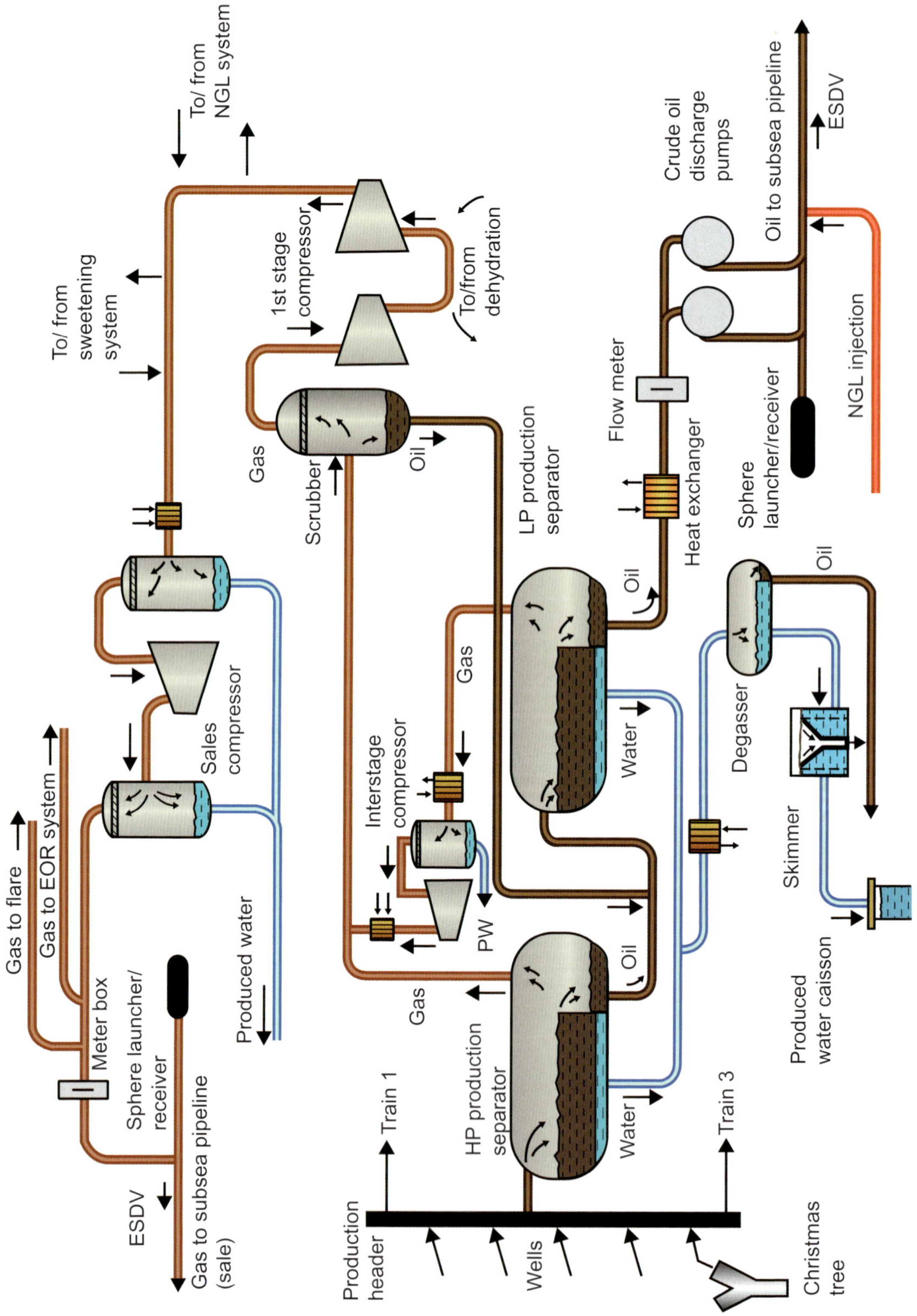

Figure 5.19 – Production of oil and associated gas

Production separators

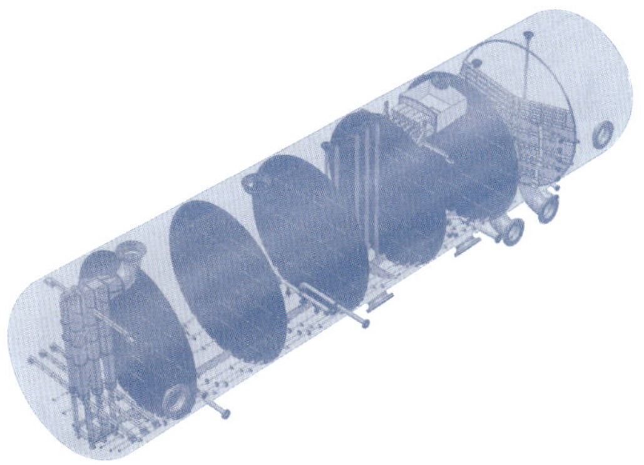

Figure 5.20 – Production separators

The number of production separators employed will be dependant on the volume and pressure of the liquids entering the system, but most installations require at least two and occasionally three per train, which operate at successively lower pressures.

The production separators facilitate the separation of oil, associated gas and water. The removal of gas is assisted by the reduction in pressure to atmospheric or near atmospheric conditions (45 psi–3 bar), while the separation of oil from the water relies on an oil/water interface. The oil may be heated prior to entry into the separator to lower its viscosity and enhance the separation process.

The water is disposed of via the produced water system while the oil is virtually ready for export. The export oil is referred to as dead crude if processing was completed at atmospheric pressure, while live crude describes oil processed at a pressure slightly above atmospheric.

Test separator
The test separator is identical to the production separator in all but size and fulfils the same function as the test separator described in the gas process.

Crude oil export
Prior to export, the crude oil will normally be subjected to cooling, metering, injection of natural gas liquids (NGL) and a significant increase in pressure by the discharge pumps.

(a) Cooling
The temperature of oil produced from a deep reservoir can be considerable, occasionally in excess of 180°C (380°F). Consequently, a number of heat exchangers may be deployed prior to, between and after the various stages of separation to change the temperature of the oil to assist in the separation of the oil from the water, and in the stabilisation of the crude prior to export.

(b) Metering
Because oil is normally transported by a subsea pipeline that is frequently shared with other installations, it is important that the quantity of oil exported is measured so that each installation may be credited. The oil is metered prior to the injection of NGL, which is measured separately.

(c) NGL injection
The disposal of NGL, by injection into the crude oil prior to export, is analogous to the re-injection of condensate encountered on gas production installations. The process is known as spiking and transforms a dead crude into a live one.

The practice of spiking crude can only be performed on pipelines that terminate at a refinery equipped to handle live oils.

(d) Pumping
Crude oil must be subjected to a considerable increase in pressure (750 psi/50 bar) if it is to be transported by subsea pipeline. The discharge pumps are generally of a two stage centrifugal type, electrically powered through a synchro-torque variable speed coupling that can be adjusted to suit fluctuating oil production. Alternatively, gas turbine driver pumps may be installed.

In common with gas producing installations, the export pipeline is fitted with a sphere launcher/receiver and an emergency shutdown valve (ESDV).

5.2.2 Improvement in Oil Production

Pour point depressant
As the crude oil flows from the reservoir on initial start-up, wax particles can precipitate from the oil as the cooling effect of the seawater is encountered. If the concentration of wax increases beyond a certain point, an interlocking structural network can form as the oil changes into a gel with solid properties. This is an undesirable situation. A pour point depressant lowers the temperature at which the precipitation of the wax commences.

Solvent
This is injected into subsea flowlines to reduce wax deposit build-up.

Asphaltene inhibitors
These are injected to prevent precipitates forming from the asphaltene fraction of crude oil. Asphaltenes are non-volatile, high molecular weight fractions in the oil and, while they are usually dissolved or dispersed within the structure of the crude oil, they can separate out, particularly under turbulent flow conditions. The stability of the asphaltene dispersions can also be disrupted when condensates and other non-compatible oils are mixed together.

Paraffin inhibitor
This is injected into the oil at the wellhead, before it has cooled to its cloud point, to prevent or minimise paraffin deposits. The effectiveness of the inhibitor is partly dependent on the composition of the oil and asphaltene as this can reduce the effectiveness of the inhibitor.

5.2.3 Associated Gas

The quantities of gas associated with crude oil production can be considerable and dictates both the processing arrangements, and the method of disposal.

5.2.3.1 The gas process

The main objective of gas processing is to produce a gas that complies with the dew point (hydrocarbon and moisture) criteria specified for the subsea pipeline to ensure the avoidance of problems associated with corrosion, hydrate formation and the build-up of condensed liquids. This involves the removal of water, natural gas liquids (NGL) and gaseous impurities such as hydrogen sulphide (H_2S) and carbon dioxide (CO_2), should they occur.

Production separator
The gas process commences at the production separator where the reduction in pressure facilitates the vaporisation of gas from the crude oil. Two stages of separation are shown in the sketch, but three successive stages of pressure reduction are frequently employed.

Gas from the HP separator passes directly to the compressor inlet scrubber, while gas liberated from the LP separator must undergo compression, cooling and scrubbing (liquid removal) before recombining with the HP gas.

Inlet scrubber
The inlet scrubber is concerned primarily with liquid removal prior to the gas entering the first full stage of compression. Liquids removed may be returned to the inlet of the LP production separator or they may be disposed of via the produced water system, depending on the design of the plant.

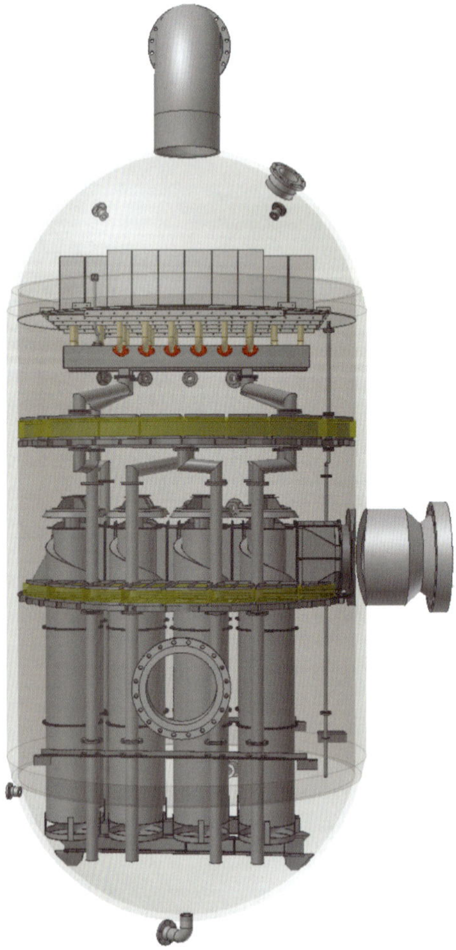

Figure 5.21 – Inlet scrubber

Compression
Gas processing involves a considerable number of compression stages. This is because associated gas is released from the crude oil at relatively low pressures and requires elevation to in excess of 1,000 psi (70 bar) for export and 6,000 psi (420 bar) if it is to be re-injected into the reservoir for enhanced oil recovery.

Figure 5.22 – Suction scrubber vessel

The inclusion of gas compressors greatly complicates the process plant as each unit must be provided with suction and discharge scrubbers and heat exchangers to avoid the dangers associated with condensed liquid carrying over into the machinery. Both electrically powered reciprocating and gas turbine driven centrifugal compressors are used extensively offshore, the centrifugal types favoured for high volume, low to medium pressure service while the reciprocating variety are more suited to low volume, high pressure duties such as gas re-injection into the reservoir.

Intermediate processing
At this stage in the process, the gas is relatively moisture free and the subsequent processing required will be dependant on the condition of the gas as it left the reservoir. While water removal and compression equipment are common to all installations, dehydration, NGL removal, stabilisation and sweetening facilities are only incorporated as and when required. These intermediate process arrangements will be discussed at the end of this section and their take off points (if required) can be seen on the Production of oil and associated gas sketch, Figure 5.19.

Export
Depending on the quantities produced, the process gas may be sold, used for enhanced oil recovery (EOR), fuel gas or flared.

(a) Sale
When substantial quantities of associated gas are produced, and it can account for up to 20% of the installation's total hydrocarbon

output, the financial returns may justify the construction of a subsea pipeline for its transportation and sale.

(b) Enhanced Oil Recovery (EOR)
Associated gas can be utilised for re-injection into the reservoir or for gas lift operations. Such is the demand for gas for EOR that installations frequently import gas through a subsea pipeline when their own reserves are inadequate.

(c) Fuel Gas
Virtually all installations are self-sufficient in fuel gas, which is used to fire regeneration boilers, gas turbine driven alternators, pumps and compressors. The fuel gas is normally produced as a by-product of processes such as NGL removal and stabilisation.

Figure 5.23 – Reciprocating gas powered generator

(d) Flaring

Figure 5.24 – Flare stack

The flare on an oil producing installation burns continuously, providing a way of disposing of unsaleable and toxic gases, and any gas surplus to requirements produced during periods of plant start-up or maintenance (Figure 5.24).

Where the gas is destined for sale it will undergo re-compression prior to being transported to an onshore reception facility by subsea pipeline. The pipeline will include a metering box, sphere/launcher receiver and an emergency shutdown valve (ESDV), identical to the arrangements employed on a gas production installation.

5.2.3.2 The dehydration process

The function of the glycol contactor, dehydration or absorption tower is to dehydrate process gas to ensure that the dewpoint temperature is below that point at which moisture condensation, and hydrate formation may occur in the subsea pipeline.

The dehydration process (Figure 5.25) is based on the strong affinity that glycols have for water. They are hygroscopic and triethylene glycol (TEG) is chosen in preference to other glycols because of its greater ability to resist the high temperatures associated with the regeneration process. At the commencement of the dehydration process, the dry glycol is referred to as being in the 'lean' condition, while on completion of water absorption it is described as being 'rich'.

Figure 5.25 shows a typical dehydration/regeneration plant.

Dehydration
The main function of the contactor is to provide conditions conducive to the water absorption process, ie the thorough mixing and agitation of the glycol and process gas. It is a tall, vertical pressure vessel that is divided into numerous sections by horizontal trays fitted with non-return or bubble valves. Lean glycol entering the top of the vessel cascades downwards, flooding each tray in turn while the process gas flows upwards, passing from section to section through the non-return valves.

The non-return valves force the gas to bubble through the glycol and this contact greatly enhances the water absorption process. Eventually, dry process gas emerges from the top of the contactor, after having first passed through a demister pad designed to remove any entrained glycol. The rich glycol accumulates in the base of the vessel prior to entry into the regeneration plant.

Throughout the dehydration process the temperature of the glycol must be maintained at approximately 2°C to 3°C above that of the process gas to prevent the condensation of liquid hydrocarbons into the glycol.

Regeneration
The dehydration process benefits from high process pressure and low process temperature, conditions that increase the water absorption capacity of the glycol. In the regeneration process, the conditions are reversed to encourage the release of water vapour and to return the glycol to the lean condition suitable for re-circulation.

The main components of the regeneration plant are the flash separator that reduces the glycol pressure to atmospheric, and the reboiler where the actual drying out or purification process takes place.

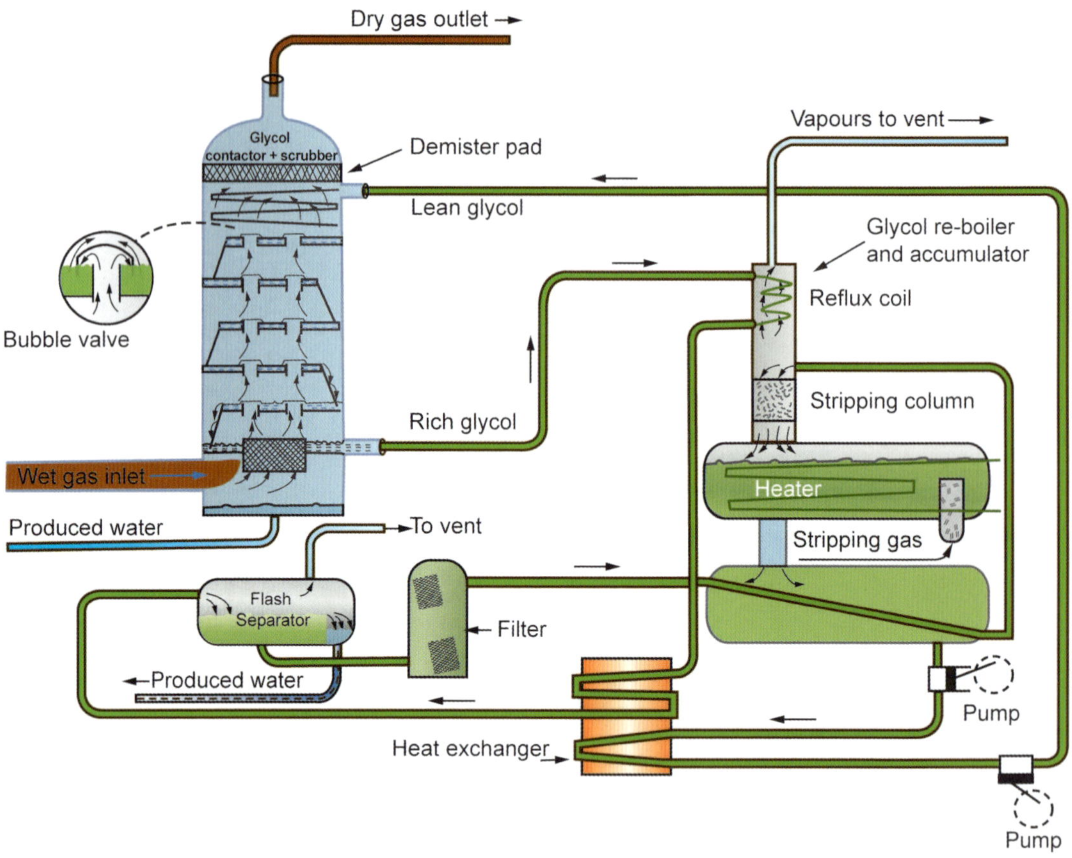

Figure 5.25 – Gas dehydration process

With reference to the diagram in Figure 5.25:

(a) The Reflux Coil
The reflux coil, or the 'reboiler vapour condenser', is located in the top of the stripping column and is a heat exchanger whose function is to pre-heat the rich glycol entering the regeneration system and to reduce glycol losses by encouraging the condensation of glycol from the hot gases and water vapours leaving the reboiler drum.

(b) Flash Separator
The flash, or condensate, separator operates at atmospheric pressure and provides three phase separation. Process gas is released to the LP vent stack, water and condensate to the produced water system and the glycol proceeds to the filtration package.

(c) Filtration Package
Glycol is a viscous liquid with a tendency to foam. Impurities, whether in liquid or solid form, accentuate this tendency. The flash separator provides for the removal of liquid impurities while the filtration package removes solid particles.

(d) Heat Exchangers
The glycol regeneration process involves a considerable transfer of heat both to and from the glycol and a number of heat exchangers are employed to increase the overall efficiency of the operation.

(e) Glycol Reboiler
The purification of rich glycol happens within a glycol reboiler or regenerator. The boiler, which operates at atmospheric pressure, employs a combination of heat and stripping gas to remove entrained water.

Rich glycol enters the reboiler drum at the base of the stripping column, where it is heated to a temperature of 200°C (418°F) by gas fired burners (or electric heaters). Ideally, a slightly higher temperature is preferred because a temperature of 200°C (418°F) will only guarantee a glycol purity of 98.5%. Unfortunately, chemical breakdown of the glycol commences at 204°C (425°F) and the oxalic and glycolic acids produced are highly corrosive. To compensate for the temperature limitations of the glycol, a stream of stripping gas (fuel gas) is admitted into the base of the reboiler drum. The gas enhances the separation effect by reducing the partial pressure of the water vapour, which permits an improvement in glycol purity to 99.9%, the value required by the dehydration process.

The lean glycol leaving the base of the reboiler drum is temporarily stored in the glycol accumulator, before cooling and subsequent transfer to the glycol contactor.

5.2.3.3 Natural Gas Liquid (NGL) removal

Natural gas is a complex mixture of hydrocarbons, primarily methane and ethane but including heavier gaseous hydrocarbon fractions such as butane and propane, which possess boiling points within the ambient temperature range. These heavier fractions readily condense, are referred to as natural gas liquids (NGL) and can account for as much as 10% of the total oil output of an oil production installation. The more volatile of these liquid fractions are referred to as condensates and tend to be associated more with gas producing installations.

The function of the NGL system is to create conditions conducive towards the condensation and subsequent separation of the heavier gaseous fractions (NGL) from the export gas to ensure that it complies with the hydrocarbon dewpoint criteria specified by the subsea pipeline.

In common with the majority of hydrocarbon process operations, the removal of NGL uses basic principles of fluid and thermodynamics, ie a reduction in temperature to encourage condensation followed by a sharp change in direction designed to throw out the liquid phases formed. Seawater cooled heat exchangers provide the means by which the initial reduction in temperature is achieved, with subsequent stages employing a throttling process or a dedicated refrigeration plant.

Throttling
The NGL removal system employs two preliminary stages of cooling and NGL separation prior to the

Throttling Process

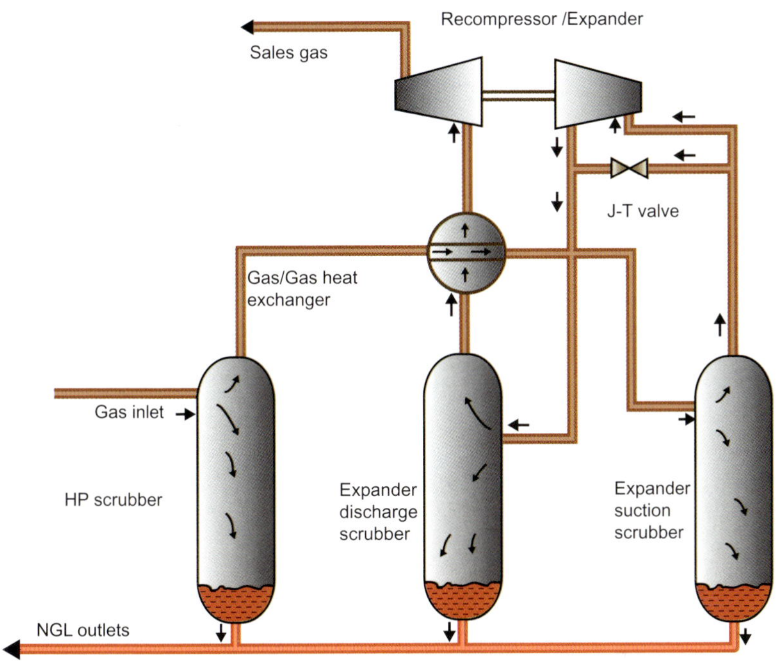

Refrigeration Process

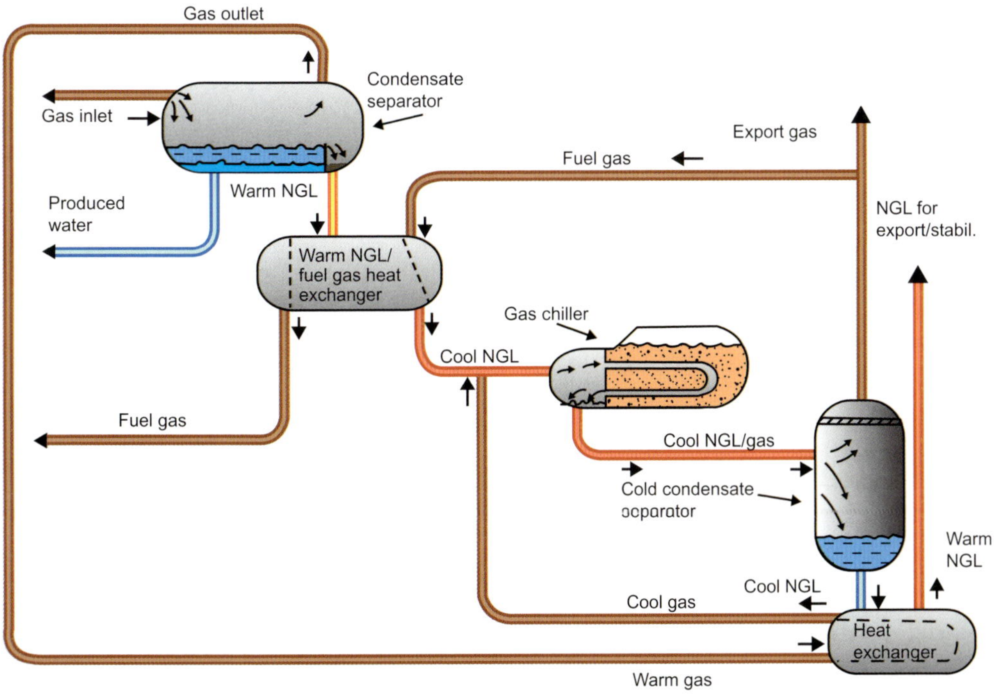

Figure 5.26 – Refrigerant plant

gas being subjected to a more drastic temperature reduction within either an expander/re-compressor, or a Joule-Thomson valve.

The Joule-Thomson (J-T) valve is a throttling valve named after the pioneering scientists James Joule (whose name was given to the SI unit of energy) and William Thomson (later Lord Kelvin after whom the absolute temperature scale was named). It is used in preference to the expander/recompressor during periods of off-peak gas production.

The expander/recompressor is an extremely efficient piece of machinery that utilises the expansion of process gas to impart rotational movement to a turbine wheel, which in turn drives the recompressor attached to the opposite end of the shaft.

Cold gas leaving the expander (minus 20°C/minus 5°F) or J-T valve (4°C/40°F) enters the discharge scrubber where any remaining NGL is deposited prior to re-compression and export.

Refrigeration

The Refrigeration Process diagram shows a simplified layout of a refrigerated NGL removal process. The majority of vessels employed are heat exchangers designed to improve the efficiency of the operation while for NGL removal in Figure 5.26, a reduction in temperature benefits the subsequent stabilisation process (if installed).

The initial separation of NGL and process gas takes place within the condensate separator. Gas leaving the separator is cooled prior to re-combination with NGL at entry to the refrigeration vessel, where the subsequent reduction in temperature (minus 30°C/minus 22°F) accelerates the condensation of any remaining NGL. The contents of the refrigeration vessel are discharged into the cold condensate separator where NGL is once again separated from the gas, the gas destined for export leaving the top of the vessel.

The disposal of NGL will be dependant on the quantity produced. The liquids may be re-injected into the crude oil export pipeline, into a dedicated NGL pipeline or into the reservoir for recovery at a later date. When transportation by subsea pipeline is intended, the NGL may be subjected to a stabilisation process to prevent problems created by the vaporisation of dissolved gases.

5.2.3.4 Stabilisation

The function of the stabilisation process is to further refine the NGL by removing the more volatile of the dissolved gases. Removal of gases likely to vaporise in the subsea pipeline will ensure trouble free transportation and operation of equipment at the onshore reception facility.

Stabilisation is a distillation process that permits the composition of the NGL to be closely controlled by the addition of heat in a reboiler. The plant shares many similarities with the gas dehydration equipment.

The NGL entering the top of the stabiliser gravitates downwards from tray to tray, while hydrocarbon gases released from within the reboiler travel upwards, creating a stripping action designed to assist in the release of dissolved gases. The NGL eventually accumulates in the base of the vessel prior to transfer to the reboiler.

The temperature of the NGL in the reboiler is set at a value that depends on which of the dissolved gases are to be removed from the NGL by vaporisation, normally methane and ethane. The gases are

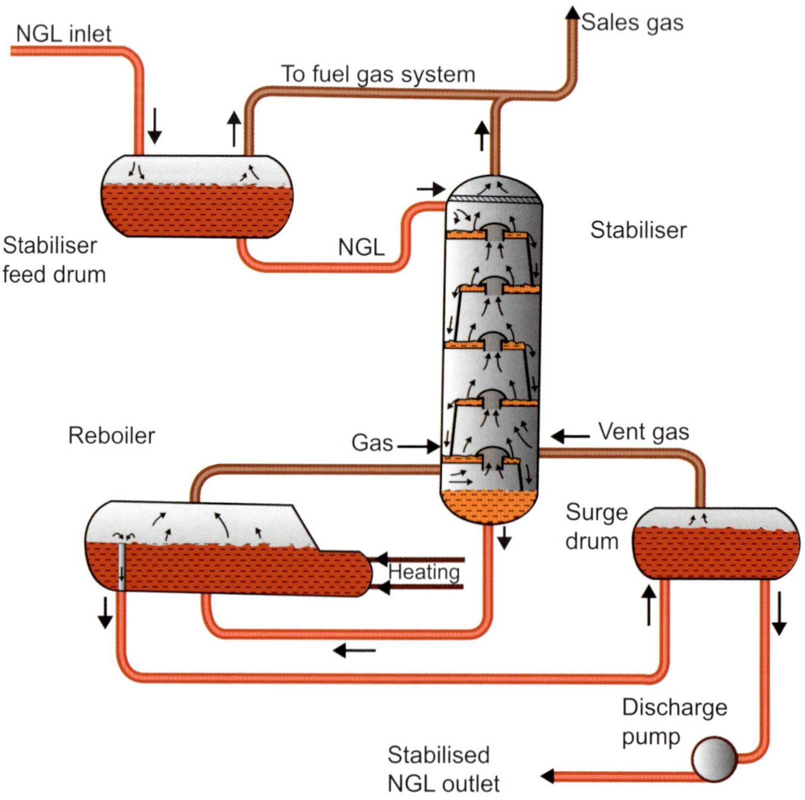

Figure 5.27 – Stabilisation process

returned to the stabiliser where they transfer heat into the NGL prior to commencing the journey up through the vessel. Gas expelled from the top of the stabiliser may be routed back to the sales compressor, or used as fuel gas.

Stabilised NGL is removed from the surface of the reboiler via a standpipe, and transferred to a surge drum for cooling prior to metering and re-injection into the crude oil export line.

5.2.3.5 The sweetening process

Two of the least desirable gases that may occur in a reservoir are carbon dioxide (CO_2) and hydrogen sulphide (H_2S), the latter creating a condition referred to as a sour gas or oil. The function of the Selexol system is to sweeten the gas, ie to remove the H_2S and to a lesser extent the CO_2 contaminants, prior to export or re-injection of the gas into the reservoir.

Selexol is a proprietary solvent, a complex glycol (dimethyl ether of polyethylene glycol), that is used in an absorption/regeneration plant that operates on similar principles to the glycol dehydration/ regeneration process.

A diagram, Figure 5.28, has been prepared to assist in the explanation of the process.

i) Absorption process

The Selexol tower contains two packing elements designed to promote contact between the process gas and selexol liquid. The upper packing element is flooded with lean (pure) selexol and removes the H_2S content of the process gas, while CO_2 is absorbed by the semi-lean selexol within the lower packing element.

The selexol accumulates in the base of the vessel in what is referred to as a rich condition, ie containing dissolved H_2S and CO_2 gases, prior to regeneration while the gas proceeds to the next stage of refinement.

ii) Regeneration process

The regeneration process relies on a combination of pressure reduction and stripping gas to purify rich selexol.

Selexol entering the regeneration system is reduced in pressure in three successive stages by a series of flash separation drums. Gases released from the HP flash drum consist primarily of the lower boiling point fractions, such as methane and ethane, which are routed back to the absorption tower for recycling or admitted to the selexol stripper as stripping gas.

Gases released from the LP flash drum and vacuum flash drums consist primarily of CO_2 and H_2S and, being of no commercial value, they are disposed of via the LP flare header. The selexol emerging from the vacuum flash drum is in a semi-lean condition and suitable for re-entry into the lower packing element of the absorption tower.

To obtain the lean selexol required by the upper packing element, approximately 5% of the semi-lean selexol is subjected to further refinement within a stripping column. The intermingling of the selexol with stripping gas completes the purification process. The H_2S contaminated stripping gas leaves the top of the vessel for disposal at the LP flare header while lean selexol accumulates in the base of the vessel prior to re-circulation in the absorption tower.

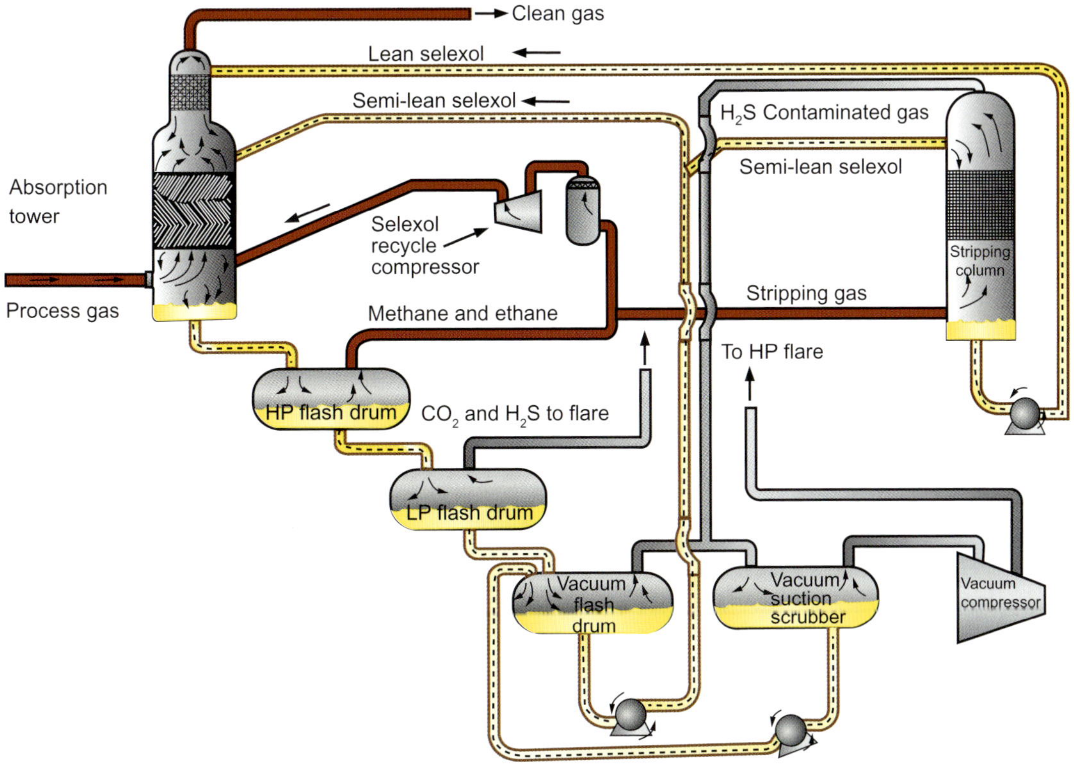

Figure 5.28 – Selexol system

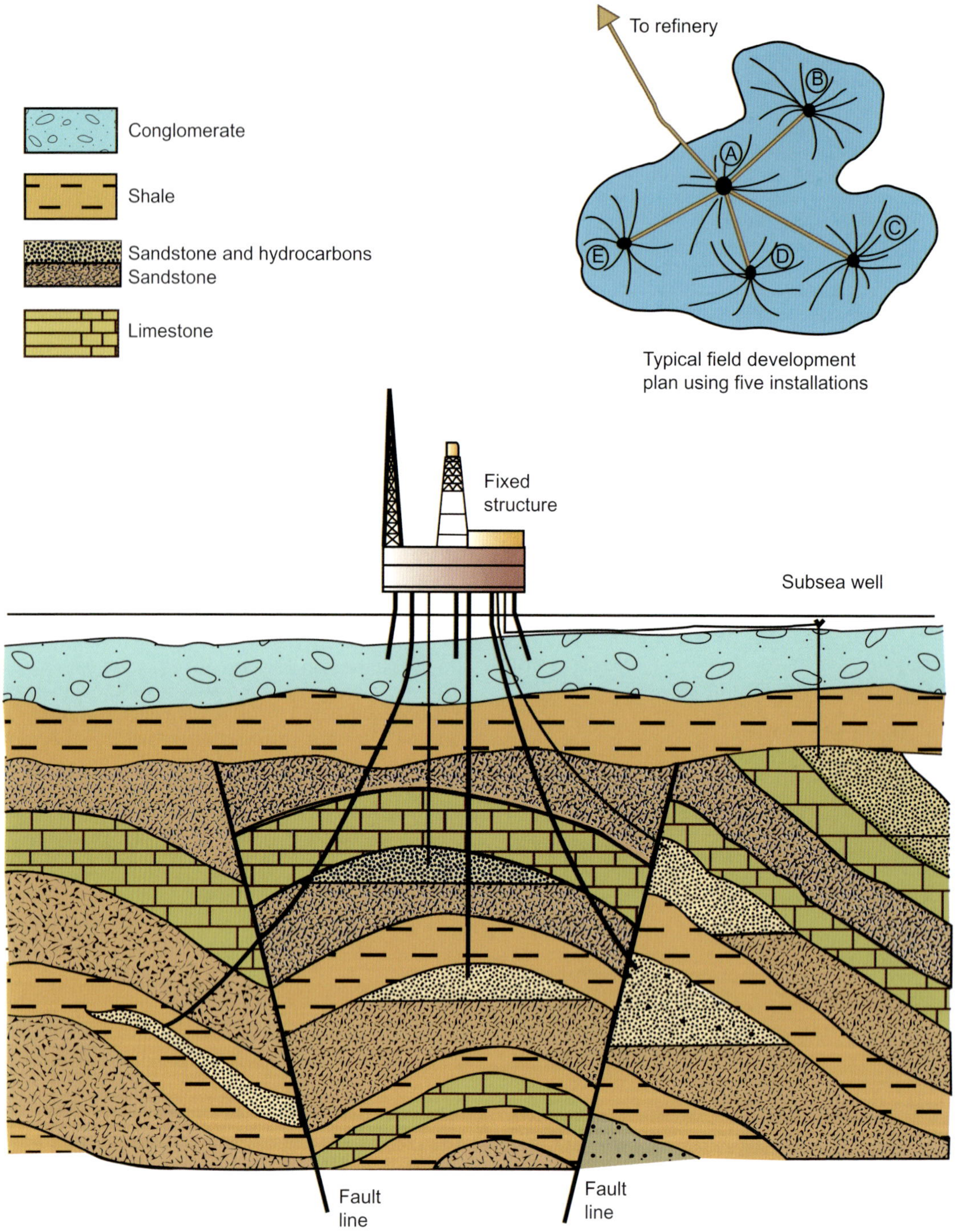

Figure 5.29 – Formation of oil and gas field schematic

5.3 Enhanced Oil Recovery and the Oil Drive Mechanism

5.3.1 The Oil Drive Mechanism

The pressure contained within a new oil field can be considerable, anything from 3,000 to 15,000 psi (204 to 1,000 bar).

3,000 to 8,000 psi (240 to 545 bar) is representative of conditions prevailing in the North Sea.

Unfortunately, being a liquid, oil is virtually incompressible and once production has commenced the natural reservoir pressure deteriorates rapidly and artificial means must be employed to enhance oil recovery if output is to be maintained at economic levels.

Enhanced oil recovery (EOR) involves the injection of either water or gas into the formation to increase the reservoir pressure, or the location of equipment within the production tubing to assist in the extraction of oil. Frequently a combination of methods are employed.

To fully understand the principles on which EOR rely, we must first familiarise ourselves with the natural mechanisms that drive the crude oil from the

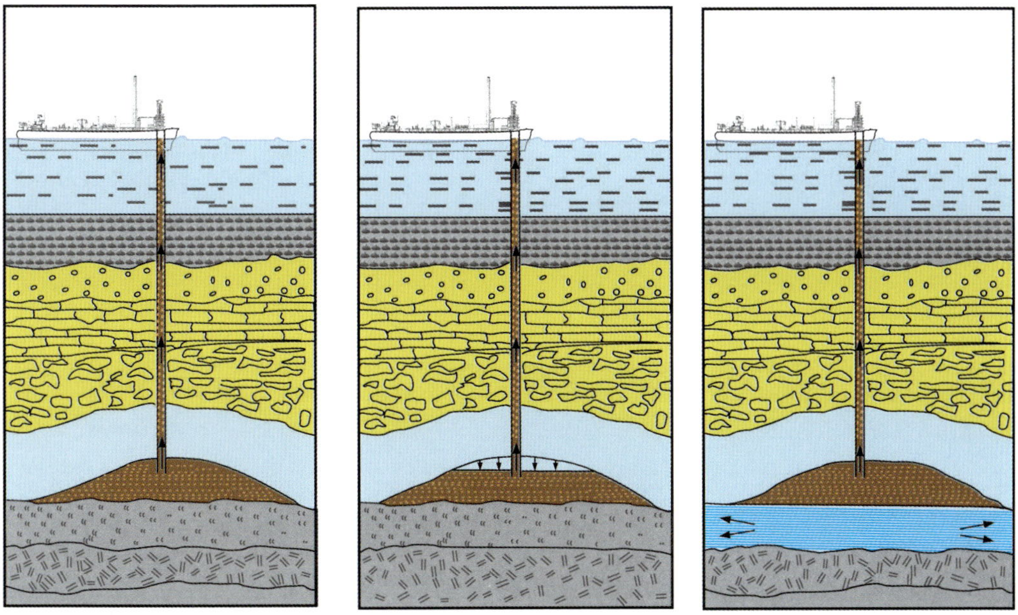

Solution-gas drive Gas-cap drive Water drive

Figure 5.30 – Reservoir drive mechanisms

reservoir. It is the reservoir pressure that provides the motive force for oil production and the reservoir pressure is created by associated gas and/or formation water.

The driving force for the production of oil may be any one of four mechanisms (Figure 5.30), depending on the rock formation surrounding the reservoir and the quantity of water and associated gas accompanying the oil.

a) **Depletion/solution drive**
Depletion drive describes the situation encountered when reservoir pressure exceeds the bubble point pressure of the hydrocarbon gas. This means that there is no free gas available as it is all dissolved within the crude oil. The gas is produced with the oil and it is the expansion of the gas that provides the driving force.

Solution drive is extremely inefficient and can only be relied on to recover 5% to 25% of available oil reserves.

b) **Gas drive**
When the reservoir pressure is less than the gas bubble point pressure, a layer or pocket of free gas exists above the oil. The gas acts like an accumulator, pushing the oil into the well bore and up to the surface, and assists in the drainage of the upper reaches of the reservoir. Gas drive permits the recovery of 20% to 40% of the oil contained within the reservoir.

c) **Water drive**
Water drive is the most efficient of the natural recovery mechanisms relying on the expansion of the underlying water zone or aquifer to provide the driving force for oil production. Water is virtually incompressible so, to be effective, the volume of the water must be considerable in relationship to the volume of oil. Theoretically, water drive can be used to recover 40% to 80% of the reservoir. Unfortunately, the large volumes of water produced with the oil from an ageing field limit the commercial viability of the process.

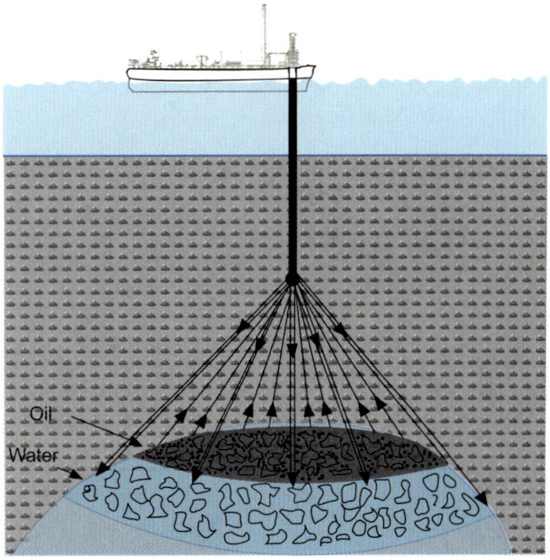

Figure 5.31 – Water injection

d) **Gravity drainage**
Gravity drainage is associated with steeply dipping reservoirs and this type of rock formation tends not to have efficient oil output.

Initially, oil production commences under the influence of depletion drive, but as pressures fall the development of an overlying layer of free gas creates conditions more associated with gas drive.

5.3.2 Enhanced Oil Recovery Systems

Enhanced oil recovery may be effected by artificially increasing the reservoir pressure to supplement the natural oil drive mechanism, or equipment may be located within the production tubing of a well to help overcome the static head created by the column of oil.

The equipment associated with EOR considerably complicates life onboard an oil production installation, with the drilling derrick involved in frequent modifications to the production tubings and in the drilling of new injection wells. For all that effort, the most efficient EOR systems can only economically recover 35% to 45% of the total oil contained within the reservoir, which is why oil field capacity is described in terms of recoverable reserves. This contrasts sharply with the exploitation of a gas field, which can be almost literally sucked dry.

The four most widely used EOR techniques are:

1. Water injection.
2. Gas injection.
3. Gas lift.
4. Submergible deepwell pump.

5.3.3 Water Injection

Water injection provides a relatively cheap and efficient way of improving oil production from a depleted field. The water is injected under pressure into the flanks of the oil bearing strata through purpose drilled wells (or redundant oil wells). Water entering the oil bearing rock displaces any remaining particles of oil and the reduction in free space increases the reservoir pressure.

Water for the injection system may be drawn from the sea or the produced water system. Produced water is water that has been removed from the crude oil during processing and is particularly suitable for re-injection because of the absence of dissolved oxygen. A mature oil field produces a considerable quantity of formation water, typically four times as much water as oil, and the disposal of up to 150,000 barrels (21,000 tonnes) a day can present problems. If the water is to be pumped into the sea, it must first be cleaned to ensure that all traces of oil and condensate are removed to avoid the risk of pollution. Re-injection greatly simplifies the problem of water disposal and provides water that is ideal for enhanced oil recovery.

Figure 5.32 shows a typical general service seawater system that provides water for reservoir injection, process and utility equipment. Relatively complex filtration and purification arrangements are employed to ensure that the water entering the reservoir is scrupulously clean. The fine pores contained within the hydrocarbon bearing rocks must be protected against impurities that could restrict the drainage of oil and reduce the output of the reservoir.

Seawater lift pumps
Multi-stage, centrifugal, submergible pumps, installed within separate caissons, provide the seawater to the system, the pump suctions being dosed with sodium hypochlorite solution to combat marine growth and discourage microbiological activity.

Filtration packages
The seawater system contains two filtration packages, the first for general filtration of all seawater prior to entry into the deaerator, and the second purely for the filtration of water destined for injection into the reservoir.

The first filtration vessel contains a graded filter bed consisting of typically anthracite, garnet and pea gravel, which will remove solid impurities greater than 5 microns (0.0002 inch) in size. The addition of a filtration aid (a positively charged polyelectrolyte) further enhances the efficiency of the operation.

The second filtration package employs a series of perforated stainless steel tubes covered by a woven fabric. Both filtration packages are periodically back-washed to clean the filtration medium of deposited solids.

Deaerator
Removal of dissolved oxygen greatly reduces the corrosive effect of seawater and will considerably increase the operating life of the plant. Reducing the quantity of oxygen in the water will also help prevent the growth of bacteria that could contaminate the reservoir and reduce oil productivity. The deaerator vessel may utilise stripping gas (fuel gas) or evacuation to encourage the release of dissolved oxygen, and the injection of an oxygen scavenger, such as ammonium bisulphate (NH_4 HSO_4), into the base of the vessel further improves the process.

Efficient deaeration can reduce the oxygen content of seawater from 9.25 parts per million to less than 30 parts per billion.

Water leaving the deaerator is subject to further conditioning by the addition of a scale inhibitor designed to prevent the formation of insoluble sulphates and carbonates that can occur when seawater eventually mixes with the natural waters in the reservoir.

Degassing
The degassing vessel provides agitation to the seawater to assist in the removal of excess hydrocarbon gas. Water leaving the vessel supplies both platform utilities and the seawater injection system.

Injection pump
The seawater is injected into the reservoir by immensely powerful (2,000 kW) gas turbine or electric motor driven centrifugal multi-stage pumps, which can generate water pressures ranging from 3,000 to 5,000 psi (205 to 340 bar).

The water injection wells are to all intent and purposes identical to production wells and include a christmas tree and mudline safety valve. They are also connected to the emergency shutdown (ESD) system.

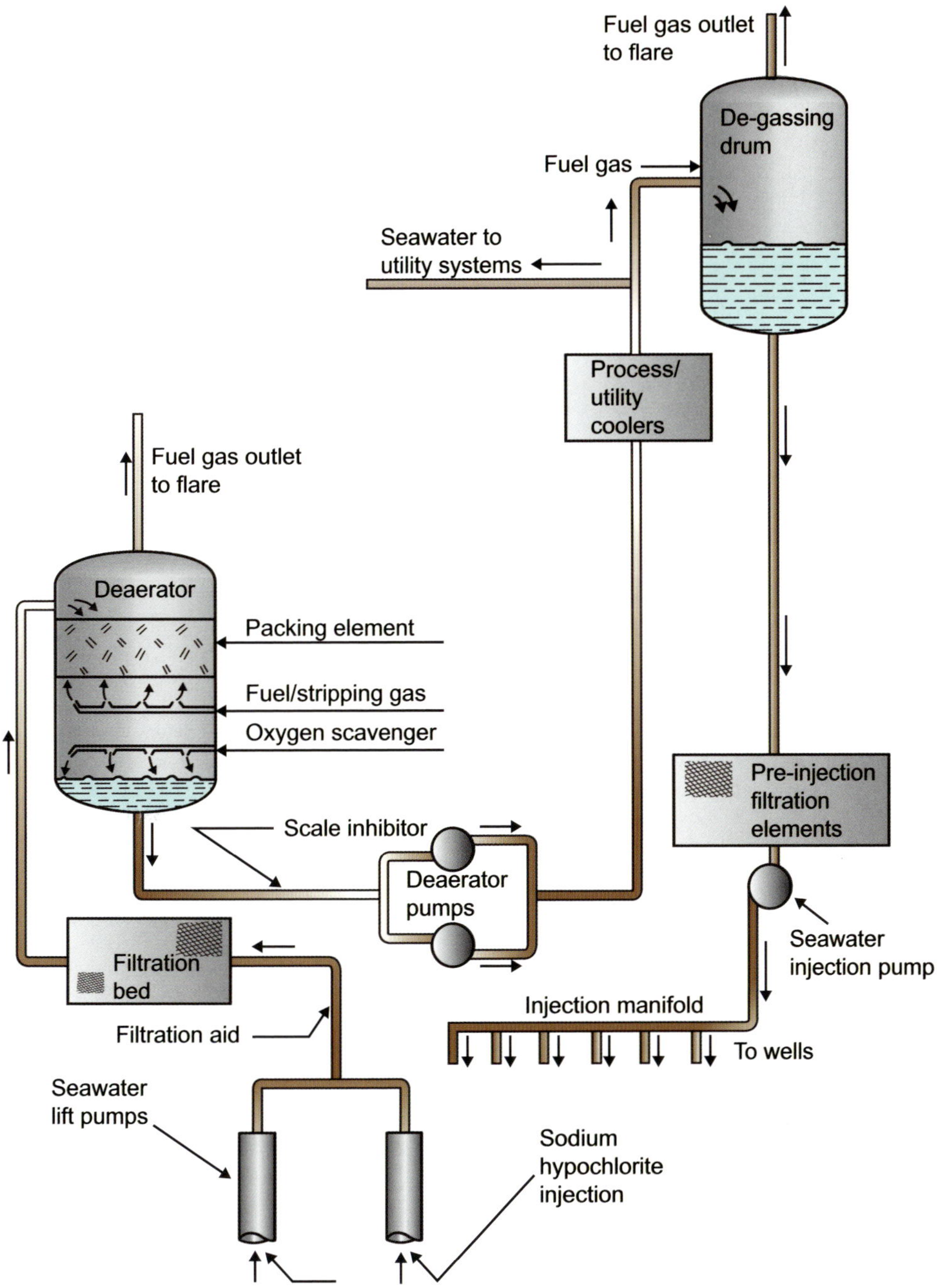

Figure 5.32 – Seawater injection schematic

5.3.4 Minox De-Oxygenation System

The Minox system is a proprietary system, manufactured by Minox Technology AS, Norway. It may be used as an alternative to the deaerator vessel described in the previous section. The process uses nitrogen gas to strip the oxygen from the seawater and it is capable of reducing the oxygen content in the seawater to as low as 10 parts per billion.

Process description
As the seawater enters the first separator, it passes through a mixing unit, where nitrogen gas is injected. Nitrogen has an affinity for oxygen and it strips the oxygen from the seawater. The nitrogen and oxygen mixture is then drawn out of the top of the separator, leaving behind seawater containing approximately 300-400 ppb of oxygen.

The nitrogen/oxygen gas mixture passes to the nitrogen regeneration unit, while the seawater enters a second separator via a static mixer, where it undergoes a process identical to the one that took place in the first separator. Water leaving the second separator contains less than 10 ppb.

Regeneration process
At process start-up, an electric heater is used to increase the temperature of the nitrogen/oxygen mixture coming from the separators to approximately 70°C. The mixture then passes through a heat exchanger into a palladium catalyst, where a small quantity of methanol is injected. Under the influence of the palladium catalyst, the methanol reacts with the oxygen in the nitrogen/oxygen mixture in an exothermic reaction that produces water, carbon dioxide and nitrogen.

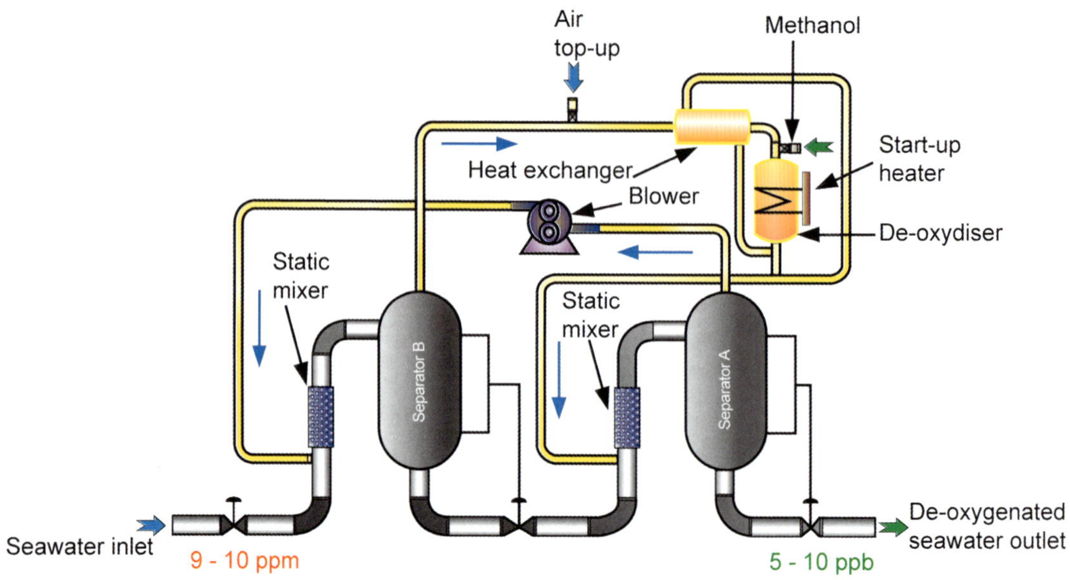

Figure 5.33 – Minox system

The water and carbon dioxide are disposed of, while the nitrogen is returned to the mixing units at the separators. Not all the nitrogen is recovered as some of it dissolves in the water that is drained from the regenerator. The nitrogen that occurs naturally in fresh air is replenished from a compressed air connection. Once the catalyst has reached its operating temperature, the electric heater can be switched off and a temperature of 120-160°C is then maintained at the inlet to the catalyst unit by routing a proportion of the hot exhaust gases from the regeneration unit back through the inlet gas heat exchanger.

5.3.5 Gas Injection

Gas injection operates on similar principles to water injection, artificially increasing the reservoir pressure. The process also provides the ideal way of disposing of the associated gas removed from crude oil when the quantities produced do not justify the installation of a subsea pipeline. The gas should not be thought of as lost because it can be recovered at a later date, on completion of oil production.

The gas injection system is a high pressure compressor and distribution manifold that delivers gas to the reservoir at pressures of up to 12,000 psi (800 bar) through purpose drilled wells. Gas destined for re-injection is subjected to the full purification process that sales gas would receive, to avoid contamination of the reservoir and injection equipment.

EOR equipment
For oil to reach the surface, the reservoir pressure must exceed the static head pressure created by the column of oil within the well bore. The pressure required can be considerable, approximately 4,000 psi (310 bar) for a well of depth 12,000 feet (3,650 metres). Frequently, the reservoir pressure produced by an ageing field is considerably less than this and the installation of a submergible pump or introduction of gas lift valves into the production tubing can greatly enhance oil output.

Gas lift valves and submergible pumps achieve the same end result by differing means and it should be remembered that both methods assist the reservoir oil drive mechanism and are not designed to produce oil from a field devoid of all natural reservoir pressure. The gas lift valves create a reduction in static head pressure by reducing the specific gravity of the oil, while the submergible pump provides the oil with a helping hand in the form of a pressure boost.

5.3.6 Gas Lift

Gas lift involves the re-injection of process gas into the production tubing to produce a foaming gas/oil mixture with a specific gravity considerably less than the naturally occurring crude oil.

Process gas is introduced into the annulus space surrounding the production tubing through a connection on the wellhead. It then passes into the production tubing through a series of gas lift non-return valves, the flow being regulated by the pressure in the annulus.

Figure 5.34 shows a general arrangement of the gas lift equipment. The non-return valves are located within eccentric tubular housings, referred to as side pocket mandrels (SPM), which are run (installed) as part of the production tubing. The SPMs ensure that an unrestricted channel is maintained within the production tubing, which permits the maximum flow of oil and the passage of the wireline tools employed to install and replace the gas lift valves.

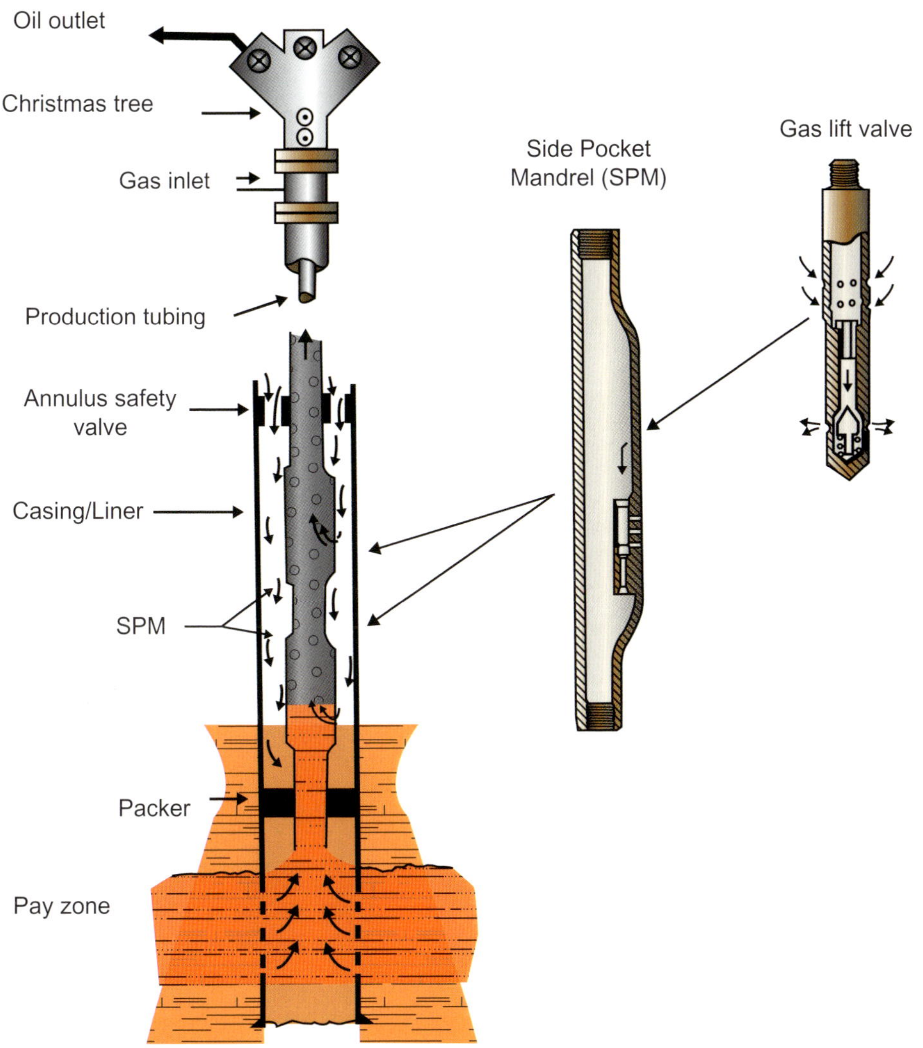

Figure 5.34 – Gas lift arrangements

5.3.7 The Deepwell/Submergible Pump

5.3.7.1 Electrical submergible pumps (ESP)

On fixed installations, electrically powered submersible pump sets are used to boost production rates beyond those achievable by natural reservoir pressure. These pumps are referred to as downhole, submergible or submersible pumps and they are installed in the well as part of the production tubing.

The pump units are individually tailored to suit the conditions of a particular well and up to 200 impellers may be used.

A typical pump designed for service in a North Sea well would comprise 100 pump stages powered by a 150 to 300 kW electric motor, output being 13,000 barrels of oil a day at a surface pressure of between 7 to 15 bar (100 and 300 psi).

Downhole pumps have also been developed that will function in wells with extended horizontal sections, a situation perhaps best illustrated by a well on BP's *Wytch Farm Offshore* development that is drilled from an onshore site. The specification is impressive, 5000 metres of power cable providing electricity to a 480 kW motor at a voltage of 4–5 kV, the pump producing 20,000 barrels of oil a day.

5.3.7.1.1 Hydraulic drive submergible pump (HSP)

In recent years a hydraulically powered pump has been developed, partly as a competitor to the ESP and partly because the manufacturers saw a hydraulic power source as providing greater potential for dealing with multiphase well fluids.

Like the ESP, the hydraulic drive downhole pump utilises a multi-stage centrifugal pump to produce the oil, but there the similarity ends. A multi-stage, axial flow power turbine is used to drive the pump in preference to an electric motor. The power fluid that drives the turbine is supplied from a charge pump unit located either on the seabed or on the reception facility. Seawater or produced water may be used as the power fluid, suitably filtered to remove impurities that could cause damage to turbine bearings and seals, delivered at a pressure of 270 bar (4000 psi).

A hydraulically powered pump operates at much higher speeds than an equivalent electrically powered unit, 7000 to 9000 rpm as compared

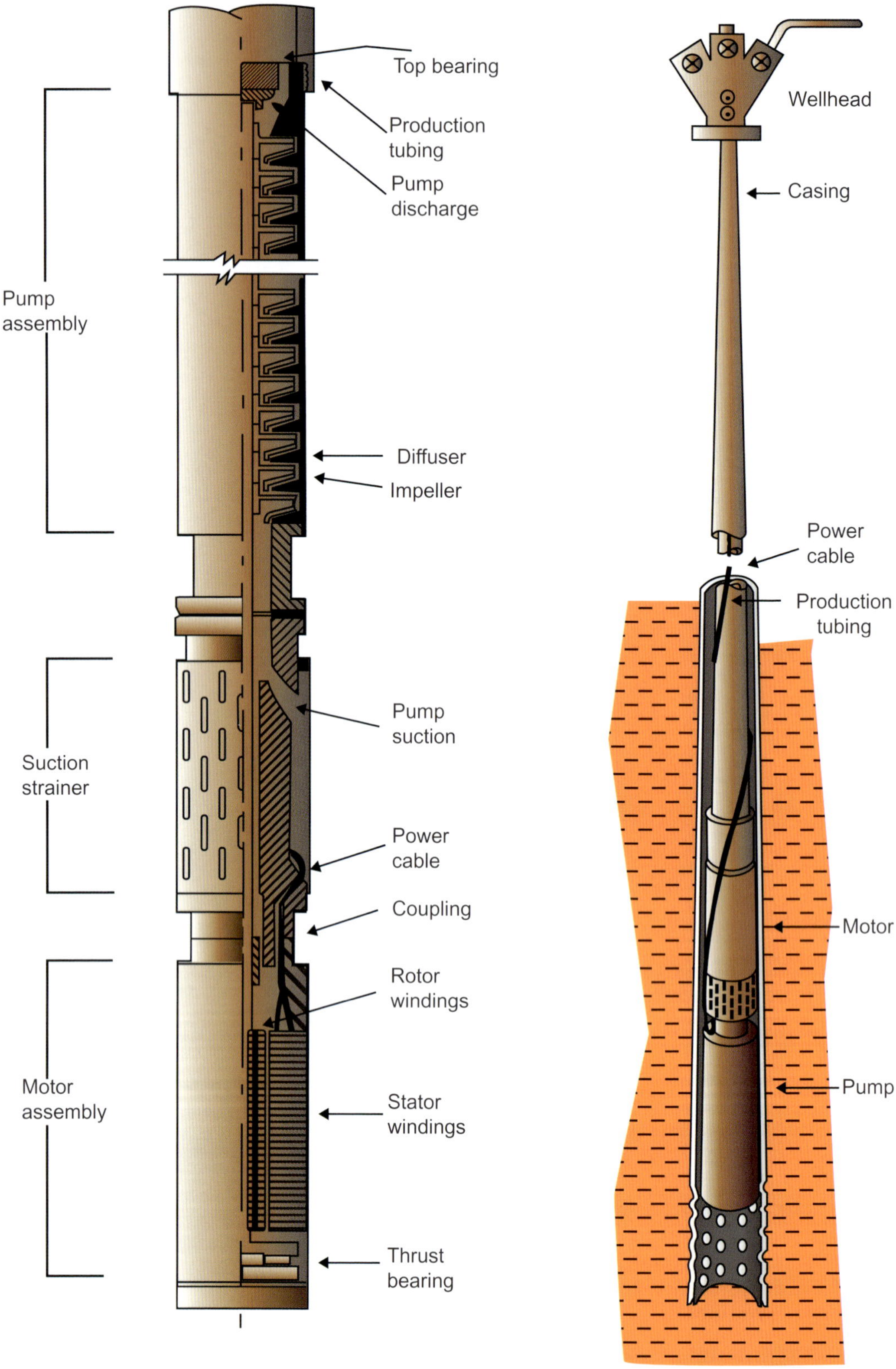

Figure 5.35 – Submergible deepwell pump

to 3600 rpm, so they require fewer pump stages, perhaps 20 compared to 90 stages for an ESP of equivalent pumping capacity. This reduction in the number of pump stages creates a more compact, rugged and lighter pump unit.

The hydraulically powered submergible pumps should not be confused with the hydraulically powered seabed booster pumps.

5.3.7.2 Submergible pump applications in subsea wells

Until recently, submergible pumps were not considered suitable for use in subsea wells, gas lift being the preferred method of enhanced oil recovery because the equipment is simple, reliable and can be installed with wireline equipment. ESPs had a reputation for being unreliable and a service life of as little as 90 days was typical, electrical equipment not being ideally suited to the aggressive operating environment encountered deep within an oil well.

The weak point of the ESP is the mechanical seal between the motor and the pump, but the failure of bearings, electrical connections and the motor itself, all tend to exacerbate the problem. Replacing an ESP is an extremely expensive business even on a fixed platform, because it necessitates a well workover to remove the christmas tree and pull the production tubing. This costs approximately $400,000 and does not include costs attributed to lost production. On a subsea well it has been estimated that the cost of replacing an ESP would be in the region of $2,000,000 due to the necessity to mobilise a jack-up or a drill ship to carry out the workover. However, claims of improved reliability and the advent of the horizontal christmas tree has encouraged operators to take a fresh look at the situation.

Pump manufacturers now claim to be in a position to guarantee a service life of two years for their equipment and, while a workover is still required to replace the pump, horizontal christmas trees can be left in situ while the production tubing is removed, which considerably simplifies the situation, the costs being reduced accordingly.

These improvements prompted Shell to install an ESP in the (1998) *Gannet* field development to tie back subsea wells a distance of 30 km (20 miles) from the platform, ultimately looking towards a tie back distance of 50 km (30 miles) for future wells. Texaco also opted to experiment with a submergible pump in their *Captain* field, a multiphase, hydraulically powered downhole pump. In conjunction with Weir Pumps of Glasgow, they have developed a 340 kW pump with a capacity of 20,000 barrels/day.

As well as equipment improvements, considerable research has been carried out on installation techniques and there are now both ESPs and HSPs available that can be installed by coiled tubing or wireline equipment. These are referred to as 'through tubing pumps', wireline installation being used for wells deviated up to 30 degrees from the vertical, while coil tubing is used beyond these angles and for horizontal wells. This greatly simplifies pump replacement because there is no need to carry out a full well workover, which saves both time and money.

5.3.7.3 Submergible multiphase pumping

Having resolved the problem of finding a reliable power source for the pump, be it electrical or hydraulic, manufacturers have also had to contend with problems with the pump impellers. The conventional radial and mixed flow impellers fitted to ESPs have difficulty in coping with high gas fraction oils. The gas gravitates to the impeller hub and eventually gas slugging or vapour locking occurs causing the pump to lose suction and surge. Damage often occurs at this point.

Gas slugging is one of the problems that plagues production from horizontal wells when high gas fraction oils are encountered, particularly when producing oil at pressures below the reservoir bubble point where free gas caps may occur. To combat these problems, axial flow pump impellers have been developed that can cope with high gas to oil ratios, where gas volumes may exceed 90%.

The axial flow impellers have proven to be particularly effective in field trials when powered by a hydraulic turbine because, being a constant power device rather than a constant speed device, the pump speed adjusts automatically to compensate for the varying product density that occurs with changing void fractions. If a slug of gas enters the pump and the pump loses suction, it accelerates to a point where pumping will resume, the gas slug is dealt with and steady state conditions return. This variable speed capability is a highly desirable feature when dealing with multiphase fluids.

6

Underwater Engineering

6 Underwater Engineering

System Summary
The vessel is fitted with a saturation diving complex, rated to 450 m depth, for 18 divers in round the clock service, with the possibility of operating two diving teams at different working depths and decompressing a third diving team simultaneously. The decompression chambers and the diving bell are equipped with a gas reclaim system. The diving bell is provided with heave-compensating main cable and guide wires, which can be disconnected when sidehauling is required. Separate sidehauling-davits with winches are installed. Heave-compensation also provided for, when sidehauling.

The diving bell is driven through the splashzone and moonpool by a cursor-system.
The moonpool (approximately 4 × 4 metres) is located at the centreline amidships where the motion-amplitude of the vessel is minimum.

The diving system fulfills classification requirements of Det Norske Veritas.

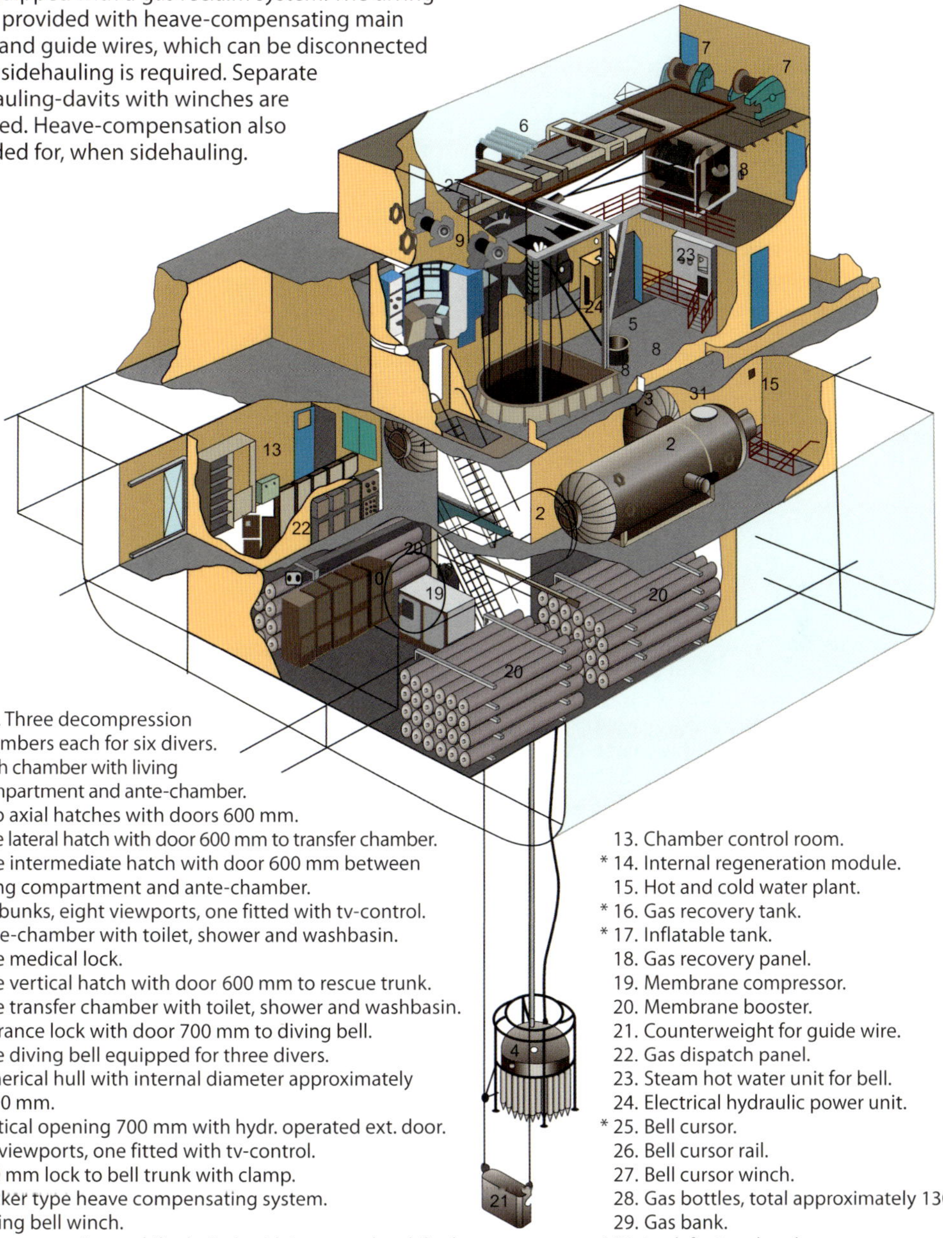

1. & 2. Three decompression chambers each for six divers.
- Each chamber with living compartment and ante-chamber.
- Two axial hatches with doors 600 mm.
- One lateral hatch with door 600 mm to transfer chamber.
- One intermediate hatch with door 600 mm between living compartment and ante-chamber.
- Six bunks, eight viewports, one fitted with tv-control.
- Ante-chamber with toilet, shower and washbasin.
- One medical lock.
- One vertical hatch with door 600 mm to rescue trunk.

3. One transfer chamber with toilet, shower and washbasin.
- Entrance lock with door 700 mm to diving bell.

4. One diving bell equipped for three divers.
- Spherical hull with internal diameter approximately 1900 mm.
- Vertical opening 700 mm with hydr. operated ext. door.
- Six viewports, one fitted with tv-control.

5. 700 mm lock to bell trunk with clamp.
6. Rucker type heave compensating system.
7. Diving bell winch.
8. Constant tension umbilical winch with integrated umbilical.
9. Constant tension winches for guide wires.
10. Diving bell trolley.
11. 1500 mm therapeutical chamber.
12. Diving control room.
13. Chamber control room.
* 14. Internal regeneration module.
15. Hot and cold water plant.
* 16. Gas recovery tank.
* 17. Inflatable tank.
18. Gas recovery panel.
19. Membrane compressor.
20. Membrane booster.
21. Counterweight for guide wire.
22. Gas dispatch panel.
23. Steam hot water unit for bell.
24. Electrical hydraulic power unit.
* 25. Bell cursor.
26. Bell cursor rail.
27. Bell cursor winch.
28. Gas bottles, total approximately 13000 m^3
29. Gas bank.
* 30. Lock for iuc chamber.
31. 600 mm lock for future rescue chamber trunk.
* 32. 600 mm lock to hyperbaric rescue boat.
* 33. Hyperbaric rescue vessel equipped for 18 divers under pressure.

* Item not shown on this picture.

Figure 6.1 – Saturation diving complex

6.1 Diving

Offshore installations are analogous to icebergs in as much as the bulk of the structure lies beneath the waves and, as such, the industry is heavily dependant on the diving fraternity for its continued well-being.

The offshore industry in the North Sea provides employment for in excess of 1,000 divers who are involved in the installation, repair, maintenance and inspection of structures and pipelines and, while the current trend is towards a reduction in diver intervention, there will always be a place for the levels of manual dexterity that only the human hand can provide.

Diving operations may be divided into two categories, which are dictated by water depth. For depths of less than 50 metres (165 feet) air diving techniques may be employed, while greater depths necessitate a full saturation diving programme. Before discussing the various aspects of air and saturation diving, we should first consider the effects that prolonged immersion in deep water have on the human body, for it is in the sensitivity of the body to environmental change that dictates the type of equipment required for life support.

At normal atmospheric pressure the tissues of the human body are in a state of equilibrium with their surroundings, ie they are saturated with the dissolved gases that constitute air, primarily nitrogen and oxygen. This state of equilibrium or saturation is disturbed the instant the diver enters the water.

As the diver descends, the hydrostatic pressure on the body increases the gas absorption capacity of the tissues. Given sufficient time at any particular depth they will once again reach a point of equilibrium or saturation. The eventual ascent from saturation depth must be carefully controlled and include in-water stops or pauses to permit the natural release of these newly dissolved gases if the diver is to avoid the bends (expansion of gas in the blood stream). This is a time consuming process and highlights the major difference between air and saturation diving. The air diver must restrict the depth and duration of his dive to prevent the body reaching the fully saturated condition.

6.1.1 Air Diving

Air diving is used primarily for dives of short duration in water depths of less than 50 metres (165 feet), where excessive decompression times can be avoided. As the name suggests, the divers breathe compressed air that is supplied through an umbilical.

Air diving is used for the vast majority of underwater inspection and repair programmes carried out in shallow waters. However, it should be appreciated that dive duration is limited dramatically with increase in depth, and at 50 metres a stay of only 10 minutes is permitted. The ascent must also include in-water stops to facilitate the body's decompression processes to proceed naturally.

Figure 6.2 – Diver

Where conditions permit, a deck decompression chamber (DDC) can be used to increase dive time by removing the need to carry out in-water decompression stops. The diver returns to the surface as quickly as possible and enters the DDC, which is then pressurised with oxygen to a pressure equivalent to the water depth at the work site. Over a period of about an hour the pressure is gradually reduced to atmospheric, the oxygen assisting in the dispersal of dissolved bodily gases.

The main advantage of air diving is that it does not require the sophisticated equipment associated with saturation diving. Air diving plant may be containerised and located on the deck of a supply boat, or a fixed installation, saving on the considerable costs incurred when chartering a fully equipped diving support vessel (DSV).

6.1.2 Saturation Diving

The main attraction of saturation diving is the increased productivity attainable from a diver due to the removal of restrictions on dive time and the need to decompress after each and every dive. Decompression takes place only on completion of the diving programme, or when the diver is due to go on leave, normally after a period of 30 days. The disadvantage of saturation diving is the cost, which can be considerable.

A diving programme commences with the transportation of the divers to the work site within the diving bell. At the site the divers enter the water through a hatch in the base of the bell, having first donned helmets/masks and attached their umbilicals to the distribution manifold on the inside of the bell.

The umbilicals provide the diver with breathing gas and a supply of hot water to heat the diving suit. Subject to favourable tidal conditions, the divers spend approximately six to eight hours at depth, during which time the diving bell is used as a habitat

Figure 6.3 – State of the art diving support vessel, the *'Deepwater 1'*, involved in survey work during the installation of a jacket. Inset – An offshore supply boat equipped for diving support duties, the *'Smit Manta'*.

where they can eat and rest. On completion of work the divers return to the bell, lock the hatch and are winched back to the diving support vessel (DSV).

Once onboard the DSV, the bell is locked onto the transfer hatch of the deck decompression chamber (DDC), where they will remain until their next shift. Throughout the entire operation the divers remain under pressure, which may be considerable, anything from 5 to 25 bar (75 to 375 psi) depending on the saturation depth of the work site. Loss of pressure at any time would cause serious injury or death. This cycle may continue for a period of up to three weeks after which time the divers commence decompression.

Decompression involves the gradual reduction of the pressure within the DDC to atmospheric to permit the natural dissipation of excess gases from the diver's body. It takes approximately one hour per metre (3.3 feet) of water depth to safely decompress, so a saturation depth of 180 metres (600 feet) requires a decompression interval of nearly a week.

6.1.3 Equipment

A saturation diving system consists of a diving bell (submersible compression chamber, SCC) and a deck decompression chamber (DCC), and this combination of pressure chambers provide the means by which a diver's body is artificially maintained at a pressure equivalent to the work site for prolonged periods.

Diving bell (SCC)
The diving bell provides transportation to and from the work site. It is secured to the diving support vessel (DSV) by a steel cable and an umbilical provides the occupants with breathing gas, heated water, electrical power and communication facilities.

The bell is housed within a substantial steel framework designed to provide protection against impact and a location for the ballast tanks and life support systems. An emergency release system permits disconnection of the steel cable and umbilical should the divers encounter a situation that necessitates isolation from the DSV, for instance a fire on the DSV or a fouled umbilical. The life support systems contain 96 hours of breathing gas and the facility to de-ballast the bell to permit a return to the surface.

Deck decompression chamber (DDC)

Deck decompression chambers are used for both air and saturation diving operations. The air diving DDC is relatively simple, consisting of a cylindrical twin chamber pressure vessel. The saturation diving DDC is considerably larger and more lavishly equipped and provides an out of water home for up to six divers for the duration of the diving programme. It contains beds, toilets, showers, messing and recreational facilities, and small air locks through which hot meals, laundry and medication can be passed to the divers. The occupants of the DDC are monitored 24 hours a day by the diving support crew and a particularly watchful eye is maintained during the decompression process.

Diving suits

Diving suits used for offshore inspection and repair programmes fall into two categories, largely dictated by the degree of thermal protection required by the diver.

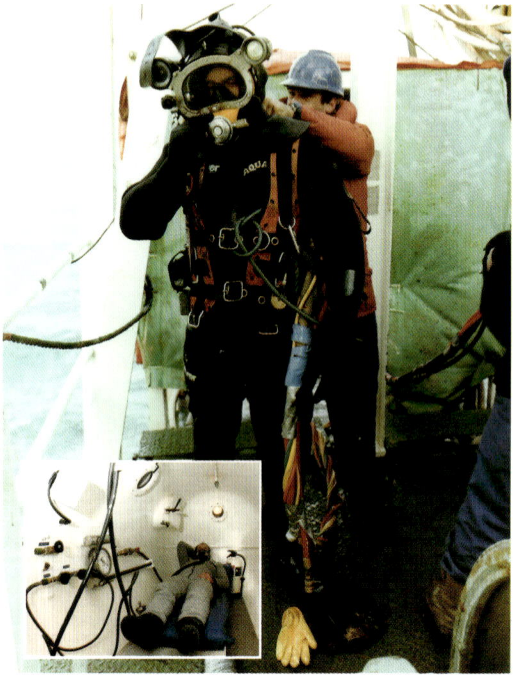

Figure 6.4 – Typical North Sea air diving dry suit and associated attire.
Inset – The spartan interior of an air diving deck decompression chamber. The diver can be seen inhaling oxygen to accelerate the decompression process.

(a) Dry suit

The dry suit consists of a watertight outer layer under which the diver wears thermal clothing to assist in the retention of body heat. In a cold water environment such as the North Sea, the dry suit is only suitable for air dives of short duration beyond which a heated wet suit is preferred.

(b) Wet suit

The prolonged periods of submersion associated with saturation diving necessitate the wearing of a heated wet suit. Hot seawater supplied via the umbilical cable is circulated around the inside of the suit prior to exit at the diver's extremities.

Heliox

Air consists primarily of oxygen and nitrogen and both these gases can prove injurious to the human body when ingested under the pressures associated with water depths greater than 50 metres (165 feet). The oxygen becomes toxic while the nitrogen leads to nitrogen narcosis or 'drunkenness of the deep'. Heliox, a specially formulated mixture of helium and oxygen, provides a solution to both these problems because it permits the quantity of oxygen to be regulated to suit the body's respiratory needs at a particular depth. Helium, like the nitrogen it replaces, is an inert gas that fulfils no useful function in the respiratory process other than to bulk out the mixture and support the desired quantity of oxygen.

While heliox may be used by divers who are not in saturation, it is a gas primarily associated with saturation diving. The divers breathe the mixture throughout the saturation period, including the time spent in the DDC in between dives. It is the period within the DDC in which another advantage of heliox emerges. The gas is much lighter than compressed air, requiring less respiratory effort, which ensures that the divers obtain restful sleep.

Heliox does have certain disadvantages; the increase in heat transfer from the body that is considerable, the 'Donald Duck' speech impediment created by the action of the gas on the vocal chords, and the cost. The cost and adverse conductivity can be moderated by incorporating gas heating and recycling equipment, while a voice synthesizer will minimise speech distortion.

6.1.4 Vessels

Diving support vessel (DSV)

Figure 6.5 – Diving support vessel *'Bourbon Emerald'*

The primary function of the DSV is to provide a stable platform from which the divers can carry out their duties. The complexity of the vessel will be determined by the type of diving operation for which it was designed, ie a saturation diving vessel will be considerably better equipped than one catering solely for air diving.

Figure 6.6 – *'Bibby Sapphire'*

The DSV can represent a diving company's single largest investment and the more divers the vessel can support, the more cost effective it becomes. The latest state of the art vessels employ up to three diving systems (three deck decompression chambers and two diving bells), which can accommodate 18 divers working around the clock in water depths ranging from 20 to 450 metres (65 to 1500 feet).

The charter rates for a modern, fully equipped saturation diving support vessel are considerable ($80,000 to $250,000 a day) and to minimise lost time due to adverse weather conditions, the vessels employ sophisticated active and passive hull stabilisation equipment and dynamic positioning (DP). The DP computer correlates information pertaining to wind speed, wave height, currents and satellite reference positions and generates command signals for the control of the main propulsion and thruster units so that the vessel can remain on station and continue working in weather conditions approaching Beaufort 8 (gale force) without the need to deploy anchors. The effects of the weather are further negated by the moonpool, a vertical shaft that runs through the centre of the ship and is open to the sea. The moonpool provides a sheltered haven for the launch and recovery of the diving bells and represents a considerable improvement on the traditional stern mounted 'A' frame launching davit.

A number of DSVs have been provided with well service equipment and can carry out intervention and workover programmes for the subsea market. This facility increases the likelihood of obtaining employment beyond the summer months, after the routine underwater inspections have been completed.

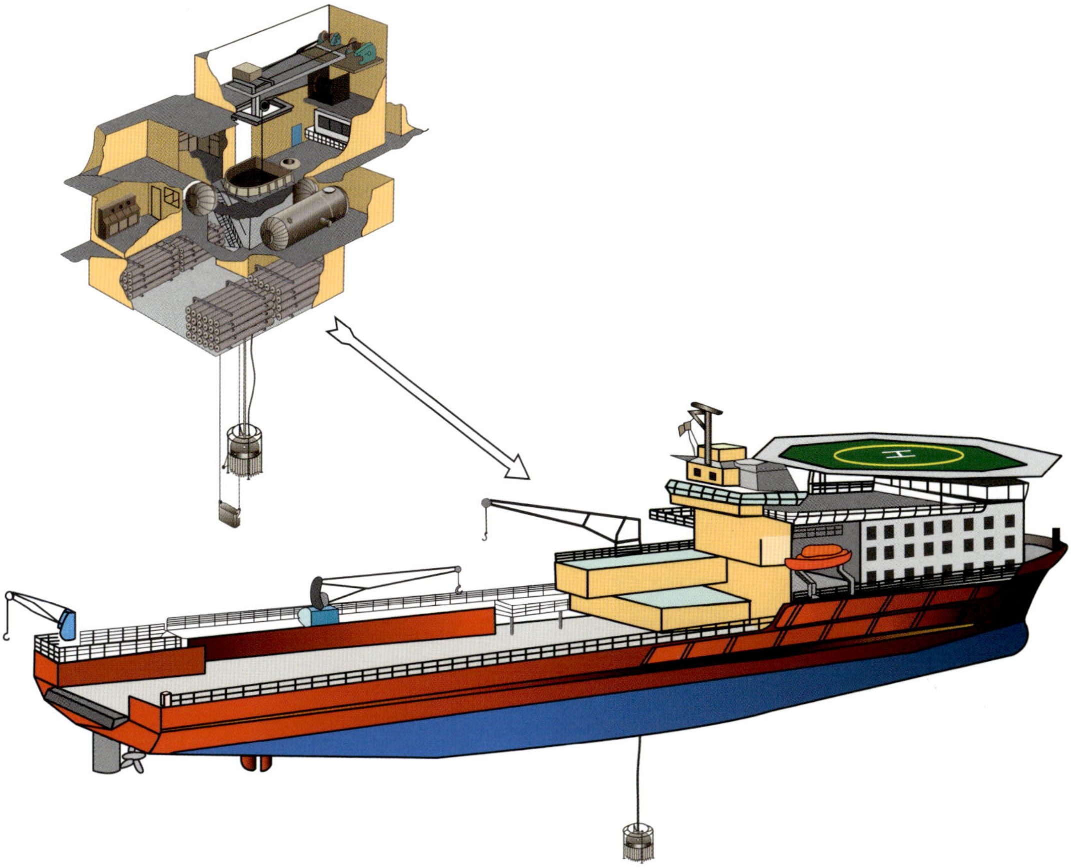

Figure 6.7 – Diving support vessel (DSV)

Remotely operated vehicles (ROV)

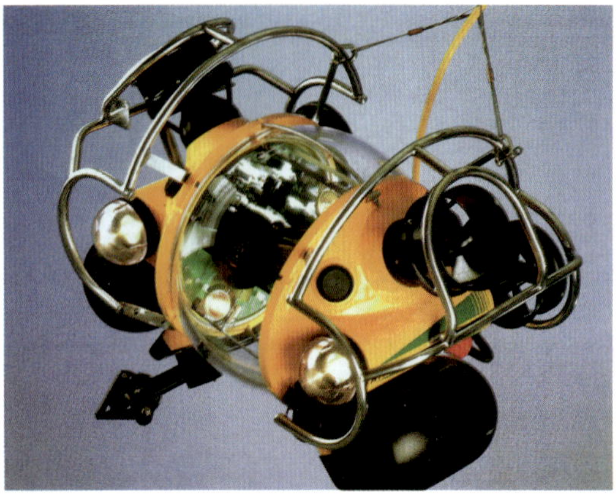

Figure 6.8 (a) – The 'Offshore Hyball', a lightweight remotely operated vehicle (ROV) equipped with video monitoring inspection equipment.

Figure 6.8 (b) – The 'Trojan' remotely operated vehicle (ROV). The propeller thruster units and articulated manipulators are clearly visible.

A remotely operated vehicle (ROV) is a mini-submarine that can be controlled from the deck of a ship or an offshore installation via an umbilical. The concept originated in the late 1970s and was developed to provide a cost effective alternative to the deep sea diver for routine inspection duties.

Figure 6.9 – 'Seaeye' ROV

The external appearance of an ROV will depend largely on the type of work it is designed to carry out. The small inspection units often employ a spherical, cylindrical glass reinforced plastic (GRP) or acrylic hull to contain the electronic control and video equipment. The larger work units tend to use aluminium alloy pods to house delicate equipment. A steel frame provides protection against impact and mounting points for ballast tanks, battery pods, manipulators (if fitted) and attachment of the umbilical. The complete assembly may weigh anything from 0.5 tonnes to 2.0 tonnes. Once in the water, the ROV is manoeuvred by a number of motor-driven ducted propellers that are electrically powered from self-contained battery packs, or from an external source supplied via the umbilical cable. Buoyancy tanks assist in the maintenance of trim and water depth.

Figure 6.10 – Subsea manifold valves equipped for ROV operation

The first generation ROVs were designed purely to look at underwater activities, being driven around the work site relaying video pictures to the command centre. They were, and still are, used extensively to carry out the visual inspections on pipelines and jacket structures associated with the annual surveys. Further developments and the addition of articulated manipulators or crab claws permit operations such as marine growth removal, ultrasonic thickness measurement and bolt tensioning to be carried out. However, it was the offshore industry's acceptance of the advantages offered by the subsea wellhead that provided impetus to ROV development during the 1980s.

The second generation ROVs were developed hand in hand with the new modular designed diverless subsea wellhead packages. These wellheads can be installed, maintained and repaired entirely by ROV and end the reliance on saturation divers, which has permitted the exploitation of reserves located in water depths beyond the range of conventional diving techniques.

A discussion on ROVs would not be complete without a brief mention of manned atmospheric diving systems (ADS) such as the Osel Mantis and Slingsby Engineering's Spider. Occasionally referred to as microsubs, they share many similarities with an ROV, with the obvious exception of the increased hull capacity required to accommodate the pilot and life support systems. The Mantis was constructed as a dual purpose vehicle and could be used in the manned and unmanned ROV mode. The main advantages of manned submersibles are the levels of control afforded by an experienced pilot compared to the remote operation of an ROV, and the fact that they can be used at depths of up to 1,000 metres (3,300 feet), nearly twice the depth at which a saturation diver can operate. The hulls are maintained at a constant pressure of one atmosphere so there is no need for the pilot to undergo decompression after the dive. Unfortunately, manned submersibles tend to be restricted to military service, their use discouraged by the leading oil companies on the grounds of safety in spite of their excellent safety record.

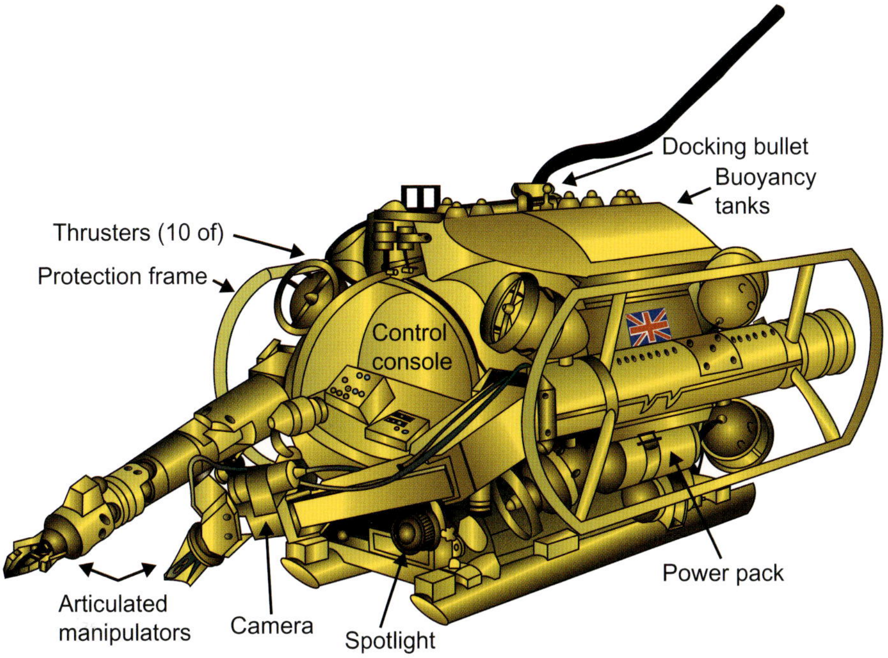

Figure 6.11 – 'OSEL MANTIS' Manned/unmanned ROV

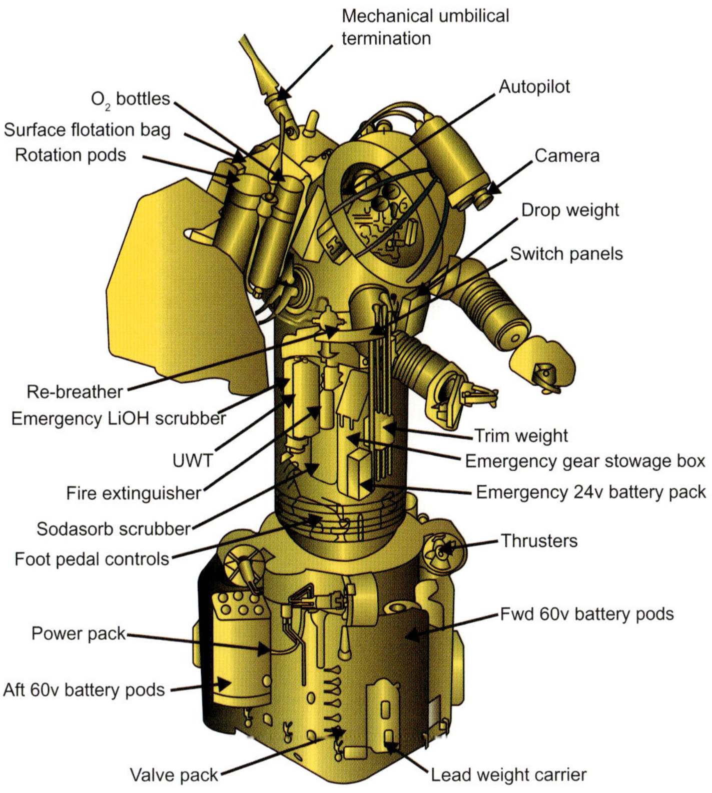

Figure 6.12 – Spider ADS system

6.2 Underwater Surveys

Specialised diving contractors are employed by the installation owners to carry out a planned inspection programme of the underwater structure and pipelines, which are designed to detect any potential defects before they develop into major problems.

Until 1998, in the UK sector of the North Sea, the scope of the planned inspection programme and its frequency were traditionally agreed with the Certifying Authority who issued the Certificate of Fitness for the installation. Under Verification, the role of the Certifying Authority has disappeared having been replaced by the Independent Competent Person (ICP), and it is the ICP who will now agree the inspection programme.

6.2.1 Splash Zone Visual Inspection

Definition: That part of the structure between the crest level of the 50 year (average) wave superimposed on the highest astronomical tide, and 3 metres below the lowest astronomical tide.

The splash zone can loosely be defined as the area of the jacket between the high and low waterlines. It is a region that is particularly prone to corrosion and erosion because wave action makes maintenance of the protective paint coatings difficult. It is also the area most likely to sustain damage from impact with supply boats.

Large installations with design lifespans in excess of 25 years are occasionally clad with copper-nickel in the vicinity of the splash zone to reduce corrosion and marine growth.

The splash zone survey will include an inspection of spider decks, boat moorings and landings, caisson clamps and risers.

Figure 6.13 – Spider deck/splash zone - fire pump caissons (red) and seawater caissons (green) clearly visible

6.2.2 Swimaround Survey

The object of the underwater swimaround examination is to detect any obvious signs of damage or distress, such as missing structural items, failed joints, impact damage, loose or missing riser and caisson clamps.

The seabed will also be examined for 'debris' that may have fallen from the rig. This frequently includes gratings, scaffold poles and an assortment of objects dropped during construction projects.

6.2.3 Non-Destructive Examination

An installation jacket consists of a considerable number of tubular members welded together. The point at which one tubular member is welded to another is referred to as a joint, or 'node'.

The planned inspection programme will identify those joints which, due to their design, are considered to be particularly highly stressed or to have low fatigue lives. A representative number of these joints will be thoroughly examined throughout the life of the jacket.

A thorough examination entails removing marine growth and scale by a powered wire brush until bright metal is exposed. The welds will then be subjected to a thorough visual examination as a minimum, supplemented with a non-destructive examination technique such as eddy current or magnetic particle inspection (MPI). Should visual examination reveal a significant defect, such as a crack, the suspect area will be subjected to a magnetic particle or ultrasonic inspection. If the crack is confirmed, some remedial grinding may be carried out. Should the defect remain after grinding, a more extensive repair programme must be considered.

It should be noted that most jacket designs include a certain number of redundant structural members, so discovery of a defect should not cause unnecessary concern. An engineering analysis will be carried out to determine the extent and the urgency of the action required.

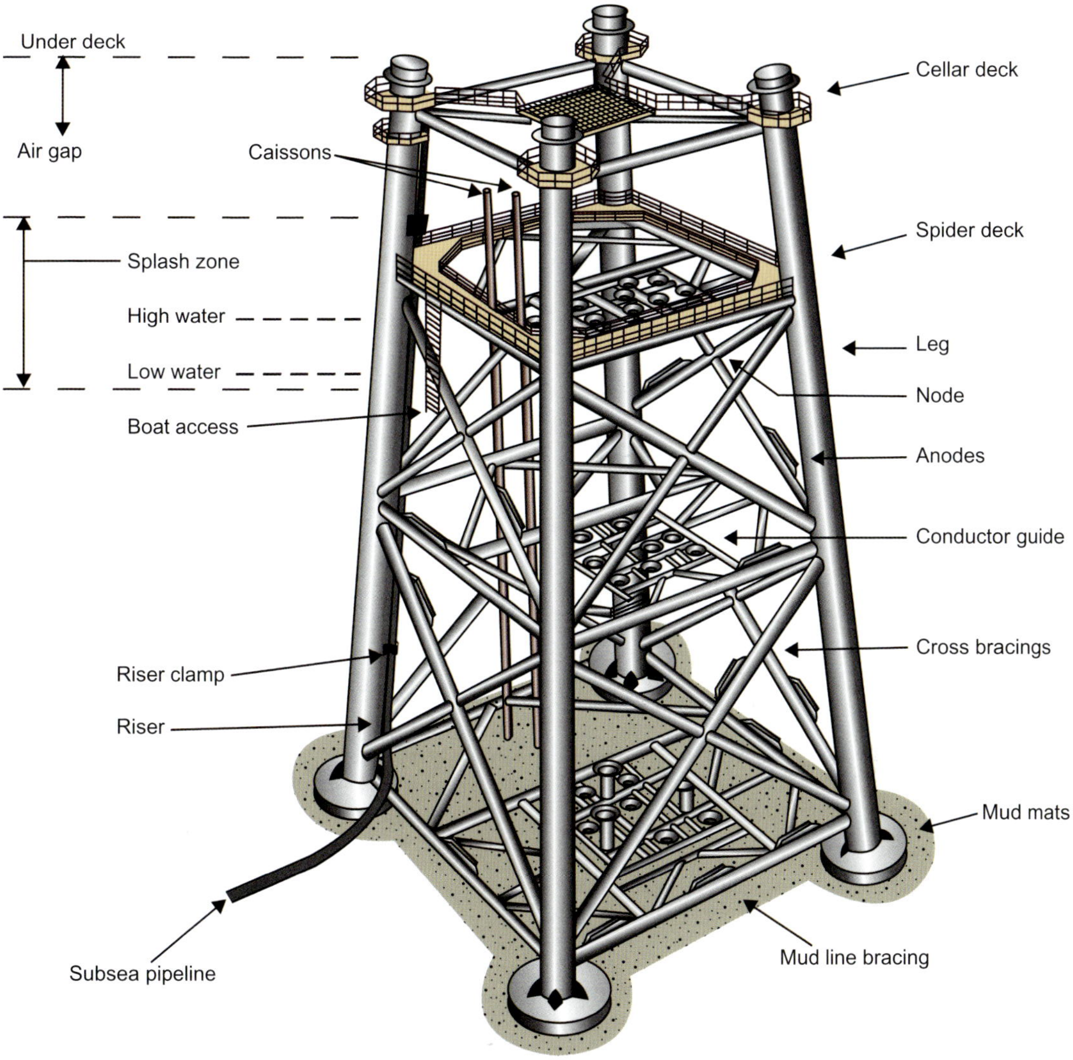

Figure 6.14 – Jacket – Component parts

6.2.4 Flooded Member Survey

This form of examination is relatively new to the industry but has gained considerable popularity, particularly during the last 10 years, due to the ease with which it can be carried out. Minimal cleaning of the structural member is required prior to attachment of an ultrasonic measuring device. The device can determine whether or not the member is full or partially full of water. The presence of water would indicate a crack or weld failure. If a flooded member is discovered, it must be thoroughly examined, using magnetic particle or ultrasonic inspection, to determine the extent of the defect.

6.2.5 Marine Growth Measurement

The additional weight imposed on an underwater structure by the presence of marine growth can be considerable. It must be carefully monitored if overloading of the structure is to be avoided.

Marine growth can be divided into two categories, hard and soft. The hard growth consists of crustacea such as mussels and the soft growth can be a combination of seaweeds and sponges. The soft weed is measured by wrapping a tape measure around the member to give a compressed growth reading.

A marine growth measurement of 50 mm to 100 mm (2 to 4 inches) in depth can normally be tolerated before a cleaning programme is required. The thickness permitted will depend on the type of growth and the design of the structure.

6.2.6 Scour Survey

The sands of the seabed have a tendency to drift in the same way as the sands of the desert. Scour is the term used to describe this subsea phenomenon. It is measured from a datum point, usually the first horizontal jacket bracing above the seabed or mud line.

It is not uncommon to find a variation in seabed levels of up to 4.5 metres (15 feet) between adjacent jacket legs and, if left unchecked, the foundations of the installation could eventually become exposed and threaten the security of the installation.

Scour can also present problems with subsea pipelines by removal of the seabed under the pipeline. The pipeline can be left suspended for considerable distances. These free spans can induce excessive strains in the steel and leave the pipeline in a position where damage may be sustained from ships' anchors or fishermen's nets.

The problem is normally rectified by dumping rocks, gravel or laying concrete mats to build up the seabed and prevent further removal of sand. Another alternative is to pin nylon scour mats to the seabed. These fibrous mats are constructed to encourage sand retention.

6.2.7 Cathodic Protection Examination

The bulk of the underwater structure, at least on very large installations, consists of bare steel and a cathode protection system is employed to inhibit corrosion. This involves the attachment of sacrificial anodes, typically zinc or an aluminium/zinc/indium alloy at strategic positions on the jacket legs and bracings. Galvanic action, promoted by the presence of seawater, decomposes the anodes in preference to the steel structure.

Sufficient anodes (anything from 7 to 70 tonnes) are attached during construction of the jacket to last the life of the installation, normally 25 years. The underwater survey consists of monitoring anode condition and measurement of the potential difference generated. The potential should exceed 0.84 volts and can be measured directly using a Bathy corrometer.

There are two further surveys associated with monitoring the performance of the installation's support structure. While not part of the underwater survey they have been included in this chapter because they are structure and foundation related.

6.2.8 Differential Settlement Survey

Differential settlement readings provide an indication as to the continuing satisfactory performance of the jacket foundation piles.

The measurements may be obtained using an instrument similar to a building surveyor's theodolite, or laser lines of sight directed at a fixed target on the shoreline. It entails measurement of the height of each corner of the installation relative to the other corners to see if any tilting has occurred.

Minor variations may be attributed to changes in deck loadings and may occur when any major items of equipment are added or removed.

6.2.9 Air Gap Measurement

An offshore platform must be designed to resist the severest weather conditions. To withstand wave action there must be sufficient air clearance under the lowest deck on the installation to accommodate what is known as a design extreme wave crest, based on a 50 year return period, and still retain 1.5 metres (5 feet) of clearance gap. The air gap must be monitored on a regular basis because it provides an indication of any settlement or sinkage that may have occurred due to reservoir depletion and seabed subsidence.

Severe problems were encountered in the Norwegian *Ekofisk* field when several installations showed signs of sinking. The problem was addressed by cutting the jacket legs, jacking up the topside structures a distance of six metres and installing spacer spools.

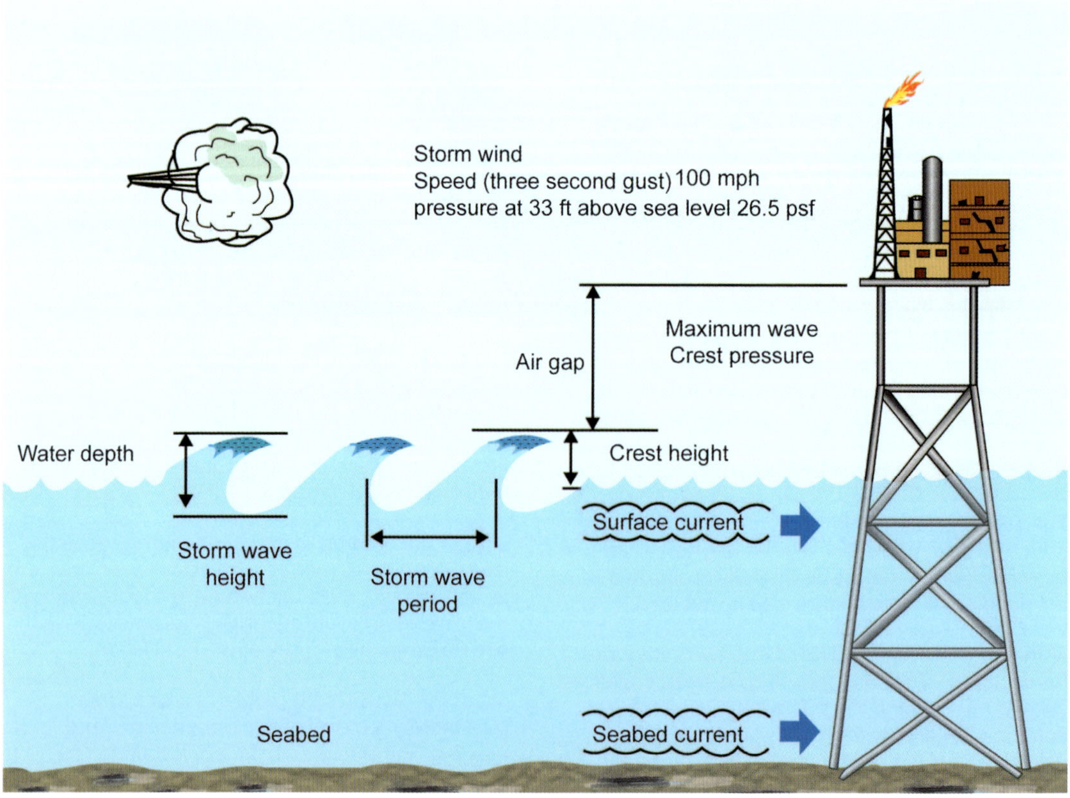

Figure 6.15 – Air gap height

6.3 Subsea Wells

Introduction

Subsea wells have been around for over 40 years and there are over 700 subsea wells operating in the UK sector of the North Sea alone. Subsea wells have two applications. They may be used in support of a fixed installation as an alternative to a satellite platform for recovering reserves located beyond the reach of the drill string, or used in conjunction with a floating production system (FPS). The combination of subsea wells and floating production systems provide a cost effective partnership for the recovery of oil from marginal fields, those containing recoverable reserves of less than 50 million barrels of oil or gas equivalent, and often the only means of developing very deep water fields.

The *Rob Roy* development illustrated (Figure 6.16) is a good example of a field development using a floating production system. Commissioned in 1989, *Rob Roy* is still actively producing oil and is likely to continue doing so into the forseeable future as new subsea wells are drilled. The British Petroleum Eastern Trough Area Project (BP-ETAP) is also illustrated (Figure 6.17), one of the largest and most innovative North Sea developments. The ETAP development is based around a conventional manned platform, the Central Processing Facility (CPF) which processes oil and gas from the Marnock reservoir and the six satellite reservoirs. A not normally manned installation (NNMI) is located over the Mungo reservoir whilst Machar and Monan are produced from subsea manifolds, as are Heron, Egret and Skua (owned by Shell), which are developed jointly as three separate subsea wellhead clusters.

The individual reservoirs that comprise the ETAP development are relatively small, but collectively they constitute some 435 million barrels of oil and NGL (natural gas liquids – propane and butane), and 1.1 trillion cubic feet of gas. At peak production, the development produced 216,000 barrels of oil a day and 553 million standard cubic feet of gas. In the North Sea, which is now considered to be a mature province, there are a considerable number of these marginal fields and extensive research has been carried out on the various options for the recovery of the oil and gas as the number of new, large field discoveries have diminished.

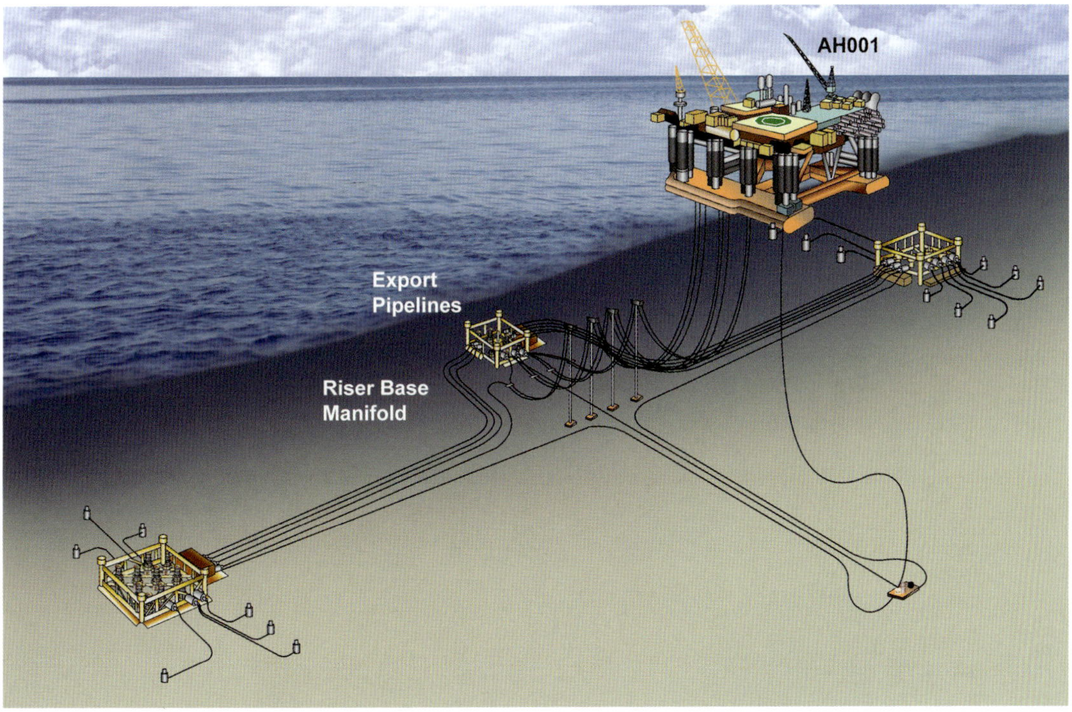

Figure 6.16 – Subsea field development

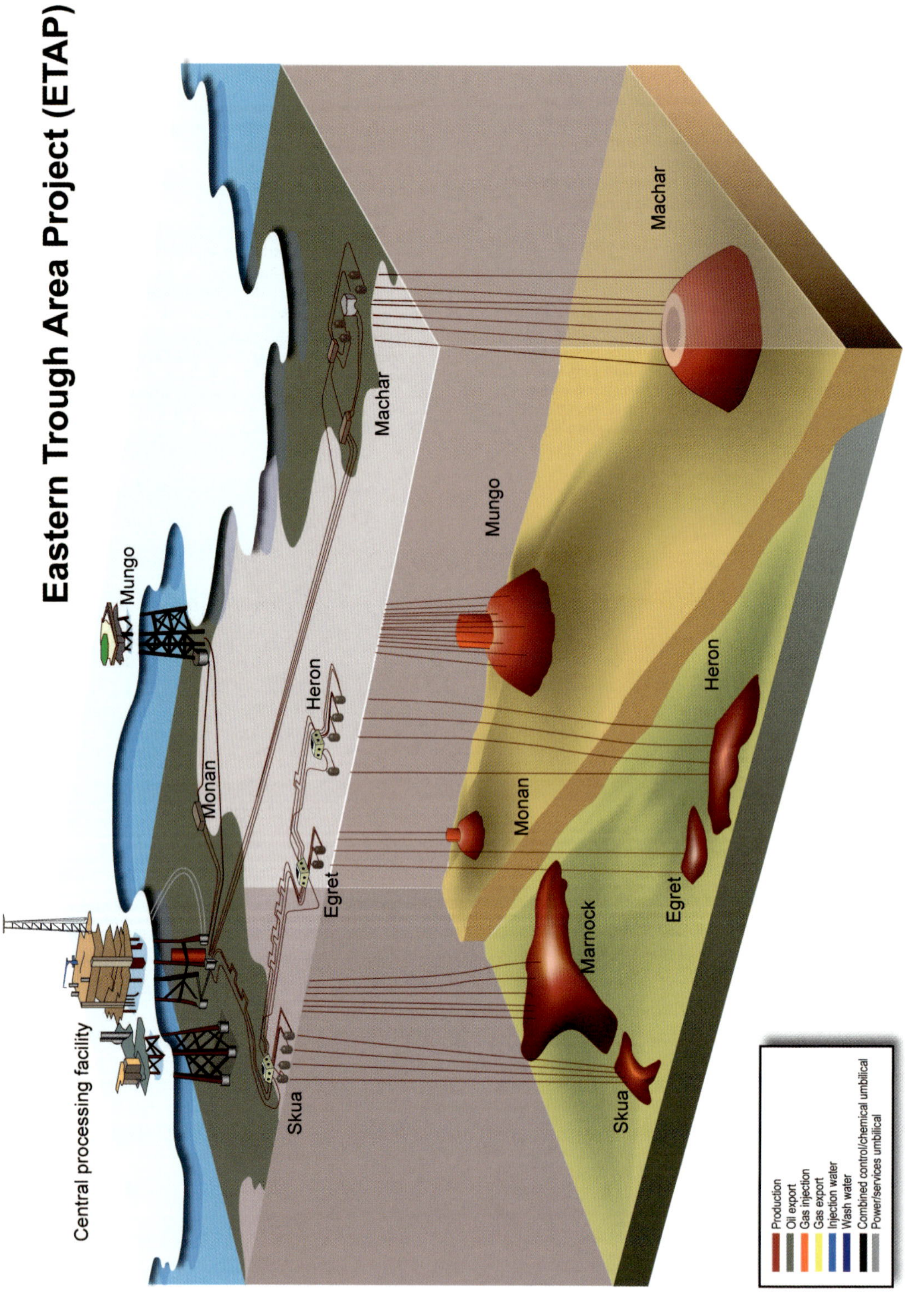

Figure 6.17 – Eastern trough area project

6.3.1 Subsea Development Options

A subsea development may consist of any number of wells. Where the development consists of more than one well, the wells may either be located on a template, in clusters feeding back to a manifold, or a subsea flowbase may be used.

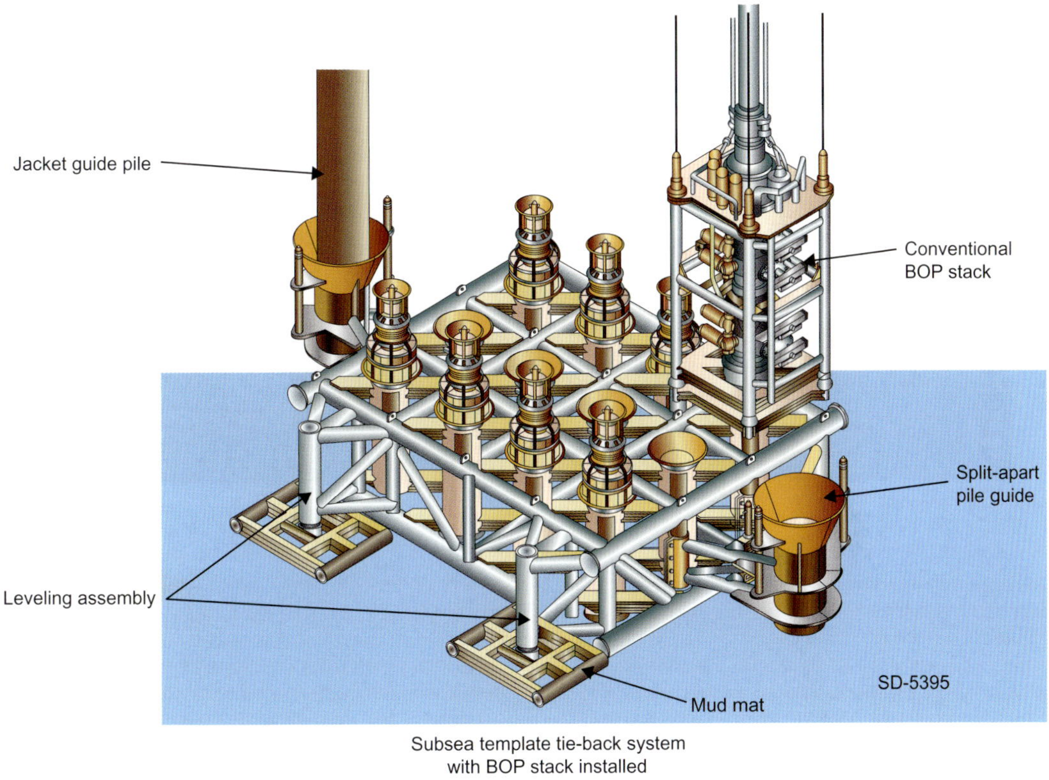

Figure 6.18 – Subsea template

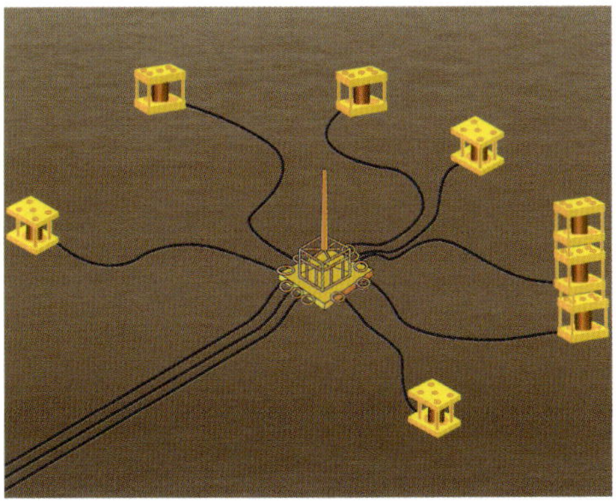

Figure 6.19 (a) – Hoist manifold centre with satellite wells

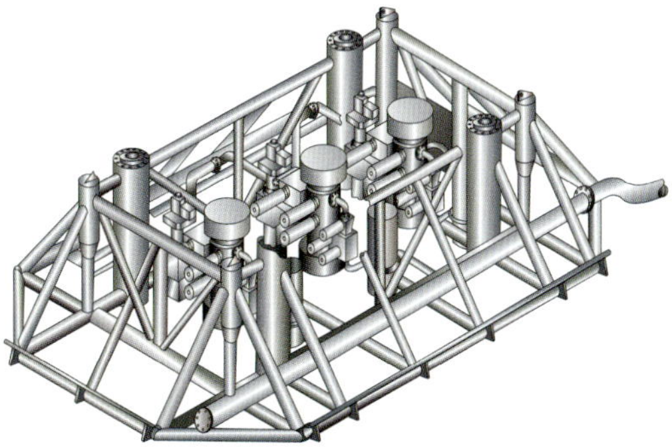

Figure 6.19 (b) – 3 Well subsea template and wellhead protection structure

6.3.1.1 Templates

A subsea template is a steel guideframe that is secured to the seabed by piles or suction anchors and through which the wells are drilled. The templates may be extremely large, heavy structures, up to 800 tonnes, which can accommodate a considerable number of wells, or they may be small and light with provision for just 2–4 wells. The template provides guides to assist in the location of the christmas trees and a manifold arrangement to connect the trees, with the subsea pipeline. A removable structure sits over the structure where over trawl protection is required.

6.3.1.2 Manifolds

At first glance, a subsea manifold looks very similar to a subsea template, the difference being that the structure contains just a manifold, the christmas trees being located around the structure rather than within it, often at locations several hundred metres distant. The manifold is a marshalling point for the flowlines and control umbilicals.

Figure 6.20 – Manifold

There are advantages and disadvantages associated with both manifolds and templates. Manifolds may reduce the amount of extended reach drilling required because the wells can be sited at the optimum field location. Also, being relatively light, the manifold structure can be installed from the drilling rig rather than a crane vessel, which considerably reduces installation costs. However, the cost of the additional subsea flowlines and control umbilicals associated with manifold structures can be prohibitive as they are extremely expensive to manufacture, install and test. Ultimately, the decision as to whether manifolds will be used in preference to templates or vice versa, will be governed partly by the reservoir characteristics and partly by installation and material costs. Some large developments may even use a combination of the two.

6.3.1.3 Subsea flowbase

A flowbase allows one well to be flowed and then co-mingled with the output from a second well in what is generally described as a daisy-chain arrangement. The use of flowbases allows the operator to drill and tie back wells over an extended period of time, the only drawback being that it reduces the flexibility for individual well testing. When reliable subsea multiphase meters become available, the need to individually test wells will no longer be necessary.

6.3.1.4 Pipelines and risers

A subsea pipeline is generally used for the transportation of the reservoir products from the manifold or template back to the host installation, and there are a number of issues to be dealt with if transportation problems are to be avoided, particularly on some of the new developments.

Due to the waxy nature of some crude oils, a system for the circulation of a warm fluid (diesel oil) has been used on some recent developments to preheat the flexible production risers, pipelines and subsea manifold to minimise the deposition of wax from the oil on start up. Other developments have dealt with similar situations by installing one pipeline within another to provide an outer jacket through which hot water can be circulated, or in which insulating materials may be housed.

Another difficulty that has to be overcome is the pigging of the pipeline, a subject explained more fully in Chapter 5. Pigging is facilitated by either installing two pipelines to the manifold or template so that a round trip pigging operation can be carried out to clean the pipeline, or by installing a loop at the blind end of the pipeline with a series of remotely operated valves, the pig can be re-directed within the loop prior to its return to the installation. Alternatively, subsea pig launcher-receivers may also be installed, but all these options are extremely expensive to put into place.

6.3.2 Subsea Wells

A subsea well consists of a christmas tree, wellhead assembly, guide frame and protective structure. The wells are drilled from a jack-up or a drill ship in a manner identical to conventional wells that terminate on a fixed platform.

6.3.2.1 Christmas trees and wellheads

The most basic subsea wells employ a diver assist christmas tree mounted over a mudline suspension system, the christmas tree being bolted to a hydraulically activated collet connector prior to immersion to facilitate attachment to the wellhead adaptor. However, as the need has arisen to produce oil from deeper waters, more specialised subsea wellheads and christmas trees have been developed.

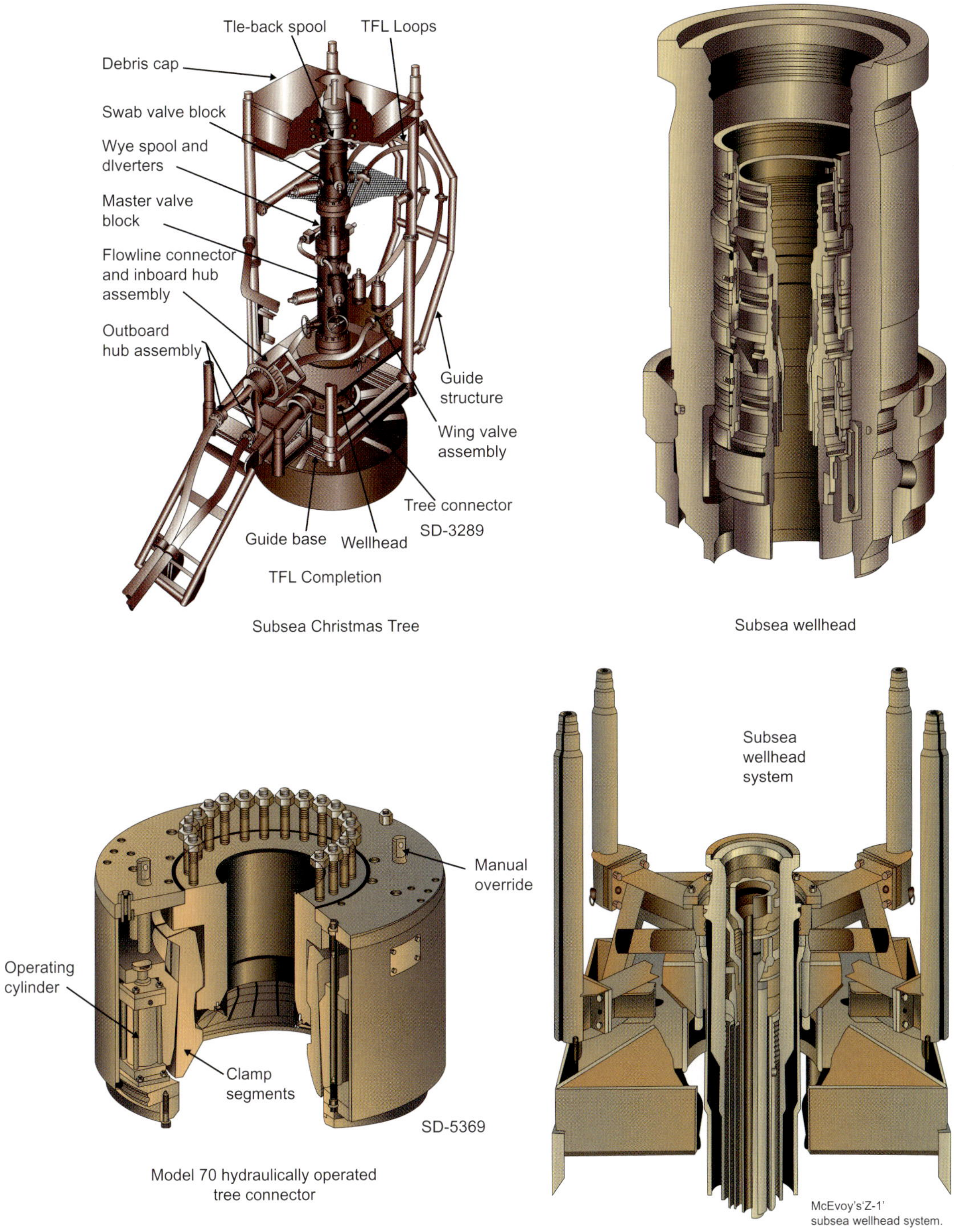

Figure 6.21 – Subsea wellhead equipment

From an installation point of view, there are four types of subsea wellhead and christmas tree, and it is the water depth that ultimately governs the type of system that will be used.

i) Mudline/simple

In waters that are regarded as shallow, less than 90 metres (300 feet) in depth, very basic systems are used that are totally reliant on divers to connect the flowline and control systems and to assist in well workover operations.

ii) Diver assist

Systems installed in waters regarded as moderate in depth, 60–210 metres (200–700 feet), will have some diverless features. Diver assistance will be required for the installation of the flowline and control systems, but workover operations will be carried out without the assistance of divers.

iii) Diverless

While divers can operate at water depths of up to 540 metres (1800 feet), the realistic limit for effective work is closer to 180 metres (600 feet). Consequently, for systems installed in deep waters in the range of 180–900 metres (600–3000 feet) and beyond, totally diverless designs are required and there are two very different approaches to solving the problems associated with developing fields in very deep waters.

The two systems are categorised as either with or without guidelines, the latter being referred to as guidelineless. The first system employs four guidelines that run from the drill ship or semi-submersible down to guide posts mounted on the subsea guidebase. The tree, flowline and control systems are designed so that they can be guided into position by the wires and, as the component parts are stacked one on top of the other, they latch together and seal.

iv) Guidelineless

Deepwater guidelineless systems are designed to be installed from vessels that are dynamically positioned, often in water depths where conventional anchors cannot be deployed. All functions, including well workovers, are performed remotely, ie without mechanical links between the ship and wellhead. The location of each component is achieved using either funnel up or funnel down designs, the operation of which is shown by the illustrations. The deepest subsea well installed to-date is in the Gulf of Mexico in a water depth of 2800 metres (9356 feet).

Installation of the mud mat using the drill string.

Assembly of the HOST onboard the drill ship.

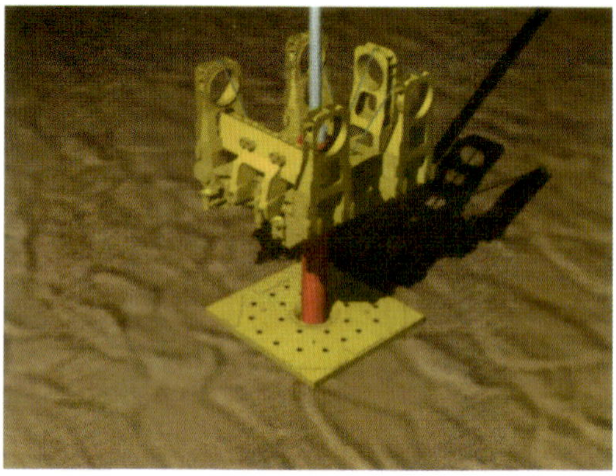

The partially assembled template is lowered through the moonpool and landed onto the mud mat.

Deployment of the wing assemblies using an ROV.

Drill the first 36 inch conductor.

The HOST GL Diverless Wellhead System.
Installation Sequence.

Install the 36 inch conductor and permanent guide base (PGB).

Drilling 26 inch intermediate casing and installation of the casing string.

Install the subsea BOP stack, drill, case and complete the first well.

Install the first christmas tree.

Install the manifold module.

Pull in the flowlines and umbilical control cables.

The completed assembly, five wells installed and connected completely diverless. (Suitable for installation in water depths up to 1,500 metres.)

Figure 6.22 – Subsea well assembly

Ending the reliance on saturation divers has opened up vast reserves up to and beyond the thousand metre (3,450 feet) depth range that were previously thought to be unattainable. While still referred to as new technology, the diverless wellhead has proven extremely reliable and is being considered for use in more moderate water depths where the high initial cost would still appear to be justified when compared with the financial savings to be made on conventional diver intervention.

6.3.2.2 Guideframes

The subsea tree is surrounded by a guideframe that acts as an installation device for the component parts of the tree and for the wellhead protection framework.

From an operational viewpoint the guideframes provide three basic functions:

i) **Interface**
The guideframes are designed specifically for a particular type of wellhead system and provide

an interface that assists in the alignment of the tree during installation.

ii) Mounting points
Attachments may be provided on the guideframe for the mounting of the flowline connector, the ROV docking station, and the control umbilical, through which the electrical or hydraulic signals are conveyed to operate the christmas tree valves.

iii) Re-entry guidance
The top of the guideframe may be fitted with guidance posts or funnels to assist in the alignment of the tree running tool and tree cap, which are used to install, retrieve and workover the tree.

6.3.2.3 Wellhead protection structure (WPS)

Subsea trees are prone to damage, the most common causes being impact damage from ships' anchors or from fishermen's trawlboards. Protective structures are normally fitted to provide some degree of protection and these will be installed over the well bay where the trees are installed on a template or built directly over a satellite or cluster tree. The protection structure may be rigidly attached to the well or it may, alternatively, be a gravity device. Either way it must be of a design that can be removed with the minimum of effort to facilitate drilling, well completion and workover operations to proceed.

Figure 6.23 – Subsea wellhead protection frame loaded onto the deck of an offshore supply boat for transportation to its final vesting place.

6.4 Subsea Developments

The use of subsea wells to recover reserves in the immediate vicinity of a host installation, be it fixed or floating, is a relatively straight forward operation. However, as the distance from the installation increases, artificial means may be required to augment the natural reservoir pressure if the flow of hydrocarbons, and oil in particular, is to continue. While water injection is often used as a way of boosting reservoir pressure, this option is only viable on relatively large oil fields, the only other alternative being the installation of gas lift equipment.

Strictly speaking, water injection and gas lift are ways of improving the flow of oil from the reservoir, a process referred to as enhanced oil recovery (EOR), and what we are about to discuss relates more to the problems caused by the distance the wells are located from the host facility, rather than the condition of the reservoir. The equipment may well be provided in addition to one of the previously mentioned forms of enhanced oil recovery, a subject more fully discussed in Chapter 5.

The transportation of unstabilised well fluids, liquids or gases, is fraught with difficulties due to the volatile nature of the component parts of the hydrocarbon mixture, a problem exacerbated by pressure drop as pipeline length increases. The flow of oil in a pipeline may be interrupted by slugs of vaporised gases, while gas pipelines are likely to suffer from slugs of condensed liquids that drop out of the gas during pressure fluctuations. Large volumes of sand and water create additional problems, particularly as the reservoir ages. Providing the equipment that can solve these problems has presented equipment designers with a number of extremely challenging tasks.

The present horizontal range for directional drilling from a fixed structure is about 8 km (5 miles) while subsea wells may be located up to 15 km (10 miles) distant from the mother platform. Extending the distance that wells may be located from the host installation has attracted considerable funding in recent years, with the research proceeding along two quite different routes, ie subsea separation, and multiphase or subsea pressure boosting, the ultimate aim being to provide a reliable system for the exploitation of oil reserves up to 50 km (30 miles) from the host facility.

Before embarking on a description of subsea separation and pressure boosting systems, it should be noted that while these options are both regarded as new technology, they have now been developed to a point where they are being considered on a number of new projects. Expenditure on this type of equipment is predicted to exceed $5 bn in the next five years, with Africa and Latin America being two key markets.

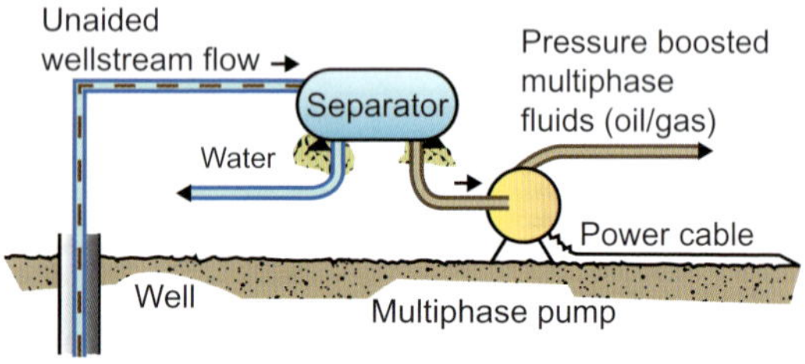

(a) The main components of a subsea separation scheme

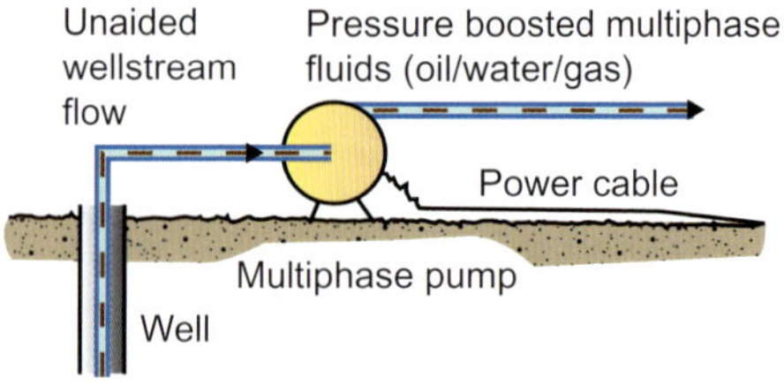

(b) The main components of a seabed multiphase pumping scheme

Figure 6.24 – Subsea developments

6.4.1 Subsea Separation

Subsea separation, or seabed processing, involves the separation of the gas from the liquids on the seabed in the vicinity of the subsea wells so that they may be transported back to the host installation through separate pipelines. This approach virtually eliminates the problems associated with the pumping and the conveyance of unstabilised reservoir products (referred to as multiphase fluids) over long distances through a single pipeline, although the injection of a hydrate inhibitor, such as methanol or glycol, into the gas pipeline may still be required in some instances.

While their use is not widespread, subsea separation systems have been available for a number of years. Recent developments, however, have centred around the potential benefits to be obtained by disposing of the separated water at source, or at least near the wellhead, the water either being re-injected into the formation, or dumped into the sea, rather than transporting the combined products back to the platform or FPSO for separation. With some new developments producing well fluids that initially contain 50–65% water, often rising to 95% water as the field ages, this option has considerable benefits and, as such, is one of the emerging technologies receiving considerable industry funding.

Perhaps the most dramatic development in subsea separation is being undertaken by Shell, who have elected to install three subsea separation units at a water depth in excess of 9,000 feet (2,700 metres) on their *Perdido* development in the Gulf of Mexico.

6.4.1.1 Subsea separation and re-injection systems (SUBSIS)

The subsea separation system sits on the seabed in the general vicinity of the template or manifold and relies primarily on a conventional gravitational separation vessel, a production separator, for the separation of the oil from the water, prior to the water either being re-injected into the formation or dumped into the sea. Single phase, constant speed, electrical powered centrifugal pumps are employed to boost the pressure of the liquids as they exit the separator and, provided that the pipeline is suitably sized to maintain the pressure above the bubble point of the oil, the subsequent transportation should proceed smoothly.

While the advantages of a subsea separation system are considerable, there are some very demanding challenges still to be overcome, such as the development of high voltage electrical connections and a reliable oil-in-water monitoring device. A water purity of 1,000 ppm (parts per million) is required for water destined for re-injection, while 40 ppm is the maximum permissible value for oil-in-water if the produced water is to be discharged directly into the sea.

The demand for electrical power will be considerable because the system will require, as a minimum, sufficient power for the pumps used to increase the oil pressure for transportation back to the host installation. Power will also be required if gas compressors and water re-injection equipment are specified. Electrical power will be fed through an umbilical, which will also contain fibre-optic control lines, chemical injection and hydraulic oil supply lines. Electrical connectors that can cope with 24 to 36 kV have been specified and the equipment will have to be extremely rugged if it is to function reliably in the harsh environment subsea, for anything from 10 to 25 years.

Operational benefits

The two main benefits of a seabed separation system relate to the saving of weight on the reception facility and the avoidance of problems associated with pumping multiphase well fluids.

Reducing the volumes of liquids being transported back to the reception facility can result in a considerable reduction in the process equipment required and, while the reduction of weight is highly desirable on a fixed installation, it can have a dramatic effect on the sea keeping properties of a floating production and off-loading installation (FPSO).

Figure 6.25 – Subsea separation and injection system (SUBSIS)

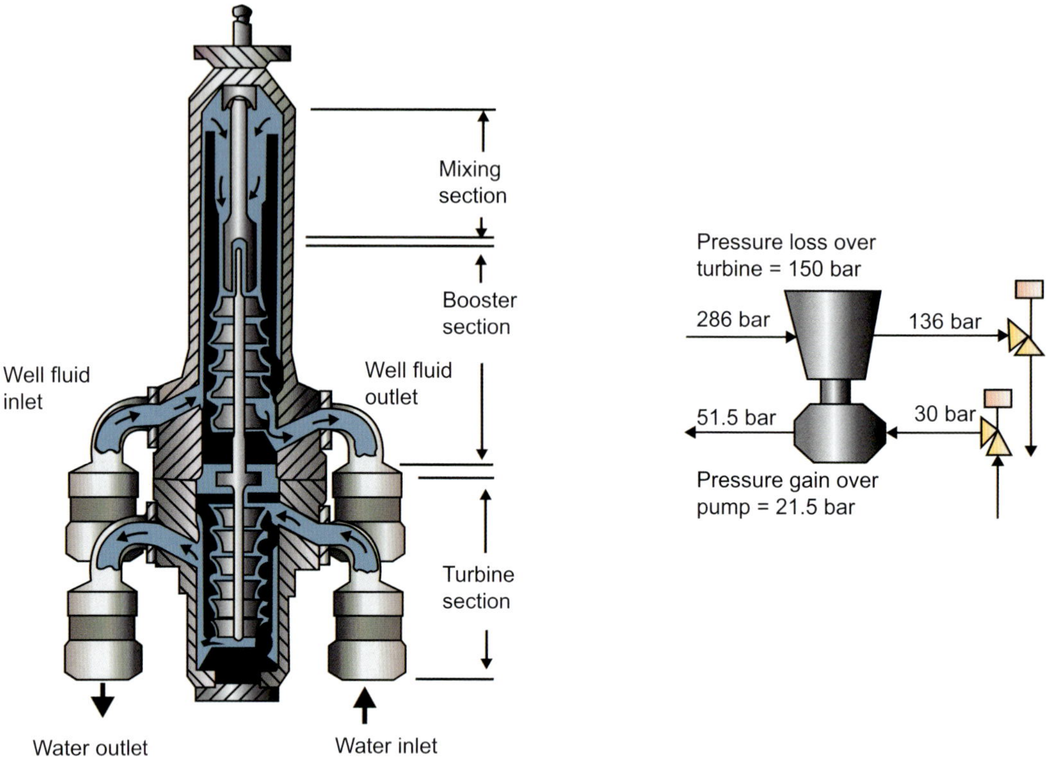

Figure 6.26 – Framo type subsea booster pump

The removal of water, virtually at source, also permits the use of much smaller diameter pipelines, flowlines and risers, resulting in a sizeable reduction in material and installation costs, particularly on fields being developed by FPSOs where there is also a limit to the number and diameter of the pipelines and flexible risers that can be installed. The pipelines and risers often appear to consume a disproportionate part of the field development budget.

6.4.2 Multiphase Pumping and Metering

Multiphase pumping and metering systems have been developed to a point where they are reliable enough to be considered as a matter of course for new developments. While initially used only for land-based applications, their use offshore has been motivated by the desire to simplify equipment and further increase the potential of unmanned installations and subsea developments.

6.4.2.1 Multiphase pumping and subsea pressure boosting

Introduction

Subsea pressure boosting is the process by which additional pressure energy is imparted to an unprocessed well stream as it exits the wellhead to prevent the formation of gas slugs in the subsea pipeline while the well fluids are transported back to the central processing installation, often up to 35 km distant. This has entailed the development of a pump that can cope with well fluids containing high percentages of gas without vapour locking, ie a multiphase pump.

The main advantage of multiphase pumping over equipment that relies on phase separation prior to pumping is the fact that the reservoir products are transported through a single pipeline. This represents a considerable saving on costs when compared with installing both oil and gas pipelines over extended distances.

In 1994, the world's first commercial subsea multiphase booster pump unit, referred to as the SMUBS concept, was installed on a spare slot on the water injection template of Shell's *Draugen* field development, and production was enhanced by 5,000 bbl/d. Following the success of this system, which was supplied by Framo Engineering AS of Norway, the BP ETAP consortium contracted Framo to provide two subsea booster pump units for the *Machar* field development for installation in 1999.

The subsea pressure boosting equipment is located on the seabed in the vicinity of the subsea wells and should not be confused with the submergible pumps that are installed downhole in the wells. The submergible pumps are used for enhanced oil recovery and, while the subsea booster pumps do improve the flow of oil from the reservoir, they are installed primarily to improve the transportation of the oil in the subsea pipeline.

Features

The pumps are installed on a separate structure as part of the water injection valve module located approximately 30 metres downstream of the production manifold. They are based on the self-regulating, helico-axial (Poseidon) principle and are driven by a hydraulic turbine powered by high pressure water supplied from the central processing platform. The water provides the motive power for the turbine prior to being re-injected into the formation for enhanced oil recovery.

Prior to the well fluids entering the pump inlet, they are thoroughly homogenized by a common flow mixer assembly, which is located immediately upstream of the pumps. Optimising the inlet conditions of the well fluids prevents slugging and broadens the operating range of the pumps. However, it should be noted that the pumps are designed to operate safely in both pure gas and pure liquid conditions.

Under normal operational conditions, the full injection water flow rate of approximately 32,000 bbl/d, at a pressure of 286 bar (4,147 psi), will be routed through each pump turbine and a pressure drop of 150 bar (1,972 psi) is expected. The pumps have been designed with reliability being of paramount importance and, consequently, no mechanical seals are used and all the bearings are lubricated by the drive water.

6.4.2.2 Multiphase metering

Accurate measurement of the quantities of oil, gas, water and solids exiting a well is an essential prerequisite to successful reservoir management. At present, it is a time consuming and labour intensive operation that involves the use of test separators to separate the flow into its constituent components, ie oil, water and gas, prior to measuring them separately using straightforward, single phase orifice metering techniques. While this approach is effective, the equipment required is considerable and somewhat impractical for subsea applications. The alternative is to provide equipment that can measure the flow of multiphase well fluids on line, ie without phase separation, a multiphase flow meter.

There are a number of multiphase meters on the market, most of which are claimed as being suitable for both topside and subsea applications. The technology varies considerably from manufacturer to manufacturer and electronic sensors, or a combination of mechanical separators and electronic sensors, are employed in various arrangements to analyse and measure the flow of hydrocarbons. Figure 6.27 shows a Framo subsea multiphase meter that was selected for the BP-ETAP development.

6.4.2.2.1 Framo multiphase flow meter

Features

The Framo multiphase flow meter is described as an inert design. The main body or receiver barrel is permanently installed on the subsea manifold structure and acts as the inlet and outlet housing, and as a guide and support during the installation of the insert cartridge, which contains the actual measuring equipment. This arrangement facilitates the removal and replacement of the operational components as one unit, either by diverless guidewire or guidelineless equipment.

Output from the flow meter is processed by a computer that provides the actual flow measurements. The equipment may be electrically interfaced to a control pod on the subsea production system with separate wires being provided for power supply and data transmission, or the flow meter may be fed from a dedicated control umbilical. Typical supply voltage will be 24 volts DC with a power consumption of approximately 20 watts.

Operating principles

The heart of the Framo multiphase meter is the flow mixer body, which is designed to provide an homogeneous flow to the measuring sensors located within the venturi. The capacity of the mixer body is such that it can accommodate sufficient fluids to ensure that the transient conditions associated with multiphase fluids are smoothed out, ie the measuring sensors are never exposed to unexpected slug flow conditions.

Well fluids entering the mixer body separate with the heavier liquid phases gravitating to the base of the chamber, while the gases migrate upwards and into the ejector via the injection pipe. The liquid and gases are reunited as they enter the ejector nozzle, which generates a turbulent shear layer and provides the venturi with a homogeneous mixture, a known flow regime that can be accurately measured.

As the homogenised mixture passes through the venturi, the fluid velocity is measured as if it were a single-phase fluid with equivalent mixture properties and standard venturi calculations can be applied. The fractions of oil, water and gas are determined by a dual gamma fraction meter, the calculations being based on the attenuation of two different gamma energy levels emitted from a radioactive Barium 133 isotope. Measurement of the detector hit rate of photons from the two different energy levels can, by physical equations, be expressed as a function of oil, water and gas volume fraction. The advantage claimed for a dual energy gamma meter is that the system is completely independent of whether the liquid is dominated by oil, water or even emulsions.

Future developments

The use of radioactive isotopes is becoming more of a common feature in measurement analysis and the latest technology under development, Nuclear Magnetic Resonance (NMR) and Pulsed Neutron Activation (PNA) devices, appear to offer direct measurement of individual phase fractions and velocities. Both developments are technically challenging and expensive to produce, particularly for subsea applications, and a great deal of further prototype testing is required before units become commercially available.

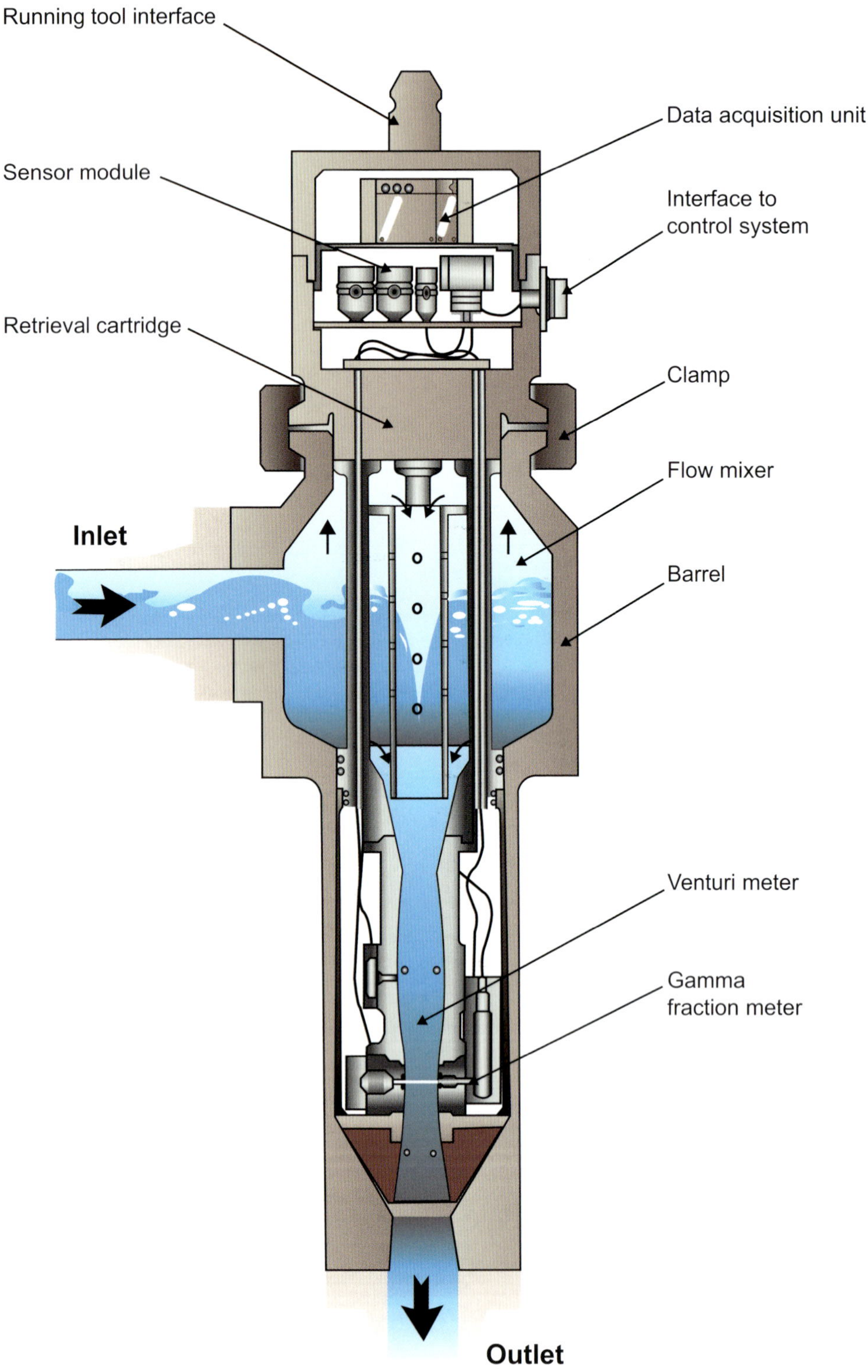

Figure 6.27 – Framo multiphase flow meter

7

Drilling

7 Drilling

7.1 Introduction

7.1.1 The Formation of Oil and Gas

Figure 7.1 – Oil field with jack-up drilling rig in the background

Oil and gas is derived almost entirely from decayed plants, algae and bacteria in an energy regeneration process, where energy from the sun, that originally fuelled the growth of the plants and bacteria, is recycled into a useful energy source in the form of coal, oil or gas.

Only a very small percentage of the vast quantities of vegetation and bacteria that are produced by nature are converted into hydrocarbon compounds because, in most cases, they decompose naturally and are absorbed by their surroundings. It is only where there is insufficient oxygen to feed the bacteria that would normally break down this material, that it remains intact and stagnates, the first stage in a unique set of circumstances required for the formation of an oil or gas field.

Where swamps, flooded forests, sheltered lakes and river estuaries occur, huge volumes of organic material can accumulate and, trapped under layers of silt and mud, the materials may be preserved in sufficient concentration to form a source rock. As the trapped deposits are further buried by younger sediments and pushed deeper towards the earth's crust, the temperature increases and a cooking process commences.

With increasing heat and pressure, the fats, waxes and oils from the algae, bacteria and spores break down into their chemical component parts, which link to form Kerogen. The cellulose and woody parts of land plants are converted to coal and woody kerogen and, as the source rock matures and is further heated, long chains of hydrogen and carbon atoms break from the kerogen and form waxy and viscous heavy oils. At higher temperatures the shorter hydrocarbon chains break away and form lighter oils and then gas, predominantly methane.

The generation of hydrocarbons continues as the source rock matures and, as the pressure increases, migration of the oil and gas commences through cracks and fissures in permeable rocks to areas of lower pressure. This migration to areas of lower pressure continues in an upward direction until, in many cases, the oil or gas escapes into the atmosphere as natural gas or tar seepages.

The upward migration of oil or gas through permeable or reservoir rocks can be halted only when they are overlain with an impermeable cap rock, such as cemented sandstones, clays or salt, all of which act as a sealing medium, salt being particularly effective as it is dense and heals itself when cracks appear due to movements in the earth's crust. Only where the reservoir and cap rocks occur in the correct formation, shapes and relative positions will oil and gas accumulate to form oil and gas reservoirs. These reservoirs should not be thought of as huge underground caverns, the hydrocarbons actually exist within the structure of porous rocks in a condition analogous to a sponge full of water.

In the North Sea, the formation of gas commenced approximately 300 million years ago during the Carboniferous period, when the area which is now the southern North Sea was covered in lush, tropical rain forests. It is thought that methane, the primary constituent of natural gas, was formed from the vegetation after it had been converted to coal bearing rock. The oil, which is found in a deep rift valley in the northern North Sea, was formed from the remains of planktonic algae and bacteria that flourished in what were the tropical waters of the Jurassic period, approximately 140 million years ago. In some parts of the North Sea, trap structures existed 125 million years ago but were not filled with oil until 100 million years later. These processes continue today. The floor of the North Sea is still sagging and new areas of source rock are reaching depths where they too are generating oil and gas.

7.1.2 Exploration

Prior to the commencement of drilling activities a licence must be obtained, which will allocate an area or block of territorial waters designated as suitable for offshore exploration. Having obtained an exploration block, a survey vessel must carry out a seismic investigation.

7.1.2.1 Seismic surveys

Seismic surveys are the way in which underground rock formations are examined to determine if potential hydrocarbon bearing reservoirs exist.

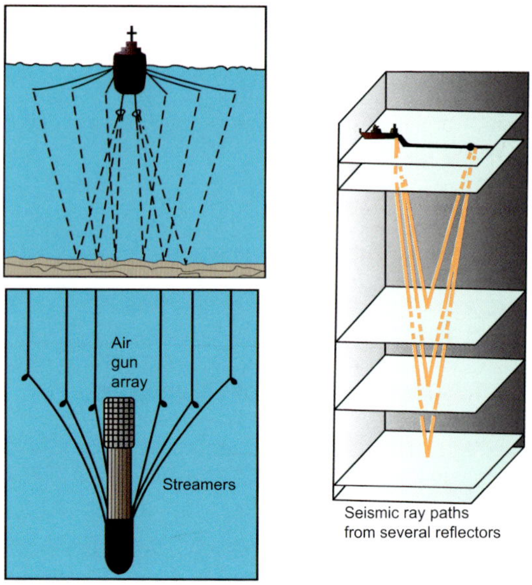

Figure 7.2 – Seismic survey emission and interpretation

Surveys are carried out by specially equipped survey vessels that traverse the area under investigation in a systematic pattern, emitting controlled bursts of sound energy from a submerged air or water gun array towed behind the vessel. The sound waves radiate through the water and into the rock formations under the seabed, where they are refracted and reflected by the different layers of rock. The return signals are absorbed by hydrophones towed behind the vessel, up to 240 contained within a streamer, several kilometres in length.

Interpreting seismic survey information takes many months. The data received by the hydrophones is used to plot graphs of sound intensity against elapsed time. Vertical seismic slices are produced which are then stacked side by side to form a seismic cross section, a two dimensional (2-D) view of the underground structure where the reflective and refractive surfaces correspond with the boundaries between rock layers. Fault lines, stratigraphic and structural traps can be identified on the seismic cross sections and, because the velocity of sound is sharply reduced in gas bearing rock, they provide some indication of where hydrocarbon deposits may be located.

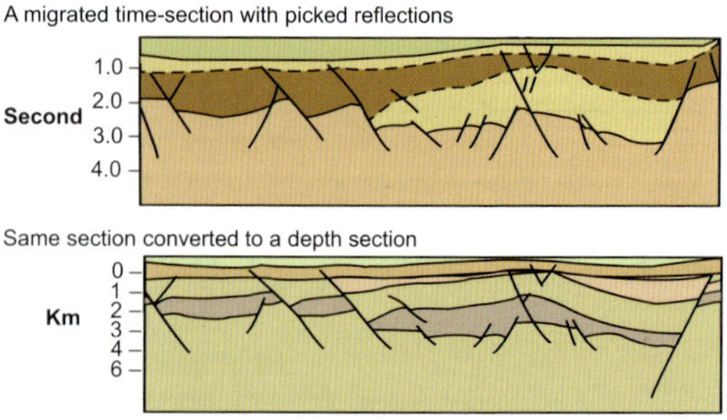

Figure 7.3 – Interpretation of seismic surveying

While 2-D seismic surveying has formed the backbone of the exploration business for many years, the recent advances in the power and speed of computers, combined with more sophisticated sound transmitting equipment deployed at much denser spacings, has permitted the development of three dimensional (3-D) seismic images. These images, comprising both vertical and horizontal sections or time slices, allow for the delineation of deeper, more subtle stratigraphic plays (the study of the depositional interrelationships of sedimentary rocks) which are a great advantage, particularly in geologically challenging areas such as West Africa.

The end result of the seismic survey investigations are the formulation of contour maps and 3-D colour and shade enhanced images that show very detailed geological structures. Visualisation software further permits the interpreters to view the 3-D data as a cube, which can be rotated and cut at any angle to view the subsurface geometry of the rock structures more closely. The results can be used to determine where further seismic surveys would prove beneficial or wildcat drilling commenced, as it is only by drilling a well that the presence of hydrocarbon deposits can be confirmed.

A 4-D survey is a time elapsed 3-D survey. By shooting over an array with a seismic vessel periodically it is possible to monitor any changes to the seismic wave character throughout the field life and to actually see seismically the oil draining from the rock, facilitating an accurate study to indicate the areas of the field where oil is not being produced efficiently.

7.1.2.2 Drilling programme

Figure 7.4 – Semi-submersible, typical of those used in wildcat drilling operations

The drilling of a wildcat or exploration test well is still the only foolproof way of confirming the existence of a hydrocarbon reservoir, and a success rate of only one in four wells gives some indication as to the difficulties encountered in determining the existence of hydrocarbon deposits. It takes approximately eight weeks and costs up to $10,000,000 ($60,000,000 in Arctic waters) to drill a test well, so it doesn't pay to have too many failures.

Figure 7.5 – The jack-up *'Galveston Key'* testing an appraisal well. The vent boom can be seen flaring the gas. The V door and pipe draw can be seen at the base of the derrick.

Figure 7.6 – The jack-up *'Al White'* drilling the production wells on the RWE *'Cavendish'* gas platform

On the occasions where drilling operations prove successful, a well test programme will be instigated to evaluate reservoir conditions and ascertain the exact nature of the hydrocarbon deposits. This entails flowing the well through a test facility located on the drilling vessel and the results will provide information as to the type of equipment that will eventually be required to produce the oil/gas and give an indication of any problems that may be encountered.

A successful well test programme will be followed by the drilling of a series of appraisal wells at locations considered by the geologists to represent the boundaries of the field. This will enable the physical dimensions of the field to be calculated and provide an indication as to its development potential.

Where exploration surveys indicate that a field represents a commercially viable proposition, a permanent installation will be located at a position where it can reach the more favourable hydrocarbon bearing formations. A field can cover several square miles and consist of several geological traps or reservoirs, all of which must be tapped individually by separate wells. Deviation drilling techniques permit the exploitation of reservoirs situated up to 8 km horizontally distant from the rig, while reserves located beyond the range of the drill string may be harnessed by subsea wells linked to the mother platform. Large fields may necessitate the installation of additional satellite platforms. The *Leman* gas field, located off the coast of East Anglia, supports more than 40 such platforms.

Frequently, fields are discovered that are considered to be marginal or borderline in terms of prospective profitability and often these must await improvements in the economic climate or available technology before they can be considered for development. Advances in subsea equipment and drilling technology now permit the exploitation of fields that were discovered up to 25 years previously and which were considered uneconomic at that time.

7.2 The Well

The majority of existing hydrocarbon reservoirs are located between the depths of 7,000 to 12,000 feet (2,100 to 3,500 metres), although discoveries of reserves located at depths of up to 15,000 feet (4,550 metres) are becoming more frequent. As one might imagine, if a hole is to be drilled to such a depth it must be drilled in several stages and reinforced to provide resistance against external pressure created by the rock formation. The process may be compared with the digging of a tunnel, as they both require additional support when passing through soft rock or sand. However, unlike a tunnel, the hole also requires reinforcing to withstand internal forces generated by reservoir pressure.

During the drilling operation, a column of dense cutting fluid (drilling mud) stabilises the hole and prevents the ingress of loose formation products. On completion of each intermediate hole, permanent support is provided by the insertion of a steel pipe referred to as a casing string. The casing strings are assembled from 30 foot (9 metre) lengths of pipe, which are screwed into tubular couplings or 'joints' as they are lowered into the hole. The casing string is then cemented into the rock formation prior to commencement of the next stage of drilling.

The process by which a hydrocarbon well evolves from a drilled hole will now be described.

7.2.1 Well Construction

i) The drilling operation commences with the installation of a large diameter steel pipe known as a conductor. Typically of 30 to 36 inch (0.75 metre) diameter, the conductor provides the foundation for subsequent drilling operations and a support mechanism for the intermediate casing strings.

Figure 7.7 – Conductors and guide frame - Spider deck level

The conductor may be installed into a pre-drilled hole and secured with cement, or it may be piled into the seabed. The latter approach involves driving the conductor to refusal, the depth at which it will go no further. The formation products are then drilled from the centre of the pipe and piling recommenced. This process is repeated until the conductor has been driven anything from 200 to 800 feet (60 to 240 metres) into the seabed. A starting head is then fixed to the top of the conductor and a diverter installed.

ii) Drilling commences in earnest with the second hole, which will be sunk to a depth of approximately 2,000 feet (600 metres). On completion of the hole, the drill string must be removed while the first intermediate casing string is lowered into the hole and cemented in place. A casing head spool is mounted above the starting head to support and seal the top of the casing and provide a temporary home for the BOP stack.

iii) The number of subsequent holes drilled, their depth and diameter, will be governed by the type of rock formation and the overall depth of the hydrocarbon reservoir. Very broadly speaking, the drill bit will be reduced in diameter every 3,000 to 6,000 feet (900 to 1,800 metres) and the hole cased off (casing inserted). However, on occasions, hole collapse or loss of drilling mud into the formation will necessitate suspension of the drilling programme and premature installation of the intermediate casing string. Having reinforced the hole, drilling may be resumed using a smaller diameter drill bit.

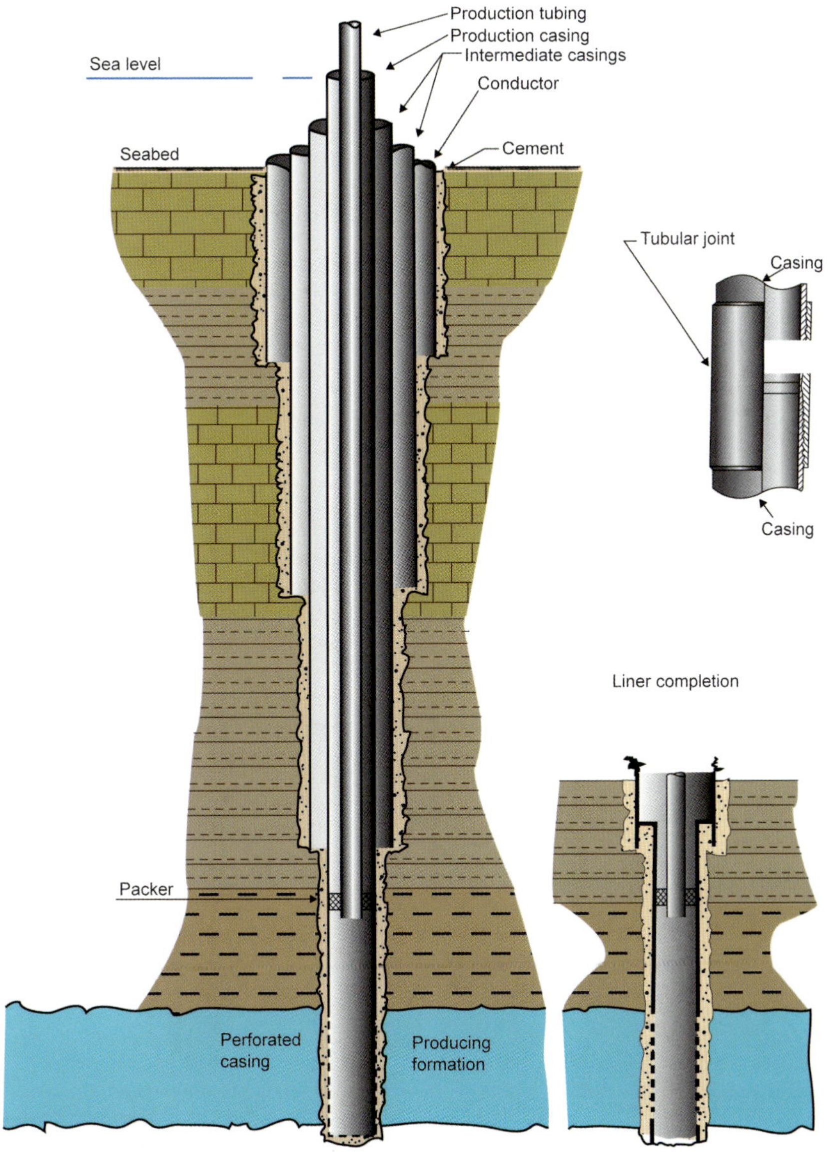

Figure 7.8 – Typical well casing arrangement

iv) Eventually the drill bit enters the hydrocarbon bearing formation. When the hole has been drilled to the required depth, the drill string must be carefully removed to permit completion of the well, the hydrostatic head of drilling mud preventing hydrocarbon products entering the hole. Completion of the well involves the insertion of the final casing string, which may be terminated using any one of three configurations.

(a) Casing completion
As shown in the main sketch in Figure 7.8, the final casing string is run to the bottom of the hole and cemented in an identical manner to the preceding casing strings.

(b) Liner completion
As shown in the detailed sketch in Figure 7.9, the casing or liner extends from the base of the previous casing and is not returned to the surface. It is installed on the end of a drill string attachment and secured by a packing device prior to cementing. Liner completions are used extensively and permit a considerable saving on casing, in excess of 3,650 metres (12,000 feet) in a typical North Sea well.

(c) Open hole completion
With open hole completion, the hole is left uncased in the vicinity of the hydrocarbon formation, the casing or liner being terminated at the top of the hole.

v) The penultimate operation entails installation of the production tubing and christmas tree in preparation for perforation and the commencement of production.

The production tubing is essentially a small diameter casing string, which is suspended from a tubing hanger located under the christmas tree and held within the lower reaches of the preceding casing by a packing device. This arrangement facilitates removal of the tubing should well modification work be required at a later date. The production tubing contains a number of machined recesses or nipples designed to accommodate the mudline safety valve, and steel plugs that are used to isolate the well during repair and maintenance operations.

With the exception of the casing located below the packer, the production tubing is the only component within the well that receives direct exposure to the reservoir pressure and it is through the production tubing that the oil and gas will flow to the surface. Once the christmas tree has been nippled up (bolted up) to the production tubing hanger, and hydrostatically tested, the perforation operation can commence.

vi) It should be remembered that prior to perforation, the well bore is isolated from the hydrocarbon bearing formation by the casing string and is still full of conditioning or well preservation fluid (normally brine or diesel oil). Perforation involves the puncturing of the final length of casing string in the vicinity of the hydrocarbon bearing formation, a distance that can extend over several hundred feet. It is achieved by lowering shaped explosive charges into the well on a wireline tool string and detonating them by remote control, either electronically or by hydraulic pressure signals. The charges blast through the steel casing and cement and permit the passage of reservoir products into the well bore.

Initial entry of the oil or gas into the production tubing is prevented by the hydrostatic head of preservation fluid and there are two main methods by which the hydrocarbon flow can be initiated and the well brought on line:

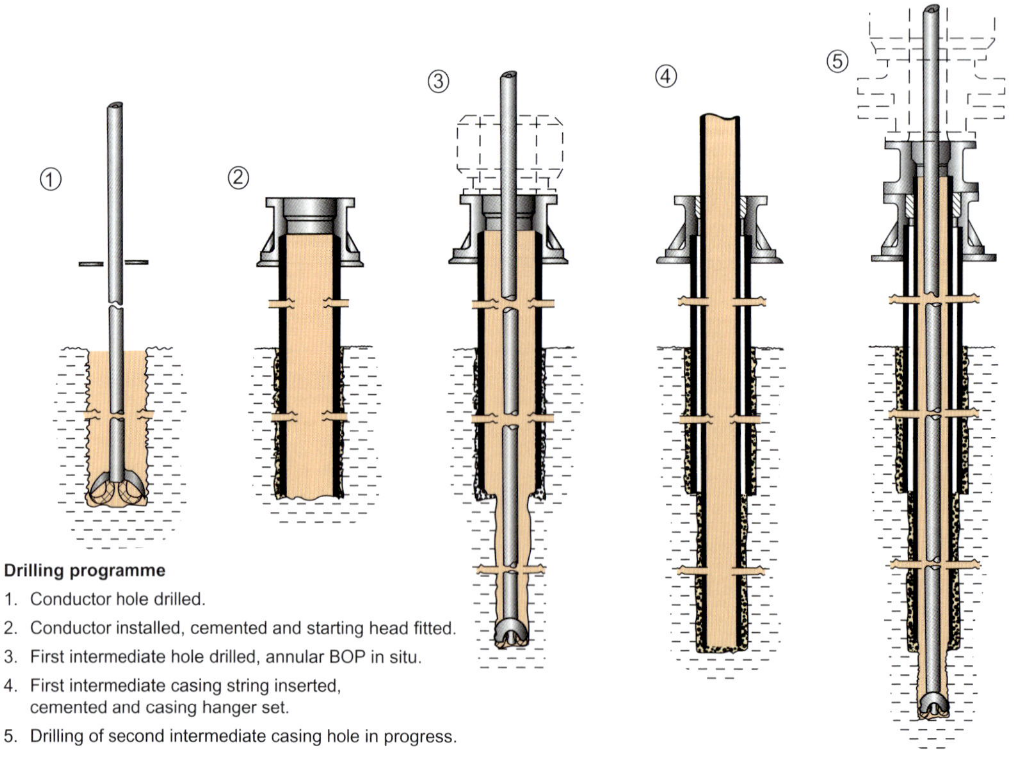

Figure 7.9 – Drilling programme

(a) **Coil tubing**
Coil tubing involves the insertion of a small diameter (1″ or 25.4 mm) steel tube through the christmas tree and down to the base of the well. High pressure nitrogen gas admitted through the tube helps displace the preservation fluid so that the flow of hydrocarbons can commence.

(b) **Blanket gas**
Prior to perforation, the hydrostatic head of preservation fluid is reduced to a value less than the reservoir pressure and a charge of nitrogen gas is then admitted to the well bore. The combined pressure created by the brine and nitrogen exceed reservoir pressure and provide the necessary safety barrier. On completion of perforation, the nitrogen gas can be vented to atmosphere and the reduction in pressure will permit the reservoir products to displace the remaining column of brine and let flow commence.

Having dealt with the drilling of the hole and the installation of the intermediate casing strings into the rock formation, we should consider how the casings and production tubing are secured on the production platform. The securing system is referred to as the wellhead and has been dealt with in Chapter 8.

7.2.2 The Cement Job

A well is drilled in stages, all of which are individually lined with a steel casing string, ie secured into the formation by cement. Cementing the casing strings produces an extremely rigid assembly and provides protection against corrosion products released by the formation.

A sketch has been prepared to assist in the explanation of the cementing process or cement job, Figure 7.10.

On completion of each stage of drilling and prior to the introduction of the casing string, the hole must be conditioned. Conditioning involves the replacement of the drilling mud with a conditioning fluid, frequently a brine solution suitably weighted to prevent the ingress of hydrocarbons into the hole. The conditioning process removes any oilyness left by the drilling mud and improves the chances of obtaining a secure bond between cement, casing and hole. With the hole suitably conditioned the installation of the casing string can commence.

The first section of each casing string lowered into the hole is fitted with an internal guide shoe and float collar, which contains a non-return valve. Subsequent casings are fitted with external sprung steel centralisers at regular intervals, to ensure that they locate centrally within the hole and provide an unrestricted channel for the passage of cement. The cementing process can proceed as soon as the final length of casing enters the hole.

The first operation entails insertion of a floating plug into the casing, which creates a barrier between the preservation fluid and the cement. A measured quantity of cement is then pumped into the casing,

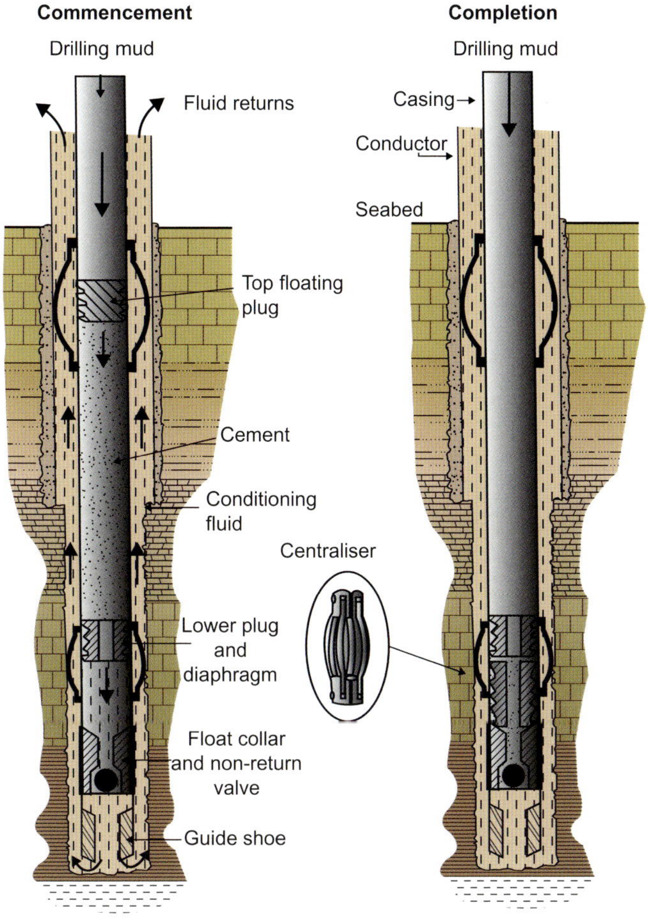

Figure 7.10 – The cement job

the volume calculated from the size of the hole and the external diameter of the casing string. A second floating plug is then installed above the cement and the drilling mud system reconnected.

The casing now resembles a giant syringe and, as pressure is applied to the drilling mud, the column of cement moves slowly down the inside of the casing until it reaches the non-return valve located in the float collar at the base of the casing. When the bottom plug contacts the non-return valve, the diaphragm within the plug ruptures, permitting the passage of cement through the plug and into the space created by the drilled hole and the external surface of the casing. Drilling mud pressure is maintained until a rapid increase in resistance indicates that the top plug has contacted the bottom plug and the cementing process is complete.

The cement takes from 6 to 24 hours to cure, depending on the downhole temperature, and during this period it is prevented from flowing back into the casing by the non-return valve in the float collar.

The diagram shows the main component parts of the drilling derrick.

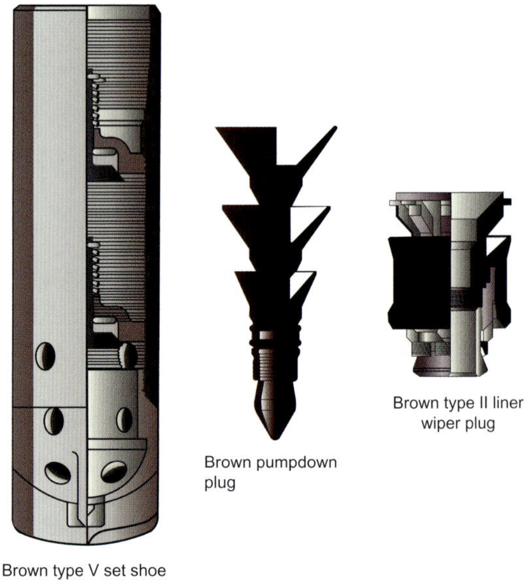

Figure 7.11 – Casing components – Cementing

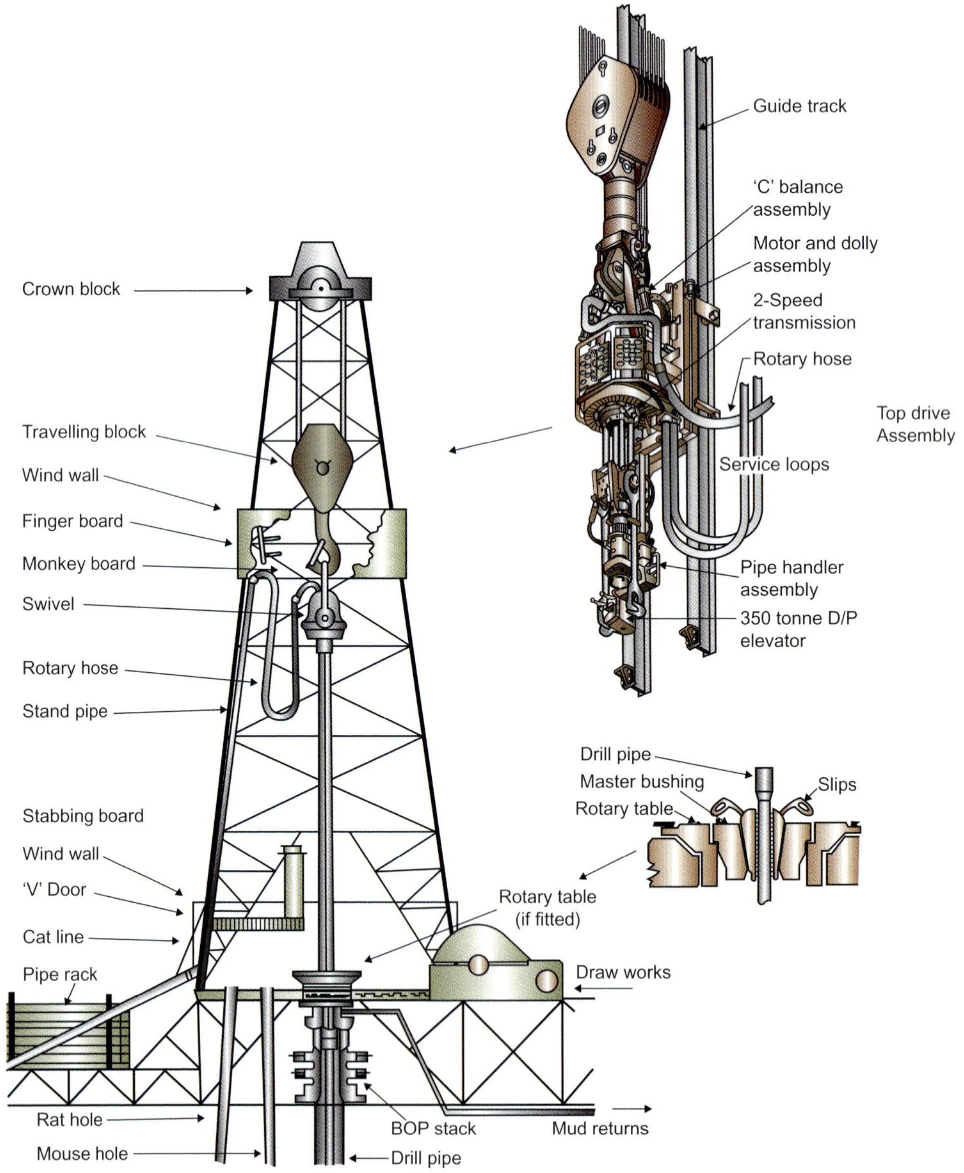

Figure 7.12 – Drilling derrick

Figure 7.13 – Upwards view of a drilling rig, showing the top drive during drilling operation

7.3 Equipment

The major items of equipment required to perform successful drilling operations from a seabed supported structure, such as a jack-up or a fixed installation, are:

1. Drilling Derrick.
2. Drill String.
3. Drilling Mud System.

7.3.1 Drilling Derrick

Figure 7.14 – Drilling derrick

The most prominent feature of an offshore installation is the drilling derrick. It consists of a steel lattice tower approximately 50 metres (165 feet) in height, which supports the Crown Block and provides temporary storage facilities for drill pipe stands.

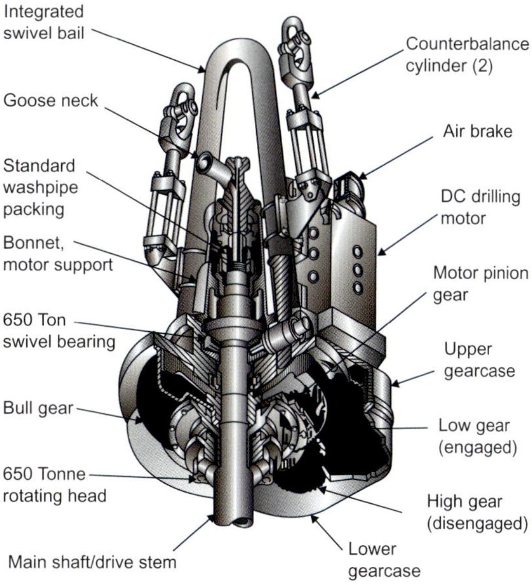

Figure 7.15 – TDS-4S top drive (Dual speed)

i) Crown block
The stationary sheaves mounted at the top of the derrick over which the wire ropes attached to the travelling block pass.

ii) Travelling block
The large heavy duty, multi-sheave lifting block that is used primarily to support the weight of the drill string during drilling operations and to hoist drill pipe and casing into and out of the hole.

iii) Top drive
The top drive assembly has all but superseded the rotary table as the means by which rotary motion is imparted to the drill string. It consists of a large electrically or hydraulically powered motor that is suspended from the travelling block and connected directly to the drill string.

iv) Monkey board
The small platform located in the top half of the drilling derrick from where the monkey, or derrick man, can manoeuvre pipe stands into and out of the finger board during a trip.

v) Finger board
The finger board resembles a large peg board and is located alongside the monkey board. The finger board provides a location for the drill pipe stands.

vi) Draw-works
The collective name given to the power control centre on the drill floor. It is a large hydraulically or electrically powered winch that provides the motive power to operate the travelling block and the rotary table (if fitted).

vii) Rotary table
Where fitted, the rotary table consists of a large casting located centrally within the drill floor and it is used to impart rotary motion to the drill string, being chain driven from the draw-works. It contains removable bushings that permit the

passage of the drill string components while providing a drive mechanism for the rotation of the square section kelly.

viii) Dog house
Tin hut or shed located at the side of the drill floor and used as a shelter by the members of the drill crew.

ix) Mouse hole
A pipe let into the drill floor in which a new section of drill pipe is inserted prior to its connection to the kelly and subsequent inclusion in the drill string.

x) Rat hole
A pipe let into the drill floor to accommodate the kelly during the making of a trip.

xi) Pipe rack
Area of deck in front of the drilling derrick where drill pipe and casing are stored prior to use.

xii) Pipe draw
Ramp leading from the pipe rack to the drill floor over which drill pipe and casing are dragged by the cat line.

xiii) 'V' door
The 'V' shaped opening in the windwall between the pipe rack and the drill floor.

xiv) Cat line (or tugger)
Small winch used to transport drill pipe and casing from pipe rack to drill floor.

7.3.2 The Drill String

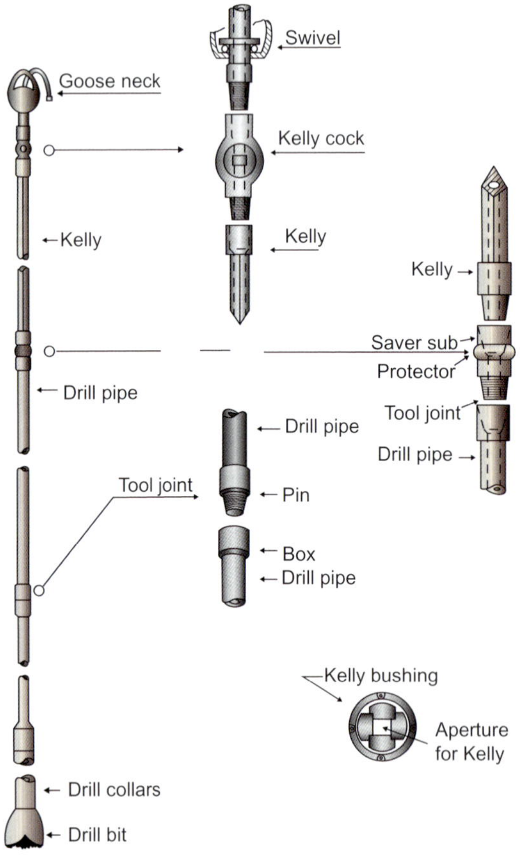

Figure 7.16 – The drill string

The drill string is the collective name that describes the assembly of components used to drill a hole. They are:

i. Drill bit.
ii. Drill collars.
iii. Drill pipe.
iv. Kelly.
v. Saver-Sub Assembly.
vi. Swivel.

i) Drill bit
The drill bit, roller cone or rock bit consists of three rotating cones that are fitted with hardened steel, carbide tipped or diamond edged teeth. The choice of tooth material will depend on the type of rock formation through which the drill bit must pass.

Figure 7.17 – Drill bit

ii) Drill collars
The drill collars are heavy, thick walled tubular couplings that connect the drill bit to the drill pipe. Anything from 2 to 30 collars may be used to provide concentrated weight in the vicinity of the drill bit to assist in the rock breaking process. The collars also provide rigidity to the drill string, which prevents the bit from wandering from the vertical during drilling operations.

iii) Drill pipe

Figure 7.18 – Drill pipe connectors and thread convertors

Drill pipe is supplied in 30 foot (9 metre) lengths and consists of 3½ or 5 inch diameter heavy wall pipe with male and female couplings welded to each end. The couplings are referred to as pins (male) or boxes (female), and permit assembly and disassembly of the drill pipe during drilling operations.

Figure 7.19 – Drill pipe thread

iv) Kelly (required only for rotary table drive)
The kelly is a long hollow, square section forging that screws into the top section of a drill pipe. It is driven by the rotary table and provides the drill pipe with both rotational movement and drilling mud.

v) Saver-Sub assembly
The saver-sub assembly is a simple male/female threaded spacer designed to protect the kelly or power take-off shaft threads from damage during the repeated make-up or break-out of the sections of drill pipe.

vi) Swivel
The swivel permits the drill string to revolve freely while suspended from the travelling block, and provides for the passage of drilling mud. The mud is supplied via the rotary hoses that are connected to the goose neck on the swivel.

7.3.3 Drilling Mud

Drilling fluid or mud, as it is universally referred to, is the lifeblood of the drilling operation. It lubricates the drill bit, removes drilling debris, stabilises the hole and provides the principal safety barrier against the ingress of hydrocarbons into the well bore.

Drilling mud is quite a sophisticated fluid consisting of bentonite, a colloidal clay dissolved in either water or fuel oil. Baryte (barium sulphate) is added as a weighting medium to permit variation of the specific gravity, and the only resemblance the final product has to mud is its consistency and the dreadful mess it creates.

Considerable care must be exercised in the preparation of the drilling mud to ensure that drilling operations progress in a smooth safe manner. The density of the mud must be continuously monitored and adjusted to ensure that the hydrostatic head it creates within the hole is always greater than the pressure likely to be encountered should the drill bit strike an unexpected pocket of oil or gas. The formation pressure is primarily a function of hole depth and can be estimated with a fair degree of accuracy. An adequate head of mud will prevent a blow out, ie a blow back of hydrocarbons into the hole and up to the rig floor. However, an excess of mud pressure must be avoided, particularly when drilling through soft formations where the mud will simply force its way into the structure of the rock and circulation will be lost. When this happens drilling must be suspended and a casing string inserted to reinforce the hole.

The majority of drilling operations are performed using a water-based mud (WBM), which is both cheaper and kinder on the environment than the oil based varieties. However, oil-based muds (OBM) are preferred for drilling operations through water sensitive shale formations and for the drilling of highly deviated wells that benefit from the additional hole stability provided by the denser mud as it cakes the well bore.

The disposal of drilling debris can create tremendous problems, particularly when one considers that the drilling of one well can liberate up to 2,000 tons of cuttings. Cuttings produced using water based muds can simply be flushed into the sea, while most countries have now prohibited the disposal of oil based mud cuttings in this manner on environmental grounds.

Operating principles
The schematic illustration in Figure 7.20 shows the basic layout of the drilling mud circulation system employed on a rotary table powered drilling rig. The system is identical to that used on the more modern top drive systems with the exception of the kelly, which is no longer required.

i) The cycle commences as high pressure reciprocating pumps deliver mud to the rotating drill string via the standpipe, flexible rotary hose and gooseneck attached to the swivel.
ii) The mud passes down the centre of the kelly and into the bore of the drill pipe, where it eventually exits around the teeth of the drill bit. The mud acts as a cutting fluid, cooling the drill bit and retaining rock chippings produced by the drilling process in a suspension that can be channelled back up through the hole and onto the rig.
iii) The mud returns are collected within the bell housing situated on top of the diverter or BOP stack, and are directed into a shale shaker prior to entry into the settling and storage tanks (or mud pits). The shale shaker consists of a series of vibrating gratings or screens that sieve the mud and remove the particles of drilling debris. Particles removed by the shale shaker are examined to provide the drill team with information pertaining to the physical characteristics and water content of the formation through which the drill bit is passing.
Where an improved level of filtration is required, centrifuges or hydrocyclones may be used in addition to the shale shakers to clean up the mud.
iv) Prior to re-circulation of the drilling mud, it is allowed to stand in the settling and storage tanks, which are open to the atmosphere. This permits the release of any gaseous hydrocarbon products dissolved during the drilling process, as will be the case when the drill bit enters a hydrocarbon bearing formation.

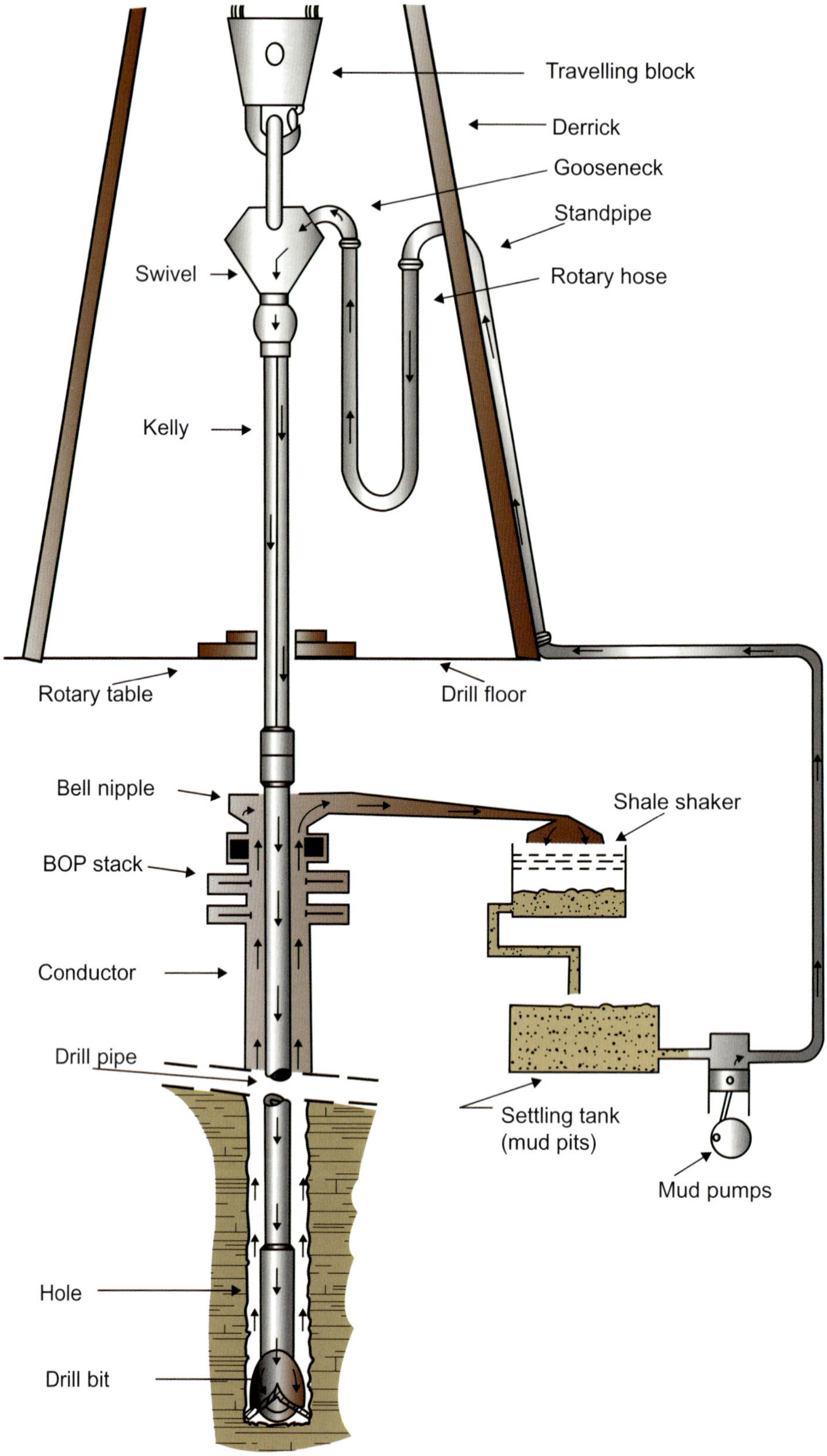

Figure 7.20 – Drilling mud system (Rotary table)

Recent Developments

Drilling muds are being continually improved and synthetic-based drilling muds (SBM) are now available as an alternative to oil-based muds (OBM). The synthetic-based muds are claimed to reduce drilling torque, the risk of the drill string sticking in the hole, and to improve penetration rates. Unlike oil-based muds, the cuttings produced with synthetic-based muds can be discharged directly into the sea, but one of the disadvantages is that it is extremely expensive, between three and five times the price of oil-based muds. Another disadvantage is that the synthetic-based muds must be displaced with a completion fluid prior to cementing if a paste with the consistency of peanut butter is to be avoided.

Underbalanced

The preceding text outlines the significance that drilling mud plays in the drilling programme, acting as the primary safety barrier against the ingress of hydrocarbons into the well bore. The process is described as drilling in the overbalanced condition. However, in recent years some companies, mostly operating onshore in Canada, have exploited the advantages of drilling in the underbalanced condition, ie where the column of drilling mud is lightened by the injection of foams or gases, the object being to keep the pressure exerted by the mud in the well above the collapse pressure of the hole but below the pressure in the formation. The well will, therefore, flow as soon as the hydrocarbon bearing formation is entered.

Underbalanced drilling is generally used when modifying an existing well, such as in the drilling of a new horizontal section, or to finish a new well in an existing reservoir where the reservoir pressure has subsided significantly. The well is drilled in the conventional overbalanced mode until the hydrocarbon bearing formation is reached and the remainder of the well is then drilled underbalanced. The main advantage of underbalanced drilling is that by preventing the well bore and surrounding formation from becoming impregnated with drilling mud, the permeability of the rock is maintained, which saves on well clean-up operations. It also greatly reduces the tendency of the drill string to get stuck in tight horizontal holes.

While there are many ways of achieving underbalance, nitrogen gas is used offshore, injected into the mud in the standpipe at a pre-calculated pressure to achieve the desired level of underbalance. The nitrogen may be supplied in tanks as a liquid, or a nitrogen generator may be used. Conventional drilling equipment is used with the addition of a rotary BOP (RBOP), which is installed on top of the conventional BOP. The inner element of the RBOP rotates with and seals the drill string and ensures that the mud returns that contain rock cuttings, nitrogen and hydrocarbons can be directed in to a four phase separator. The gases are flared, the hydrocarbon liquids deposited into storage tanks and the cuttings disposed of.

7.4 Operations

Figure 7.21 – Two roughnecks working on an oil rig platform drill floor

Drilling a hole is a particularly time consuming operation. Drilling can proceed at speeds of up to 360 feet (108 metres) an hour or down to 1 foot (0.3 metres) depending on the type of rock formation. A typical well takes approximately eight weeks to drill and consumes five miles of casing and tubing.

7.4.1 The Drill Crew

The drill team consists of the following personnel:

i) **Toolpusher**
The toolpusher or drilling supervisor is in overall charge of the drilling programme.

ii) **Driller**
The driller is in charge of all operations carried out on the drill floor, in effect the drill floor foreman.

iii) **Assistant driller**
The assistant driller's main duties involve the preparation of drill floor machinery.

iv) **Derrick man**
The derrick man works from the monkey board and assists in the installation and removal of drill pipe during the making of a trip, guiding the stands into and out of the finger board.

v) **Roughnecks**
Drill floor labourers.

vi) **Roustabouts**
General rig labourers.

Figure 7.22 – Driller at the control console during drilling operations

Figure 7.23 – Drill crew hard at work on the drill floor.
Inset – the drill crew are dwarfed by the travelling block hook, the gooseneck and rotary hose, the swivel and drive motor - note the drill pipe stands.

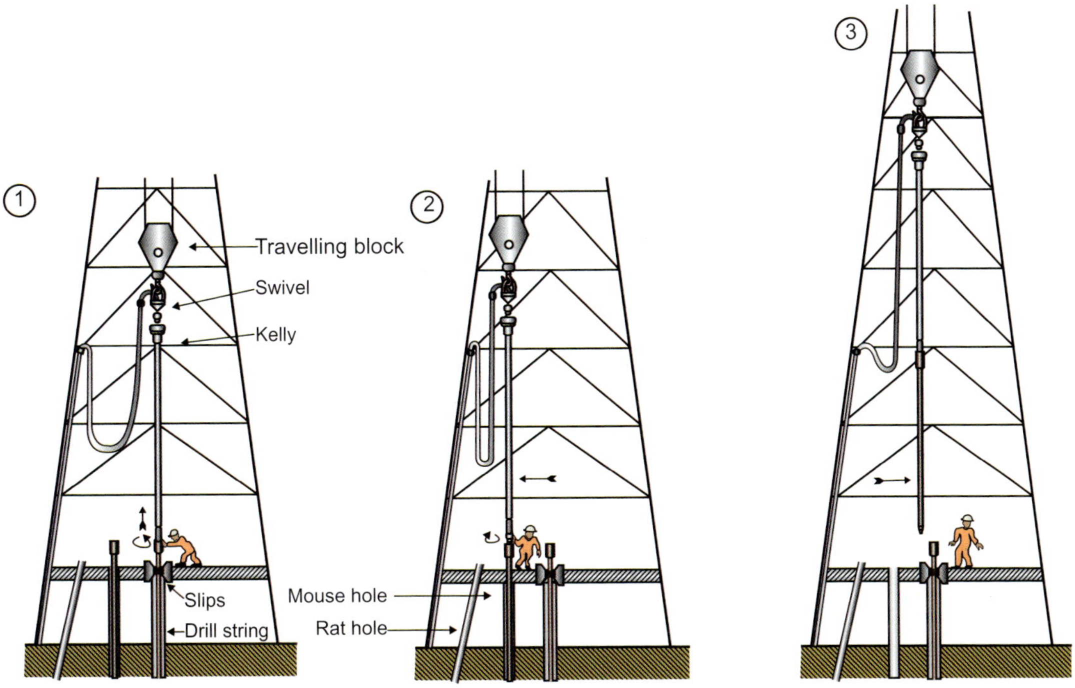

Figure 7.24 – **Making a connection**

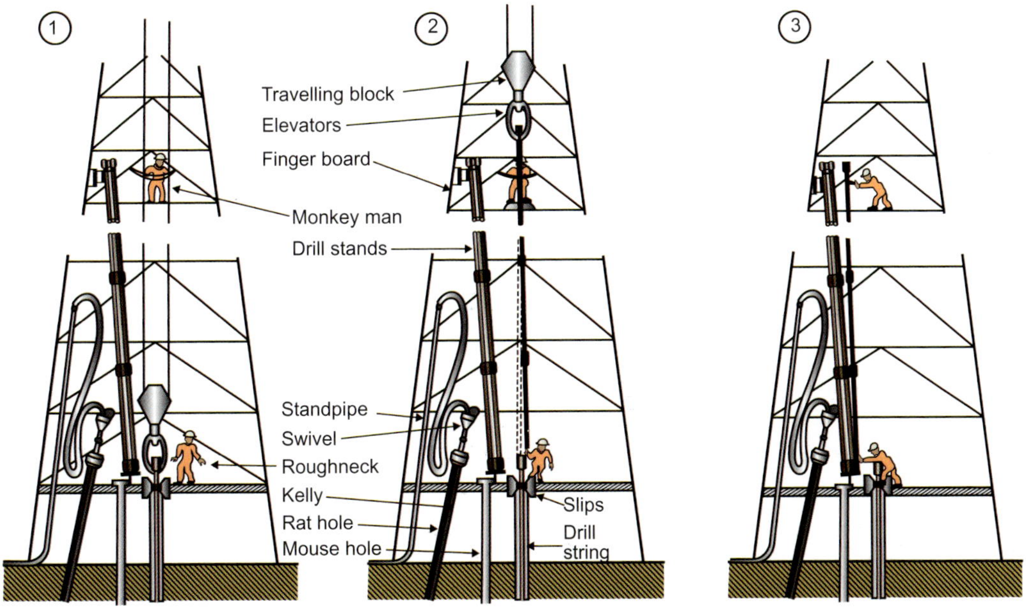

Figure 7.25 – **Tripping out**

7.4.2 Making Hole

While the rotary table has largely been superseded by the top drive assembly, an account of drilling operations using both types of equipment will be provided as the rotary table may still be encountered on older installations.

i) Rotary table drive

Rotational movement is imparted to the drill string by the rotary table, via the kelly bushings. As the hole progresses, the kelly slides through the bushings until approximately 30 feet (9 metres) of hole have been drilled and a new section of drill pipe is required. The addition of drill pipe is referred to as making a connection, and the process may be compared with the way a chimney sweep increases the length of his brush as it proceeds up the chimney.

To make a connection the rotary table must be stopped and the flow of drilling mud interrupted. The drill string is then lifted by the travelling block until the kelly is clear of the rotary table. The kelly bushings are then replaced by slips, a pair of serrated steel wedges that prevent the drill string falling back down the hole when the base of the kelly is disconnected. The kelly is connected to a new section of drill pipe located in the mouse hole and the complete assembly re-positioned over the drill string. The drill string joints are tightened using a pair of hydraulically powered pipe grips or tongs, after which the drill string is ready for re-entry into the hole. The drill string weight is once again taken by the travelling block while the slips are removed and the drill string is lowered back into the hole. Once the kelly bushings have been replaced, drilling may continue.

Figure 7.26 – **Rotary table and kelly**

ii) Top drive

Figure 7.27 – Top drive drilling motor undergoing an examination. The motor guide rails and drill pipe stands are clearly visible.

The top drive assembly employs a large, hydraulically or electrically powered motor to impart rotary motion to the drill string. The motor is suspended from the travelling block and 90 feet (27 metres) of hole can be drilled before additional drill pipe is required. The facility to drill three times as much hole as the rotary table between connections represents a considerable saving in time and is the main reason for the success of top drive systems.

To make a connection with a top drive assembly, the motor must first be stopped and the flow of drilling mud suspended. The travelling block briefly supports the entire weight of the drill string while the slips are inserted into the drill floor. The motor can then be disconnected from the drill string and hoisted into position over a new section of drill pipe.

The drill pipe awaiting inclusion into the drill string is re-assembled into stands consisting of three sections of pipe standing vertically inside the derrick finger board. The drive motor is connected to the top of a stand and the travelling block is used to lift it into position over the drill string while the two are coupled together. Drilling may be resumed once the slips have been removed and a further 90 feet of hole drilled.

While the preceding text provides an outline of the way in which a hole is progressed, it will be appreciated that occasions arise that require the removal of the drill string in its entirety, the operation being referred to as 'making a trip' or 'tripping out'. It is a time consuming operation that may require the removal of up to 12,000 feet (3,650 metres) of drill pipe during the latter stages of drilling.

The making of a trip will be required prior to the installation of a casing string and whenever the drill bit requires replacement. Drill bit life will depend on the type of formation being drilled, but a life of 10 to 72 hours can be expected unless specialised diamond tipped cutters are used.

The tripping operation commences with the suspension of the drill string from the slips in the drill floor. The procedure is identical for top drive and rotary table powered equipment once the kelly and swivel have been removed. The swivel is replaced with drill pipe elevators, a large clamp arrangement that grips the drill string below the uppermost drill joint. The travelling block is then raised until three lengths of drill pipe (a stand) are exposed and the slips are re-set in the drill floor. The stand is then disconnected and stood vertically inside the derrick, and held in the finger board until required for further use. The process is repeated until all the drill pipe has been retrieved from the hole. The procedure is reversed once the new drill bit or casing has been installed.

Description:
The Varco Pick-up/Lay-down System (PLS), is a new generation pick-up/lay-down system that, together with a rig floor pipe racking system, drastically reduces the manual handling of pipe on the rig floor and pipe deck. The PLS automatically picks up pipe from special pipe containers which may be loaded onshore or at the pipe yard. The pipe is horizontally transferred to the rig floor where a pipe pick-up boom rotates the pipe vertically and presents it to the arm of the pipe racking system. This eliminates the need to manually latch elevators on the pipe as well as removing the tugger or crane handling of each joint. The danger associated with tailing pipe into the V door and the handling of the pipe on the pipe deck is completely eliminated.

The PLS system consists of two major components, a pipe elevator and a pick-up/lay-down boom. The pipe elevator is designed to raise pipe from off the pipe deck to the rig floor level. The elevator assembly sits on top of the pipe deck next to the catwalk. The pipe containers are positioned on either side of the elevator and automatically feed the drill pipe, collars or casing to the lifting clamps on either side of the elevator. On top of the elevator is a series of V rollers that feed pipe to and from the pick-up boom. The pipe boom is designed to pick-up the pipe from its horizontal position by rotating it 90 degrees, permiting the upper end of the pipe to pass through the existing V door opening in the derrick. The PLS is controlled from the same control system used to control the pipe racking system. A control station for manual operation is located on a work platform near the rig floor. Semi-automatic sequencing is controlled by the pipe racking system operator from his console.

Specifications:
Elevator High Speed: 2 ft per sec
Elevator Low Speed: 6 inches per sec
Boom Elevation Time: 18 sec
Pipe Sizes High Range: 9-3/4 to 20 inches
Pipe Sizes Low Range: 3-1/2 to 9-3/4 inches

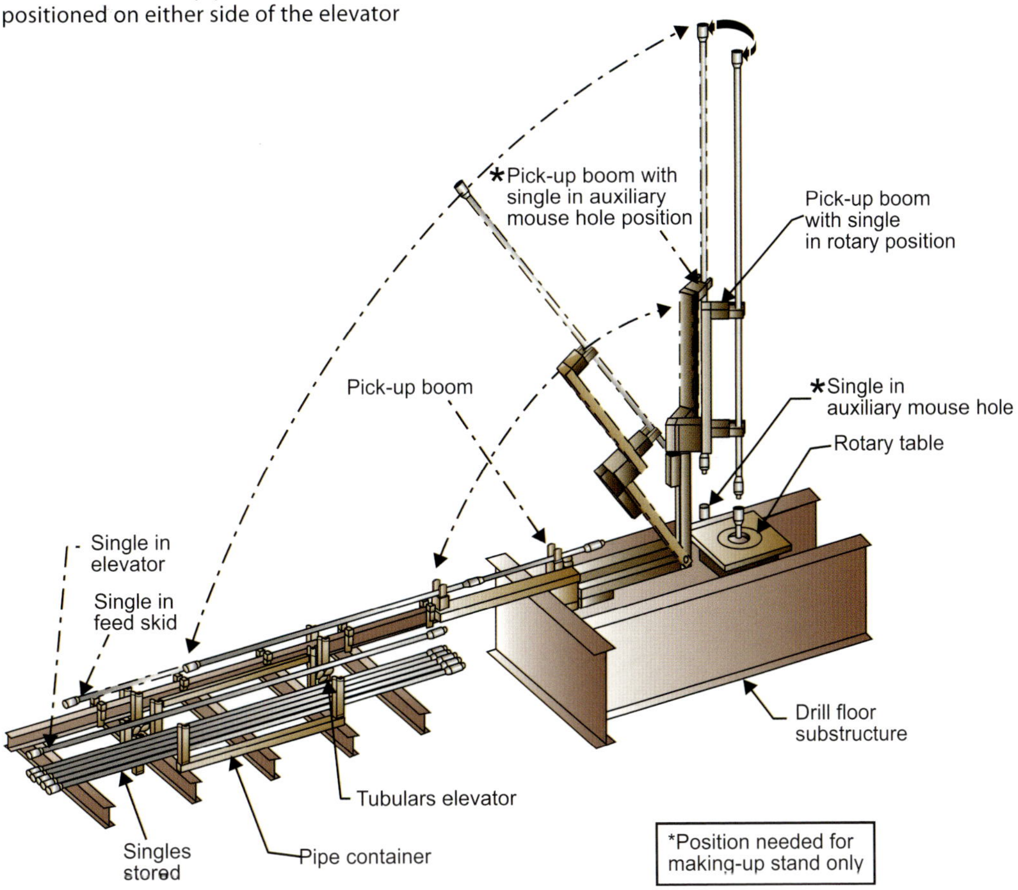

Figure 7.28 – **Automated drill pipe handling systems**

7.5 Well Control Equipment

Diverters and BOPs are the last line of defence and provide the drill crew with a breathing space to take corrective action or abandon the installation when control of the well is lost.

The initial stages of drilling a well are without doubt the most dangerous due to the ever present risk of penetrating shallow gas bearing sands, a hazard that has resulted in numerous fatalities and the complete destruction of both drill ships and jack-ups. The diverter is used in preference to a BOP to provide a degree of protection against these hazards, because attempting to confine hydrocarbons within a shallow conductor may lead to a blow-out of the formation around the outside of the conductor, which would create an unrecoverable situation. Once the hole depth has been substantially increased and the first intermediate casing string cemented in place, the diverter can be replaced with a BOP stack.

7.5.1 The Diverter

As the name suggests, the diverter provides a means of diverting an unexpected release of well fluids, primarily gas and occasionally solids, to a location at the extremities of the rig where they can be discharged safely.

The diverter is situated on top of the conductor and must permit the passage of the drill bit, while still being capable of effecting a seal around the drill pipe or kelly. During normal drilling operations, the diverter vents are closed and the drilling mud returns flow upwards and into the bell housing from where they are channelled into the shale shaker. Activation of the diverter results in the closure of the packing element around the drill string and sequential opening of the vents, providing an unrestricted passage to atmosphere for well fluids.

The diverter may consist of a proprietary item or it may be assembled from a fabricated manifold and a conventional annular blow-out preventer (BOP). The illustrations in Figure 7.29 show a diverter in which the constriction of the packing element, the opening of the vent port and the closure of the flowline port are combined into a single operation controlled by the annular piston, an inherently more reliable arrangement than can be achieved with built up diverters.

The drilling of wells from a floating rig (semi-sub or drill ship) necessitates location of the BOP stack on the ocean floor. In these cases, a diverter is normally mounted on top of the marine riser as a precautionary measure throughout the drilling programme.

The most noteworthy development in diverter technology in recent years has been diversion at the wellhead rather than on the rig. This is accomplished by locating a diverter on the seabed in addition to the diverter on top of the drilling riser and permits the venting of well fluids into the sea. The risks associated with a possible release of gas on the drill floor during the early stages of drilling are avoided.

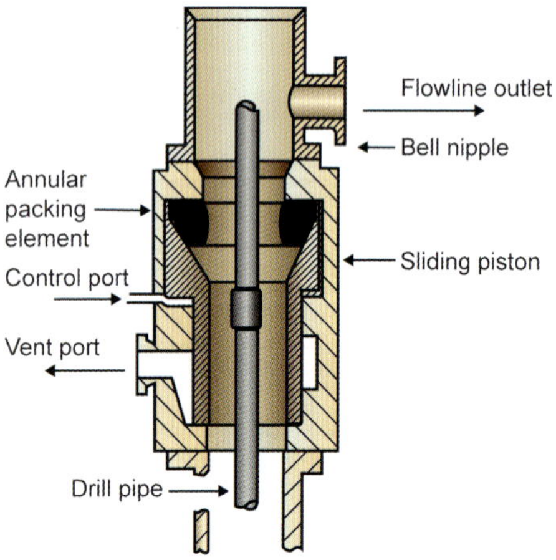

For Floating Offshore Drilling Rigs

The FS 21-500 is a diverter with a 21 inch bore and 500 psi working pressure rating. Drillships and semi-submersibles employ a diverter for venting gas flows encountered while drilling top hole through the 30 inch casing. The diverter also serves as the support for the upper flex joint and the inner barrel of the riser telescopic joint, so the diverter is in place whenever the riser is in service. On a floating rig, the diverter is never used as a BOP, so its pressure rating is 500 psi. Its bore accommodates the riser bore.

The FS diverter is fundamentally simple. Its design avoids the functional complexities which increase chance for human error

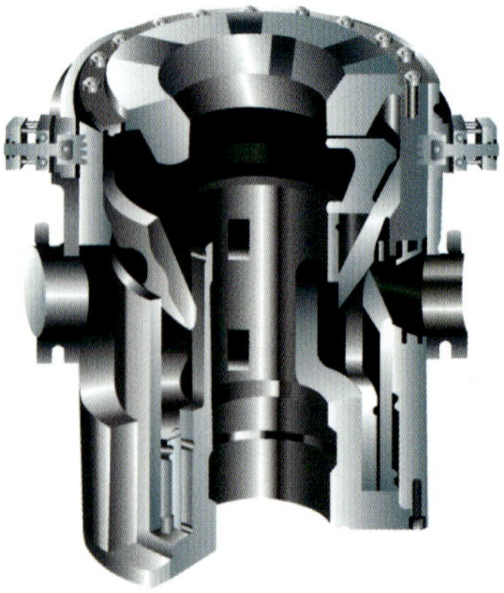

FS 21-500 Marine Riser Diverter, internal view. *The FS is simple and safe. The piston moves up and closes the annular packing unit, stopping upward flow. The sleeve opens the vent and closes the flowline. The FS can be fitted with two vent outlets.*

Figure 7.29 – Marine riser diverter

7.5.2 The Blowout Preventer

Figure 7.30 – The BOP stack is one of the critical components on a drilling rig to prevent blow-out

The primary function of the blow-out preventer is to confine well fluids within the well bore. It operates as part of a BOP system, which includes the BOP stack, choke and kill valves, choke manifold and a hydraulically powered control unit.

There are two basic types of blow-out preventer, examples of which are shown in the illustrations in Figure 7.31.

i) Annular BOPs

The annular BOP relies on the constriction of a reinforced elastomeric packing element to effect a seal around the drill pipe or kelly. The flexible packing element can accommodate considerable changes in diameter and cross section, but it typically cannot withstand pressures as high as the ram type BOPs. Consequently, it is used in conjunction with ram type BOPs, which are better suited to retaining extremes of reservoir pressure. In addition to containing well bore pressure, annular BOPs are also used for stripping and snubbing operations, which involve the vertical movement of drill pipe and casing into or out of the well under well pressure.

ii) Ram Type BOP

Ram type BOPs are used in combination with an annular BOP and may feature any one of four ram configurations:

(a) Pipe rams
These are supplied in various sizes and are designed to close and seal around the drill pipe.

(b) Variable rams
These are capable of closing and sealing on a limited range of pipe sizes.

(c) Blind rams
These seal against each other and are designed to seal the well bore in the absence of the drill pipe.

(d) Blind/shear rams
These are primarily used during subsea drilling operations where weather conditions may necessitate the rapid disconnection of the rig from the well. The blind rams incorporate a cutting edge capable of severing the drill string and sealing high pressure, high temperature well fluids within the well bore for an extended period of time.

A hydraulically controlled, fail safe power unit typically provides the fluid power for operation of the rams, and the annular piston used in the annular BOPs.

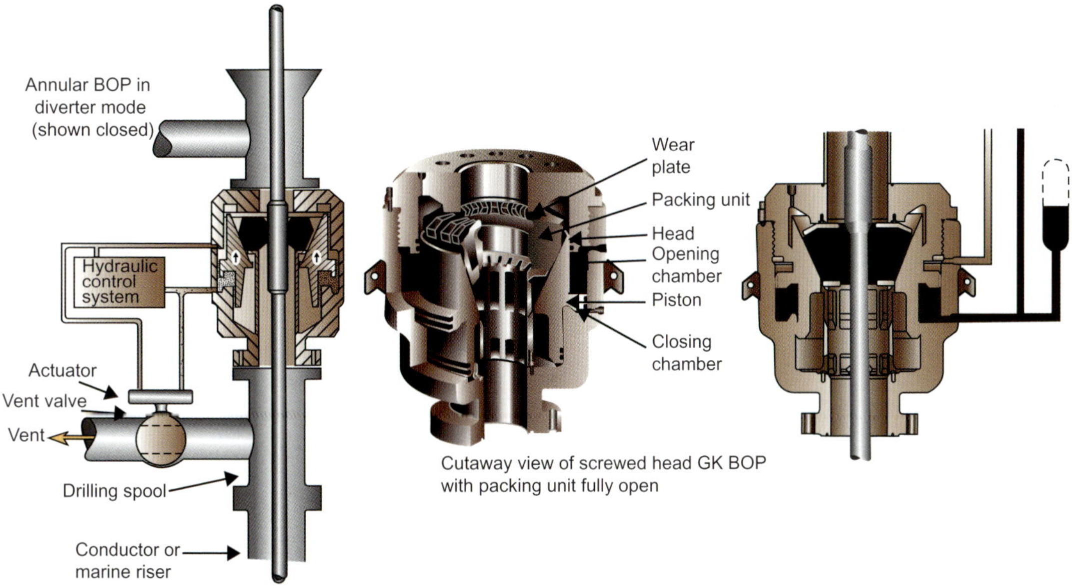

Figure 7.31 – Annular blow-out preventer

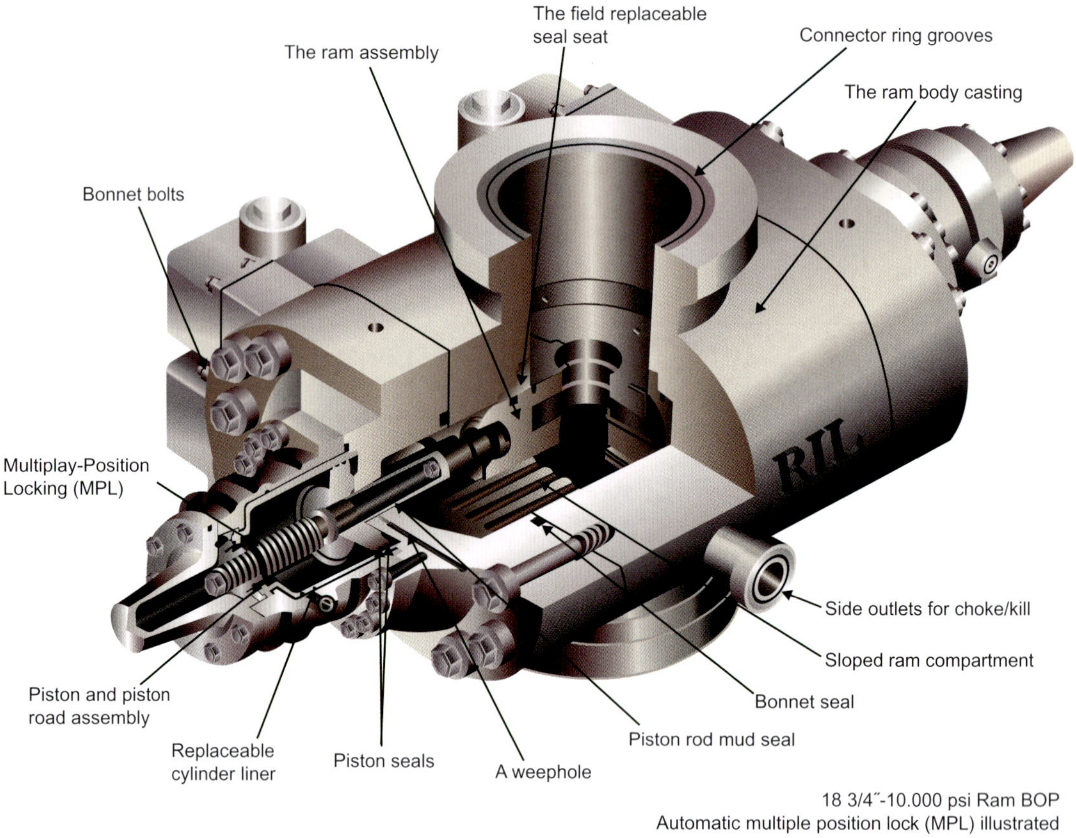

Figure 7.32 – **Ram blow-out preventer**

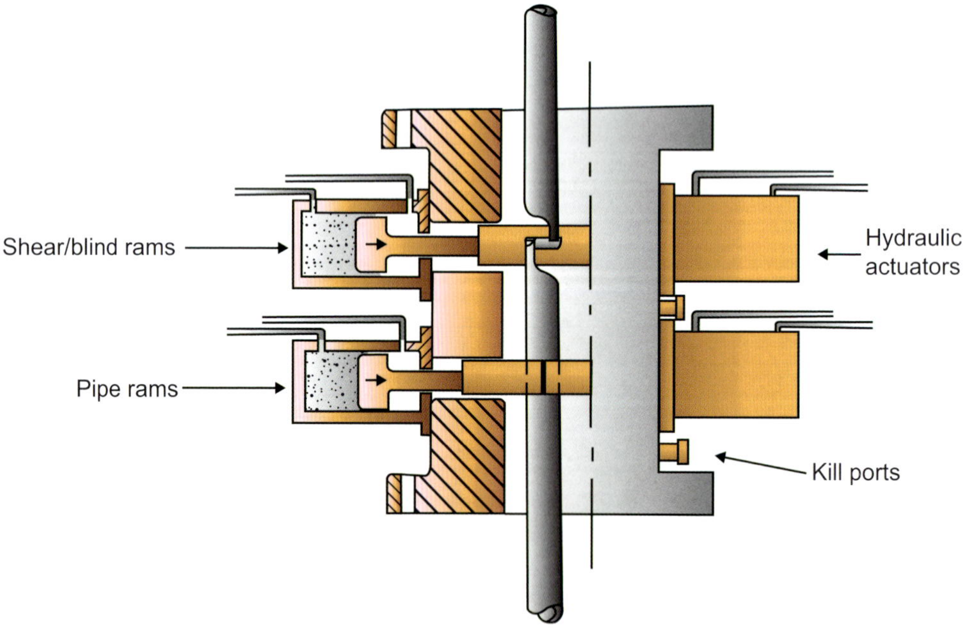

Figure 7.33 – **Ram type BOP schematic**

7.5.3 BOP Operations

Figure 7.34 – Blow-out preventer (BOP) stack piped up and ready for action. The stack consists of an annular BOP twin and single ram BOPs.
Inset – the tool pusher's console, from where all the drilling activities are controlled.

Drilling operations proceed through the centre of a BOP stack that is mounted on top of the casing head. The stack consists of a drilling spool incorporating choke and kill connections, an annular and at least one ram type BOP. As anticipated well pressures increase, additional sets of rams are installed up to a maximum of three sets. A typical surface mounted, high pressure well BOP stack would consist of one annular BOP, two sets of pipe rams and one set of blind/shear rams. A subsea BOP stack normally contains an additional set of pipe rams and two annular BOPs.

During normal drilling operations, both ram and annular BOPs are held in the open position, there being no hydrocarbon pressure within the hole. Should the drill bit penetrate an unexpected pocket of oil or gas, the hydrostatic head created by the drilling mud should prevent well fluids from entering the hole.

When the drill bit enters a hydrocarbon bearing formation and the hydrostatic head exerted by the column of drilling mud is insufficient to cope with the formation pressure, the well fluids will commence displacement of the drilling mud, which will be indicated by an increase in the volume of mud returning to the mud tanks. This situation is referred to as 'a kick' and must be dealt with immediately if it is not to escalate into a full blown blow-out.

To control a kick, the drill crew must first close the BOP pipe rams and the kelly cock to contain the formation pressure within the well bore. Eventually, the pressure will stabilise at which time a specially prepared mixture of dense drilling mud can be pumped into the well via the kill connections located on the drilling spool or BOP. The well pressure is neutralised by circulating the mud back through the choke line to the choke manifold. Having successfully killed the formation pressure with the hydrostatic head of mud, drilling may be resumed.

Should it be necessary to reduce the pressure within the well bore at any time during the well kill operation, this may be effected via the choke manifold connected to the drilling spool. The choke contains a variable orifice through which well pressure may be vented to atmosphere.

A blow-out occurs when a rapid increase in reservoir pressure displaces the drilling mud from the hole, causing a release of oil, gas or solids. This situation is virtually unrecoverable and should be avoided at all costs. When all else fails, activation of the shear rams should cut and seal the drill pipe and provide valuable time in

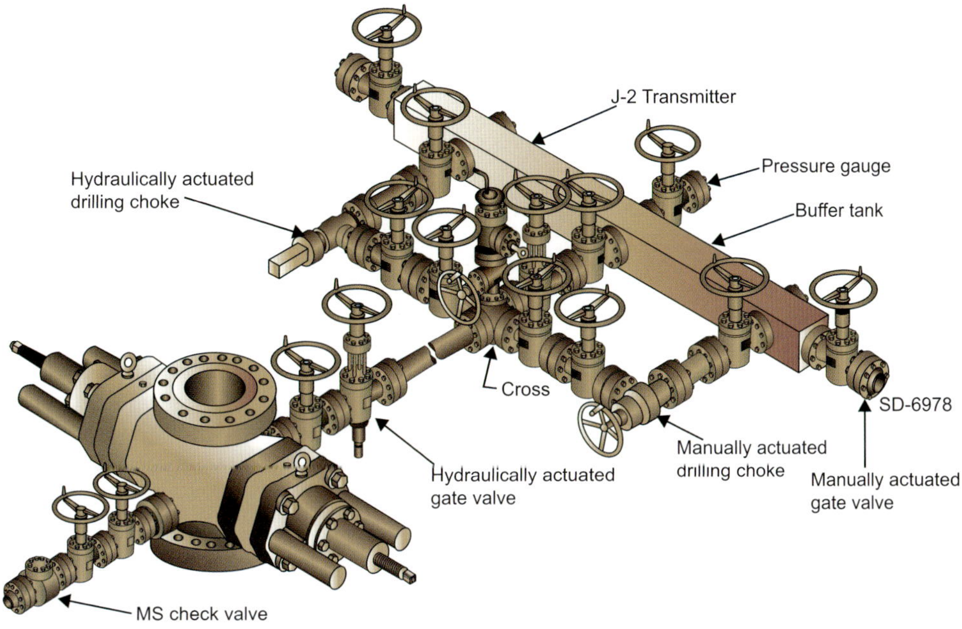

Figure 7.35 – Choke and kill manifold

which corrective action, or abandonment of the installation, may be effected.

Further information pertaining to blow-out prevention equipment can be found in the following documents which are available from the American Petroleum Institute (API):

i) **API RP 53,** Recommended Practice for Blow-out Prevention Equipment Systems for Drilling Wells

ii) **API RP 64,** Recommended Practice for Diverter Systems Equipment and Operations

iii) **API RP 16E,** Recommended Practice for the Design of Control Systems for Drilling Well Control Equipment (withdrawn).

7.6 Drill Ship Equipment

Figure 7.36 – **Drill ship**

The procedure for drilling a well is relatively unaffected by the type of vessel employed, be it fixed or floating. The major changes occur in the selection of equipment used to occupy the space between the seabed and the drill floor. This equipment must include a motion, or heave compensation, system that will accommodate vessel movement relative to the seabed.

i) Guide base

The first operation entails installation of a guide base on the seabed, through which drilling operations may proceed. Wires extending from the drill ship to the guide base assist in the location of the drill string and wellhead equipment.

Prior to commencement of drilling, the conductor pipe must be spudded (piled) through the guide base and into the seabed. The conductor stabilises the soft seabed, provides a passage for the disposal of drill cuttings and incorporates a suspension system for the support of the intermediate casing strings. The top of the suspension system is fitted with a machined collar designed for attachment to a wellhead coupling.

ii) Blow-out Preventer (BOP) Stack

When drilling operations are carried out from a floating vessel, the BOP is mounted on the seabed. This ensures that, should an emergency situation necessitate suspension of the drilling programme and departure of the vessel, the well may be left in a secure condition.

The BOP stack consists of a combination of annular and ram type BOPs housed within a tubular framework. The framework provides protection during transportation and incorporates location devices to assist in the attachment of the stack to the guide base. Remotely operated hydraulic couplings connect the base of the BOP stack to the conductor, and the top of the stack to the riser. The uppermost connector is frequently referred to as the emergency disconnect package (EDP).

iii) Marine riser flexible joint

A flexible ball joint located between the EDP and the marine riser can tolerate a certain amount of lateral movement of the drill ship relative to the wellhead. A second ball joint is attached to the top of the riser, frequently incorporated within the diverter.

iv) Marine drilling riser

The length of steel pipe extending from the BOP stack to the drill ship is referred to as the marine drilling riser, and it provides a return passage for the drilling mud. The riser incorporates a telescopic joint, the top-half of which is attached to a tensioning system.

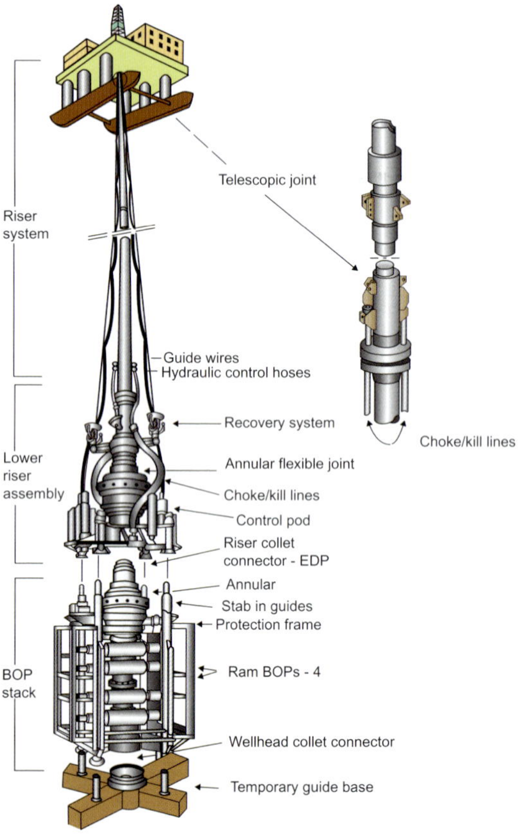

Figure 7.37 – **Drill ship equipment**

v) Telescopic joint

The telescopic joint permits relative vertical movement between the stationary lower section of drilling riser, which is attached to the seabed, and the upper section of the drilling riser, which is connected to the drilling derrick. It consists of a hydraulic or pneumatically activated packing element that is located in the top section of the drilling riser.

vi) The riser tensioning system

The riser tensioning system, or heave compensation system, is designed to keep the drilling riser in tension and to minimise the effects of hull movement. It can accommodate up to 15 metres (50 feet) of vertical wave-induced motion or heave. Wires emanating from the upper section of the drilling riser are connected to hydraulic/pneumatic cylinders which act like shock absorbers, extending as the vessel rides up and compressing as the vessel moves down.

vii) Drill string tensioning system

The travelling block is fitted with a heave compensation system that is similar to, if somewhat less sophisticated than, that employed to tension the riser. The hook is separated from the travelling block and mounted within a pneumatic/hydraulically activated piston assembly, which ensures that a constant load is maintained on the drill string at all times.

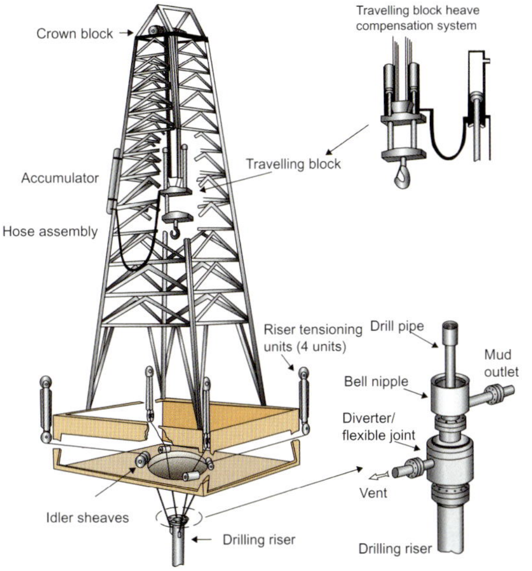

Figure 7.38 – **Drilling derrick assembly**

viii) Diverter

The diverter sits on top of the drilling riser and provides a way of controlling an influx of hydrocarbons into the well bore during the early stages of drilling.

ix) Choke and kill lines

Subsea choke and kill line connections are arranged slightly differently to those on a surface well as they are manifolded to permit pressure release, or the pumping in of drilling mud, through either connection.

They are run down the outside of the marine drilling riser in hard steel pipe with flexible couplings, providing the means of attachment to the BOP stack. The choke and kill line connections to the vessel are through flexible hoses, suitably dimensioned to absorb wave-induced motion.

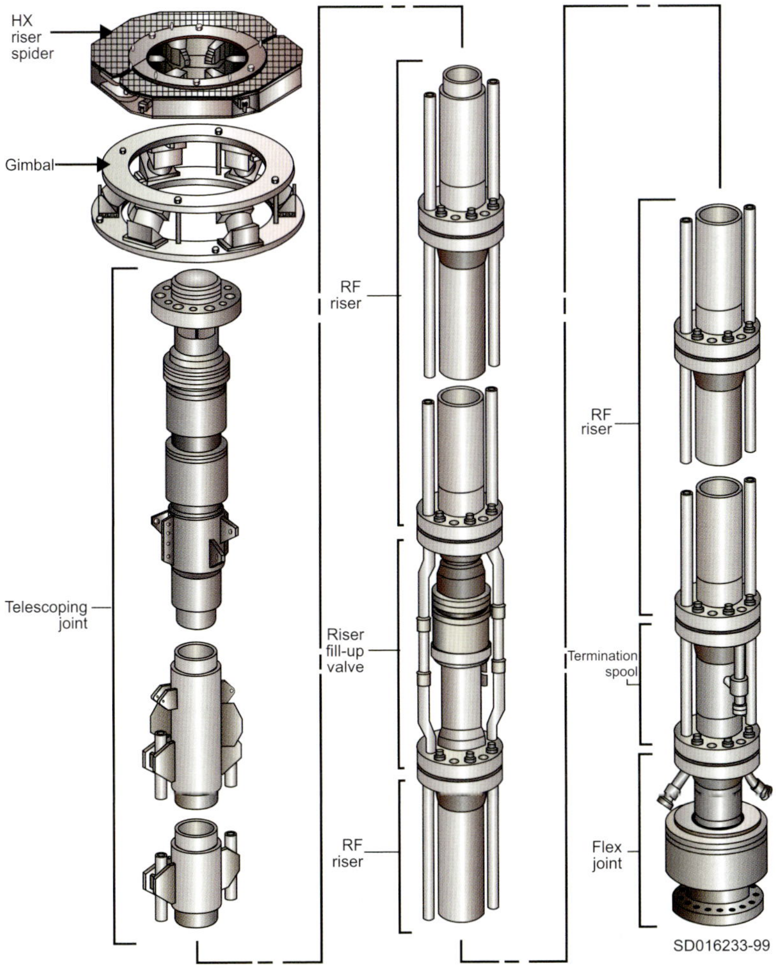

Figure 7.39 – **Subsea riser system**

Varco BJ DRILLING SYSTEMS

Description:
The Type V Racking System was designed primarily to improve the efficiency and safety of drilling operations on drillships and semi-submersibles where vessel motion makes handling tubulars dangerous and inefficient.

The Type V Racker System consists of three pipe racker assemblies normally located on the same side of the derrick as the setback area. Each assembly consists of a racker arm capable of moving in and out (towards and away from the rig centreline) and a carriage that moves laterally across the derrick. A racker head assembly on each arm is used to grip the pipe. In a typical tripping operation, the upper racker arm guides the upper end of the stand between the well centre and the fingerboard while the intermediate racker arm lifts and guides the lower end of the stand. The operation of the upper and intermediate racker arms are controlled from the derrickman's and assistant driller's consoles respectively.

The lower racker arm is used to assist in the handling of tubulars at the rig floor. The operation of handling the kelly, casing, collars and marine risers can now be easily and safely performed. The lifting force and vertical motion of the intermediate racker lifting head is provided by a stand lift cylinder assembly controlled from the assistant driller's console. In a variation on the basic three arm Type V Racker System, the lower racker arm is replaced with the hydraulically powered telescoping arm which is normally mounted on a pedestal on the rig floor. The substitution of the lower racker arm can provide more operating flexibility in some cases.

For maximum efficiency and reduction in trip times, the system should be combined with a remotely operated Varco pneumatically powered fingerboard, air operated elevators and a retract system on the top drive/travelling block.

Specifications:
Pipe Size (inches): 3-1/2 to 9-1/2
Lifting Capacity: 20,000 lbs
Max Reach: Customer specified
Max Push: 4,000 lbs

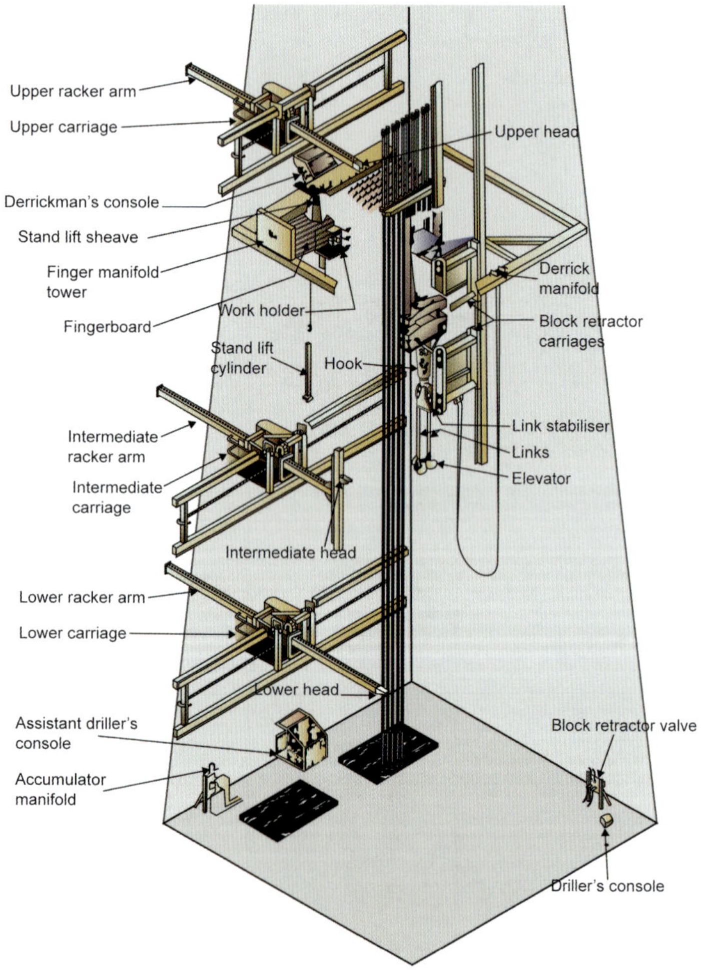

Varco BJ Type V Three-arm racker system

Figure 7.40 – **Automated drill floor equipment**

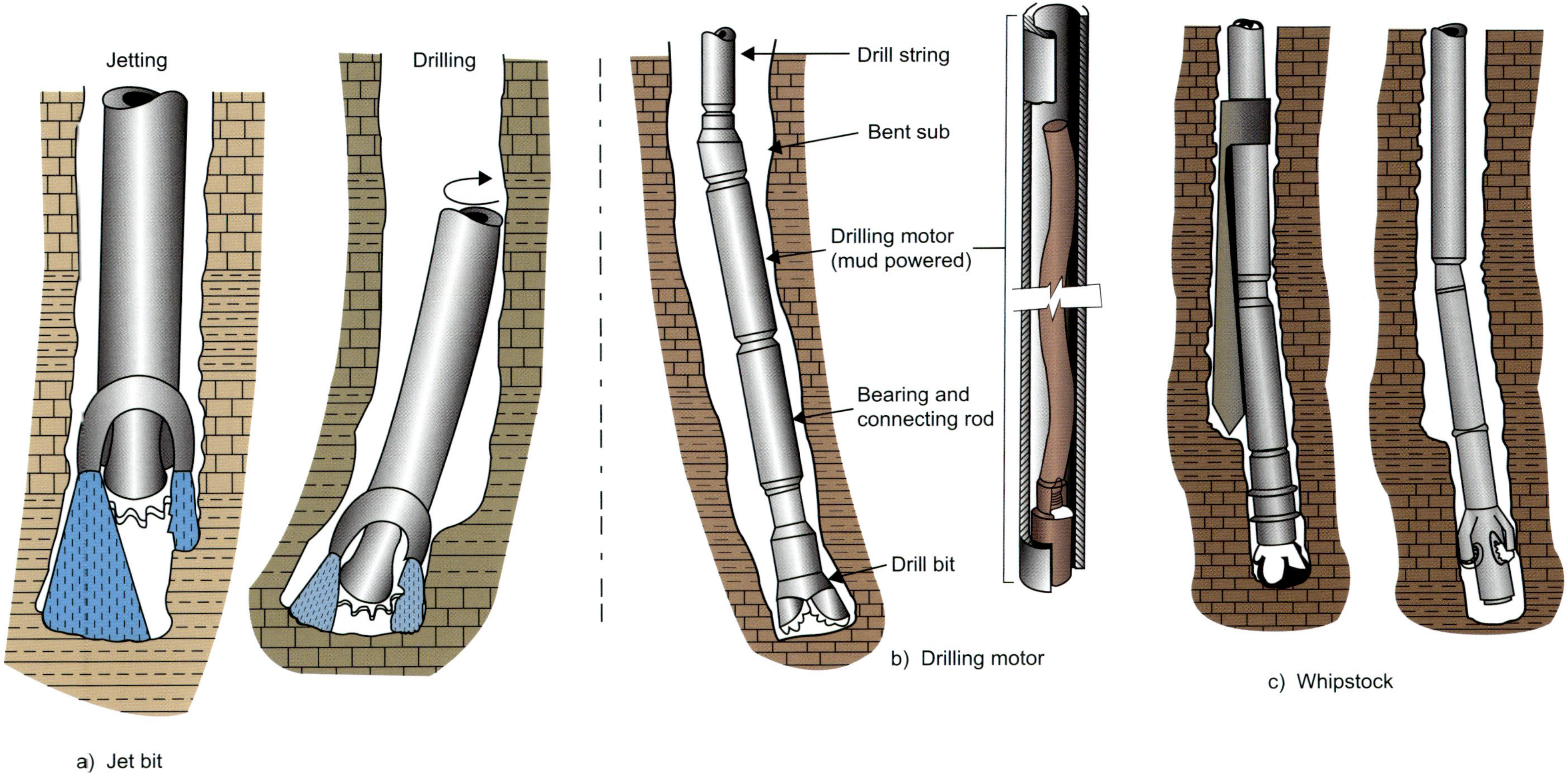

Figure 7.41 – Directional drilling

7.7 Deviated Wells

7.7.1 Directional Drilling Techniques

Directional drilling was primarily developed to permit the exploitation of hydrocarbon bearing formations at locations horizontally distant from fixed installations that would otherwise necessitate an additional platform or subsea well.

The most common techniques employed to drill a deviated hole are:

i) Jet bit
The jet deflection processes utilise a triple jet drill bit in which one of the drilling mud outlet connections or jets is considerably larger than the other two. The large jet is positioned in the desired direction of deviation while high pressure drilling mud is pumped through the stationary drill bit. The impact force provided by the mud erodes a hole into which the drill string will track on resumption of drilling. The process is simple and effective when the formation is soft, but it has largely been superseded by more modern methods.

ii) Whipstock
The whipstock has been used since the early days of the oil industry as a way of changing the direction of the drill bit. Made of steel, it is shaped with a concave tapered groove set at an angle of 2–3° and is approximately 1.5–3 metres (6–12 feet) in length. One of the disadvantages of the whipstock is that it produces an under-gauge hole, which must subsequently be reamed out. Largely superseded by more sophisticated directional drilling tools, a whipstock is still used to kick-off multi-lateral wells.

iii) Drilling motor
The vast majority of directional drilling operations are carried out using a drilling motor to rotate the drill bit and a bent-sub assembly to steer the drill string, although more recently adjustable motors have been developed that combine the functions of both pieces of equipment.

The drilling motor comprises a hydraulic turbine or a positive displacement hydraulic motor, with drilling mud used as the power fluid. The drill string does not rotate, it simply provides a conduit for the supply of drilling mud to the drilling motor. The well is steered to the target by means of the bent-sub, an offset attachment supplied in angles ranging from l°-3° positioned between the drilling motor and the drill string. This technique is referred to as 'drilling in sliding mode', the drill string sliding after the drill bit as the hole is progressed.

7.7.2 Horizontal Wells

Directional drilling technology has developed radically since the early 1990s and wells can now be drilled that extend a distance of 8 kilometres (5 miles) from the wellhead, and which have horizontal sections that exceed 2,500 metres in distance. These wells are referred to as horizontal wells, the main advantage being that they expose a larger and often more favourable section of the reservoir to the well bore. This improves reservoir drainage and enhances production, particularly from thin hydrocarbon bearing formations, and can result in a considerable reduction in the number of wells required, often by up to 25%.

Drilling highly deviated and horizontal wells
The drilling of a highly deviated or horizontal well is carefully planned by first calculating the geometric co-ordinates of the reservoir from seismic maps. A well bore trajectory is then prepared to take the drill bit to the target.

The drilling programme commences with the vertical section of the well being drilled to a predetermined depth with a conventional rotary table or top drive motor being used to rotate the drill string, the equipment being referred to as a rotary bottomhole assembly (BHA). When the desired vertical depth has been reached, a bent-sub and drilling motor are inserted between the drill string and the drill bit to provide the means by which the hole will be kicked off from the vertical plane and steered to the target.

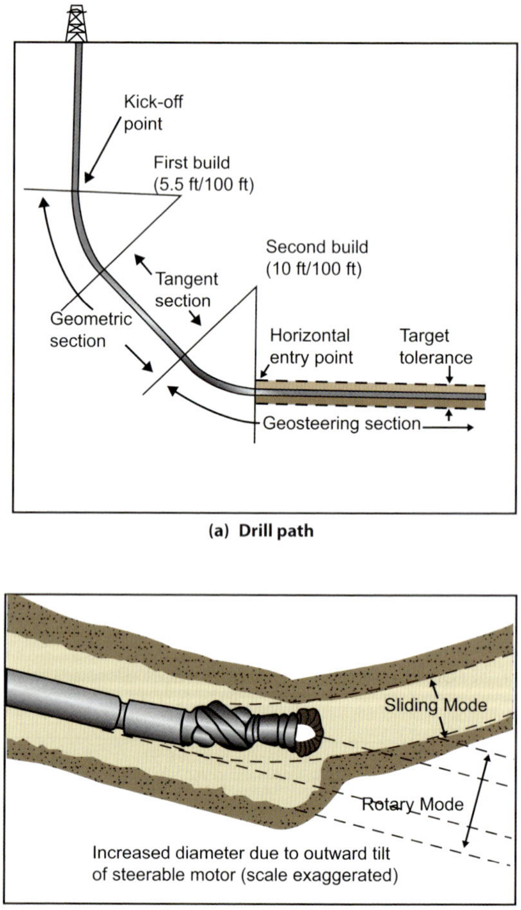

(a) Drill path

(b) Movement of drill bit

Figure 7.42 – Horizontal well plan

In rotary mode, both the bit and drill string rotate and the bit cuts a straight path parallel to the axis of the drill string above the bent-sub. In sliding mode, only the bit rotates and the hole follows the axis of the bent housing below the bend, changing the trajectory.

As the hole is progressed, the position of the drill bit is plotted continuously to ensure that it follows the desired trajectory. The position of the drill bit at any particular time can be calculated geometrically from the compass heading, and the length and inclination of the drill string.

The length of drill string can be established by counting the lengths of drill pipe in the hole, while measurement-while-drilling (MWD) tools measure inclination (angle from the vertical) and azimuth (compass direction). Accelerometers measure inclination and magnetometers measure azimuth, responding to the earth's magnetic field. The measurements are electronically converted into a binary code and are transmitted to the surface by mud-pulse telemetry, a system whereby pressure signals are sent through the drilling mud in what can best be described as a form of Morse code.

Numerous changes to the direction of the drill bit will be required during the course of the drilling programme and these are effected by means of the bent-sub, the acuteness of the angle determining the severity of the trajectory of the hole. When large changes in direction are required, drilling is suspended and the drill bit is lifted from the bottom of the hole while the drill string is rotated slightly to re-orientate the bent-sub. Drilling is then resumed.

Once sufficient inclination has been achieved, the horizontal section of the hole is normally drilled with the steerable motor while rotating the drill string. When compared to operation in the sliding mode, rotational drilling is quicker and reduces the chance of the drill string sticking in the hole. When further changes in direction are required, rotation is suspended while the drilling motor is used to change direction. Rotational drilling is then resumed.

Deviation of the hole from the vertical to the horizontal position is a very gradual process and, once drilled, the hole may be completed in a similar manner to a vertical well. Conventional casing is run in the hole although it is not cemented in the horizontal section. The casing can be pre-perforated prior to installation and sand screens may be fitted to prevent the formation invading the well bore.

As previously stated, one of the main advantages of horizontal wells is that they permit the exploitation of small, thin reservoirs by effectively providing an extended drainage channel that far exceeds what could be achieved with a vertical well bore. However, drilling horizontal wells is fraught with problems, some of which we will now consider before providing a brief outline of some of the technology and equipment that has been developed to overcome these problems.

Overcoming limitations in horizontal drilling
In recent years, the information provided by enhanced 3-D seismic survey techniques has enabled geologists to determine the location of potential oil and gas bearing formations with a far greater degree of confidence than was previously the case and small, intricate reservoirs are now being defined that would previously have gone undetected. Unfortunately, there are limitations in the accuracy of the geometric co-ordinates produced from seismic surveys and a variation of 5% in sound velocities can lead to errors in calculating the depth of the target by several hundred feet. These limitations are less of a problem when drilling into large deposits but small, complex reservoirs can be missed altogether if there are significant discrepancies in the depth and lateral co-ordinates.

The technique of geological steering or geosteering was developed to ensure higher rates of success in reaching these smaller and less clearly defined targets.

7.7.3 Geosteering

Geosteering is real-time steering of horizontal and high-angle wells using while-drilling formation evaluation. It enables the driller to guide the drill bit and the well to the optimum geological destination, rather than directionally steering the well to a pre-determined, possibly non-optimum, geometric location.

At the heart of the new equipment is a comprehensive array of measurement-while-drilling (MWD) and logging-while-drilling (LWD) tools linked to an interactive graphical interface. The logging and measurement tools are incorporated in the drill string so that the drilling team can determine, with a very high degree of accuracy, the location of the drill bit in the reservoir.

Planning
Before beginning operations, a computer generated model of the formation is constructed based on a combination of seismic survey information and on wireline logs taken from similar wells located in the immediate vicinity of the area under investigation. From the reservoir model, a simulation of the predicted response of the formation is prepared and this is compared with the actual resistivity measurements recorded while drilling.

Resistivity, which describes the response a material has to electromagnetic induction, is one of the most reliable techniques for determining the geology of the formation. Different formations have differing resistivities and they are further affected by the presence of liquids, either oil or water. By comparing the actual results with the predicted results, the position of the drill bit can be plotted accurately on the geological model of the reservoir. Should any discrepancies occur between the predicted values and the measured values, the model can be updated and the drilling co-ordinates revised. Any inaccuracies in the initial geometric co-ordinates are compensated for and the chances of the drill bit being lead astray are reduced significantly, or at least corrective action can be taken prior to the tool wandering off course by a significant distance. In this manner, the drill bit can be navigated into the pay zone along geological boundaries.

Geosteering equipment
There is nothing radical in the design of the drilling hardware, ie the drilling motor and the surface adjustable bent housing (or the adjustable motor). It is the sensing and measuring equipment, and the telemetry systems that relay the measurements back to the drill floor, where the greatest advances have taken place.

To steer a well more efficiently, with fewer unplanned lateral and vertical undulations, tools must be provided that can:

- Survey the formation
- determine the inclination of the drill string
- determine the compass direction of the drill bit
- relay this information to the drill floor.

To achieve the desired speed of response, sensing tools must be placed as near to the drill bit as possible if changes in direction are to be made with the minimum delay.

Measurement systems

i) Bit resistivity

A 1500-Hz alternating current is forced into the formation via electrical coils mounted on the drill collar and bit, the drill collar, formation and coils effectively acting like a large transformer. The flow of current is governed by the resistivity of the formation and by comparing this with the voltage induced in a toriod coil positioned approximately 1.5 metres (5 feet) above the bit face, the resistivity of the formation can be computed. The magnitude of the induced current varies inversely with the resistivity of the formation.

ii) Azimuthal resistivity and gamma ray

Azimuthal resistivity and gamma ray readings assist in determining where the formation boundaries are in relation to the drill bit. This can make the difference between drilling through the pay zone and missing it, or landing in the pay zone and drilling along it. By being able to identify the shale-sand interface and changing the direction of the drill bit as soon as a variation from the predicted formation is measured, a more productive well can be achieved.

iii) Inclination

Inclination measurements taken both at the drill bit and 18 metres (60 feet) behind the bit remove any uncertainties as to where the drill bit is pointing. They provide the driller with a feel for what the bottom hole assembly is doing and enable him to determine whether or not the drill bit is drifting in response to formation conditions.

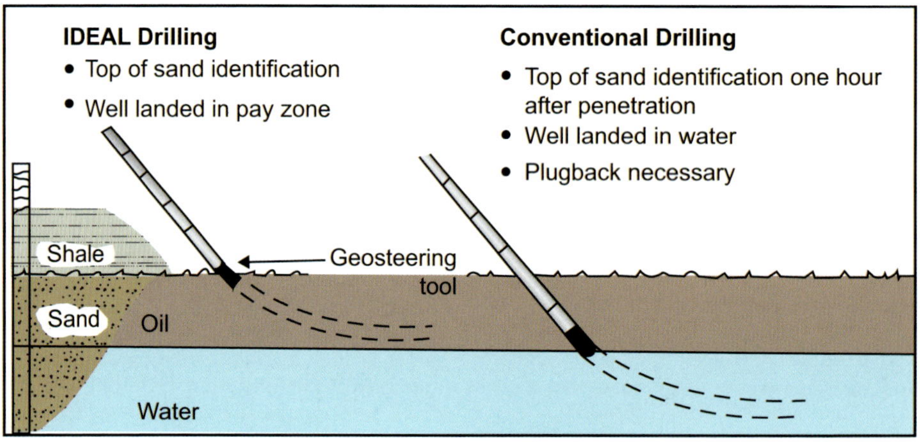

Figure 7.43 – Ideal and conventional drilling

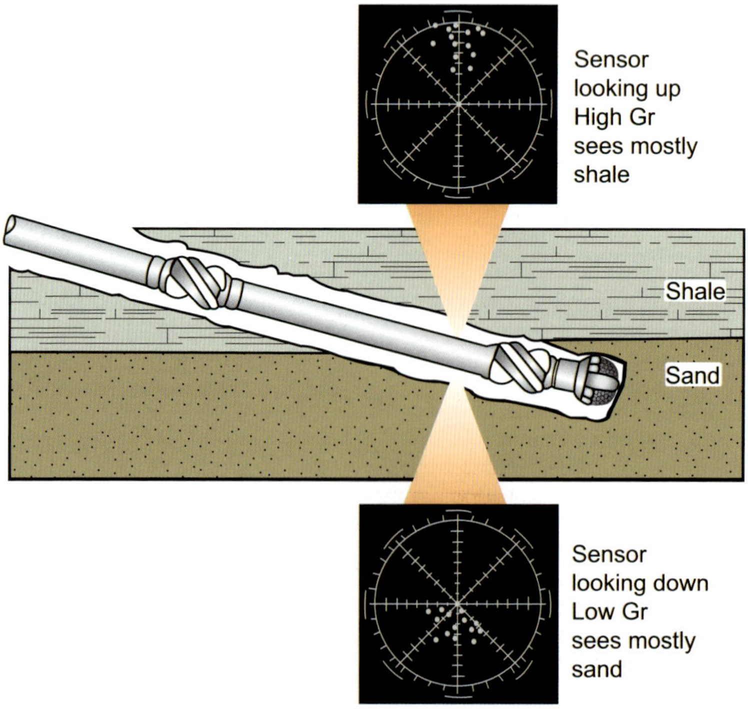

Figure 7.44 – Geosteering sensors

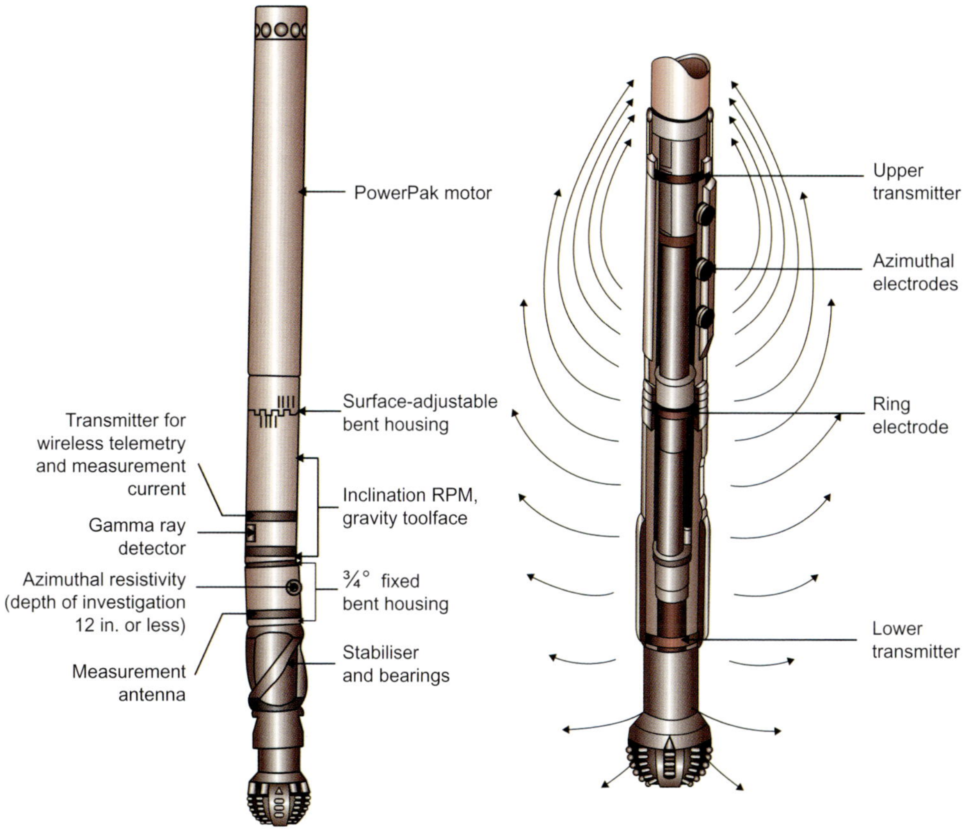

Figure 7.45 – **Geosteering and rab tools**

iv) RWOB

In addition to the tools described above, a receiver weight on bit (RWOB) tool may be included, which measures downhole weight and torque on the drill bit. This allows the driller to optimise the power curve, which improves penetration, reduces the chance of the drill pipe sticking and provides an early indication of problems with the drill bit.

Telemetry systems

The technical leap that allows measurements to be made at the drill bit is the wireless telemetry system.

The measurement tools and sensors located near the drill bit are battery powered, self-contained units. A wireless telemetry link sends data from the sensors to the MWD tool, which may be located up to 200 metres behind the bit, a path that bypasses the intervention drilling tools such as the steering motor. The MWD system records and transposes the signals and then sends the data to the surface using mud-pulse telemetry.

7.7.4 Multi-Lateral Wells

The multi-lateral well is an arrangement whereby two or more wells will share the same well bore.

There are two main reasons for opting to drill a multi-lateral well. The first is cost, because using an existing well bore saves on drilling time and materials and savings of up to 25% can be made compared to drilling a completely new well. The second reason is to improve reservoir drainage, a successfully drilled multi-lateral can improve production by up to 500% in ideal cases.

A multi-lateral is where multiple bore holes are drilled from an existing single bore well. The new bore hole may be horizontal or deviated to reach the desired bottomhole location, a 'branch' being the term used to describe a lateral drilled from a horizontal lateral in the horizontal plane while a splay is a lateral that has been drilled from a horizontal lateral in the vertical plane.

Multi-laterals are formed by cutting a hole through the side of the casing of an existing well with a milling cutter attached to the end of a drill string. A 'whipstock', a hardened steel wedge, is used to deflect the milling cutter at the desired location. The angle of the whipstock is between 1.5-3.0°, so the deflection of the multi-lateral is extremely gentle and, once the junction is completed, the well is conventional in all other respects and may be either cased or uncased.

There are six types of junction and a standard classification system has been prepared by the industry depending on the complexity of the junction. Cased junctions involve inserting a tubular casing string through the junction point and they are used to support a weak or unstable formation, to simplify re-entry of the well during workovers and when pressure isolation is required.

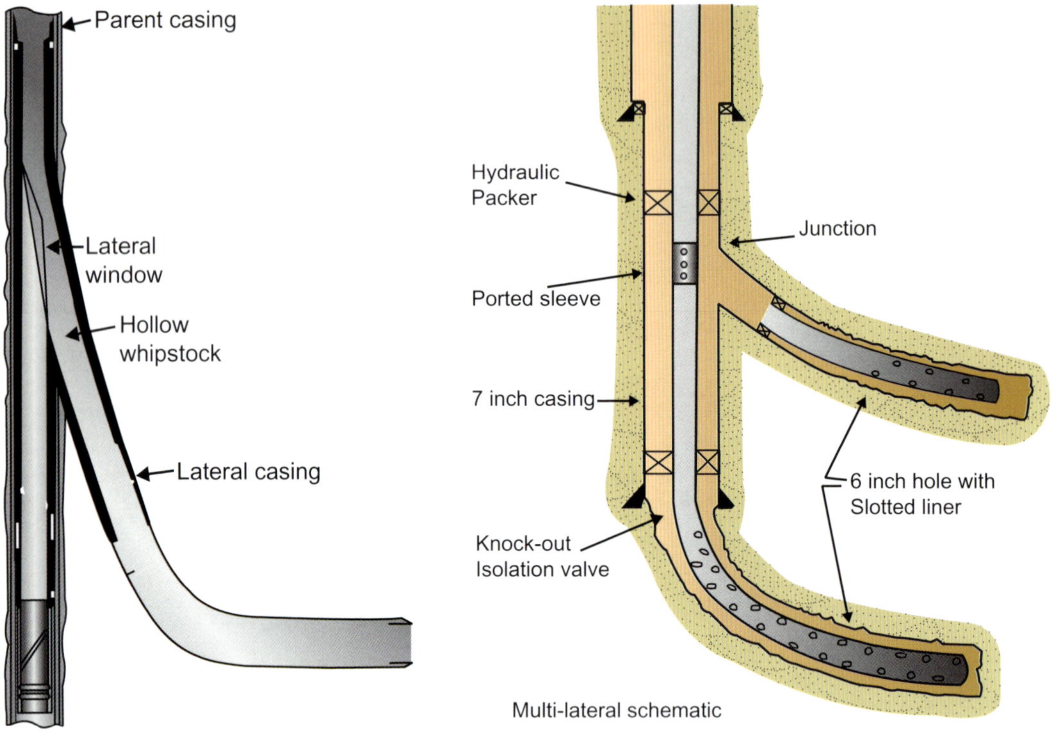

Figure 7.46 – **Multi-lateral wells**

Land Drilling System Components

Cameron BOP Stack

1. Annular BOP
2. Double U BOP
3. Drilling Spool
4. Manual FL Gate Valve
5. Hydraulic FL Gate Valve
6. R Check Valve
7. Single U BOP
8. Casing Head Spool
9. Casing Head Housing

Figure 7.47 – **Land drilling rig**

8
Component Parts

8 Component Parts

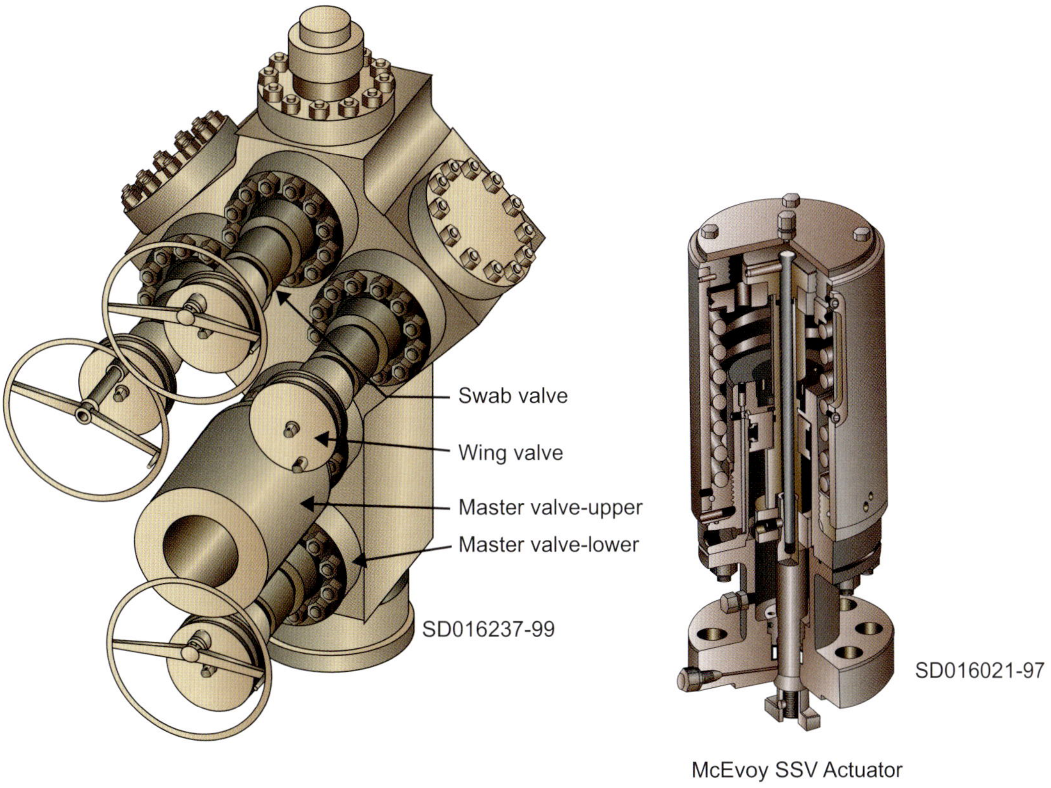

Figure 8.1 – Solid block Christmas tree and actuator

8.1 The Christmas Tree

The christmas tree sits on top of the wellhead and provides the controlling mechanism for the isolation of individual wells, and for the regulation of the flow of hydrocarbons to the production and process facilities. The christmas trees play an essential role in the emergency shutdown (ESD) system.

The christmas tree is essentially an assembly of gate valves and the term 'christmas tree' is thought to have been used to describe the onshore trees that comprise a number of single valves bolted together. Solid block trees are preferred for offshore service because they provide fewer potential leak paths being manufactured from a single alloy steel or stainless steel forging or casting into which the valve chests are machined. Depending on the composition of the reservoir products, the bores and sealing surfaces may be protected against corrosion and erosion by cladding (over welding) with a material such as Inconel.

Occasionally, a combination of solid block and individual valves are used.

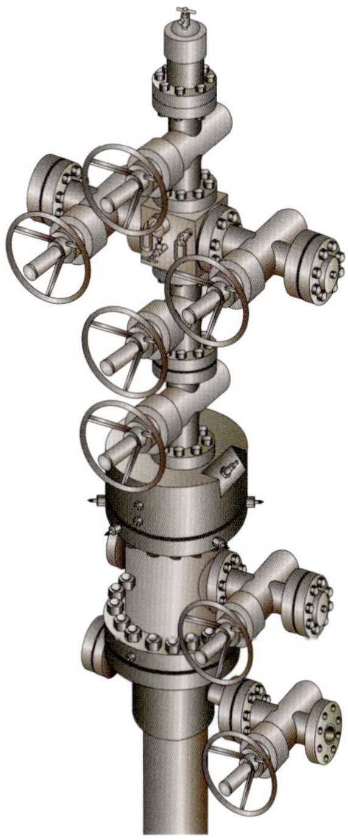

Figure 8.2 – Built-up tree and wellhead (Onshore type)

8.1.1 Solid Block Christmas Tree

A solid block christmas tree will normally comprise the following valves:

8.1.1.1 Master Valve(s)

Most offshore christmas trees are provided with two master valves which are referred to as the upper master and lower master valves. The valves are mounted one above the other in the vertical plane and provide a double block arrangement for the isolation of the well.

The upper master valve is opened by a hydraulic or pneumatically powered actuator which is controlled from the wellhead control panel. The lower master valve is almost always a manual valve which is only closed during prolonged periods of shutdown such as during maintenance programmes. The lower master valve is opened first and closed last. This ensures minimal flow of hydrocarbons over the valve seats, providing protection from abrasive particles and ensuring a good seal is maintained.

Modern christmas trees are fitted with a wire cutting master valve. Should an emergency situation arise when wireline operations are in progress, the valve will close completely, cutting the wireline in the process. The wire and tool string will fall into the well and can be recovered by fishing at a later date.

Wing valve
Christmas trees may be supplied with one or two wing valves, the valves being offset from the vertical plane so that a clear entry into the well is maintained through the swab valve for wireline work. One valve, the production wing, is permanently connected to the hydrocarbon process system and is normally fitted with a hydraulic or pneumatically powered actuator. The other valve, the service wing is operated manually and permits the injection of chemicals or gases into the well without disturbing production pipework.

The flow of gas from the well is normally regulated by a choke valve, which is fitted either above the wing valve, or further downstream in the vicinity of the production header.

8.1.1.2 Swab valve

The swab valve is the uppermost valve on the christmas tree and is positioned directly above the master valves. It facilitates the entry of wireline equipment into the well.

The valve is manually operated and the top connection is fitted with a swab cap to protect the threads and prevent the accumulation of debris.

8.1.2 Wellhead Control Panel

A wellhead control panel or wellhead skid controls the operation of the christmas trees and mudline or surface controlled subsurface safety valves (MLSV, SSSV or SCSSV). The skid permits valves to be operated locally, remotely or via the ESD system and timing mechanisms provide a means of controlling the speed and sequence of valve operation. For valves fitted with actuators the sequence would normally be:

To Close
I) Production wing valve.
ii) Upper master valve.
iii) Mudline safety valve.

During an emergency shutdown (ESD) situation, complete closure of the christmas tree valves should be effected within approximately 45 seconds to comply with API RP 14H, the only document which provides guidance on this particular aspect.

To Open
I) Mudline safety valve.
ii) Upper master valve.
iii) Wing valve.

8.1.3 Gate Valves and Actuators

8.1.3.1 Gate valves

Gate valves are the preferred choice of valve type for christmas trees because they are robust and the sealing surfaces can be easily replaced should damage occur. Gate valves are most effective when used in either the fully open or fully closed position and are not generally used to regulate flow. This is carried out by a choke valve located downstream of the production wing valve if required.

8.1.3.2 Valve gates

Two types of gate valve may be encountered, the slab gate and the split or segmented gate. The slab gates rely solely on well pressure to push the gate against the seat to establish a seal and are simpler and cheaper to produce than the split gates. Consequently, they are most effective in high pressure wells and may only seal securely in one direction, the direction of flow.

The segmented gate valve employs two gates held apart by a spring or expander and it is the expander that forces the gates against the seats as the valve closes. The resulting seal is effective in both directions and, because it does not rely on well pressure, it works well at both high and low pressures.

8.1.3.3 Valve seats

As previously stated, both the gates and the seats of a gate valve can be easily replaced. Once the valve bonnet has been removed, the gates are simply unscrewed from the valve stem while the seats can be jacked out of the valve chest. Fixed gates tend to be used with segmented gates, ie they are a metal to metal interference fit in the valve chest. Floating seats are generally used with slab gates, an elastomeric seal sealing the seat in the valve body, the seats being free to move slightly to assist in the forming of an effective seal against the gate.

8.1.3.4 Valve actuators

All the valve actuators are fail-safe in operation. The valves are held open by oil or air pressure against a compressed coil spring. Venting of the power medium permits the spring to push the valve plate into the closed position. The actuators are normally designed so that they can be replaced, should the need arise, without disturbing the gate valve. Position indicators may also be fitted with either visual or electrical proximity switches to show when the valve is open or closed.

Illustrations of gate valves and actuators can be found in Chapter 4.

8.1.4 Horizontal Trees

Unlike conventional christmas trees, the horizontal or spool tree is a through bore design with the valves mounted external to the vertical bore of the well. The well is produced from the side of the tubing hanger, the reservoir products flowing horizontally through the master valve.

The development of horizontal christmas trees has taken place over the same period of time as the development of horizontal wells but there is no direct association between the two, they are developments that stand on their own merits and one is not reliant on the other.

While horizontal trees are available for installation on conventional offshore platforms, the market has grown considerably because of the interest in deepwater developments using subsea wells.

The main advantage of the horizontal production tree is that it permits complete access to the well bore without removing the tree because there are no valves in the vertical plane above the production tubing. Major workovers can, therefore, be carried out without the need to recover the subsea tree, the BOP stack being landed directly on top of the christmas tree. They are ideal for use in wells where frequent tubing-downhole completion recoveries are anticipated, such as when electrical submersible pumps (ESP) are used. There are also economic benefits because larger diameter production tubings can be installed that permit a greater throughput of reservoir products.

The only perceived disadvantage of the horizontal tree is that with the absence of valves in the well bore, double block valve isolation of the well is not provided. However, two barriers are provided in the form of plugs which are set in the top of the tree.

Guidance notes and reference standards

(i)	API Spec. 6A	Specification for Wellhead and Christmas Tree Equipment.
(ii)	API RP 6AR	Repair and Remanufacture of Wellhead and Christmas Tree Equipment.
(iii)	API Spec. 14D	Recommended Practice for Wellhead Surface Safety Valves and Underwater Safety Valves for Offshore Service.
(iv)	API RP 14H	Recommended Practice for Installation, Maintenance and Repair of Surface Safety Valves and Underwater Safety Valves Offshore.

(v) The Department of Energy Guidance Notes on Offshore Installations: Guidance on Design, Construction and Certification, Part II, Section 43.

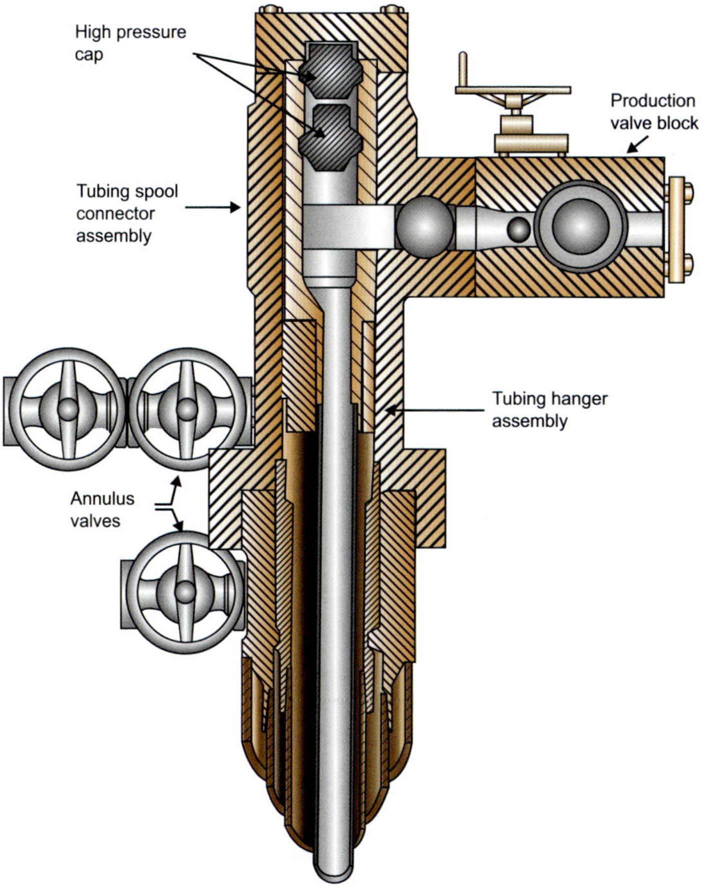

Figure 8.3 – Spool tree wellhead and christmas tree assembly

8.2 The Surface Wellhead

The wellhead assembly on a fixed installation consists of a collection of components designed to support and seal the intermediate casings, production tubing and christmas tree. It also provides a base for the location of the blowout preventer (BOP) during drilling operations.

Assembly of the wellhead commences with the installation of the conductor pipe and the various stages can best be described with the aid of a diagram, Figure 8.4.

8.2.1 Base Plate/Landing Ring

Once the conductor has been driven to the desired depth, the base plate or landing ring can be positioned and welded. Landing rings are used to help distribute wellhead loads through the conductor and into the formation. It should be noted that these components are designed specifically to support the casing strings and are not required to be pressure retaining. At this stage in the drilling operation the formation should be of a type that cannot support the existence of hydrocarbons.

8.2.2 Starting Head

Depending on the design, the starting head may be screwed onto, or welded to the top of the first intermediate casing string, and it may incorporate a landing ring. The function of the starting head is to transmit the loads created by subsequent casing strings through to the conductor, and to provide the initial location for the blowout preventer.

The BOP will remain in place for the remainder of the drilling operation, being moved onto each subsequent casing head in turn.

8.2.3 Casing Head Spools

Following the installation of the starting head, all subsequent casing strings will be terminated within a casing head spool. The spool is designed

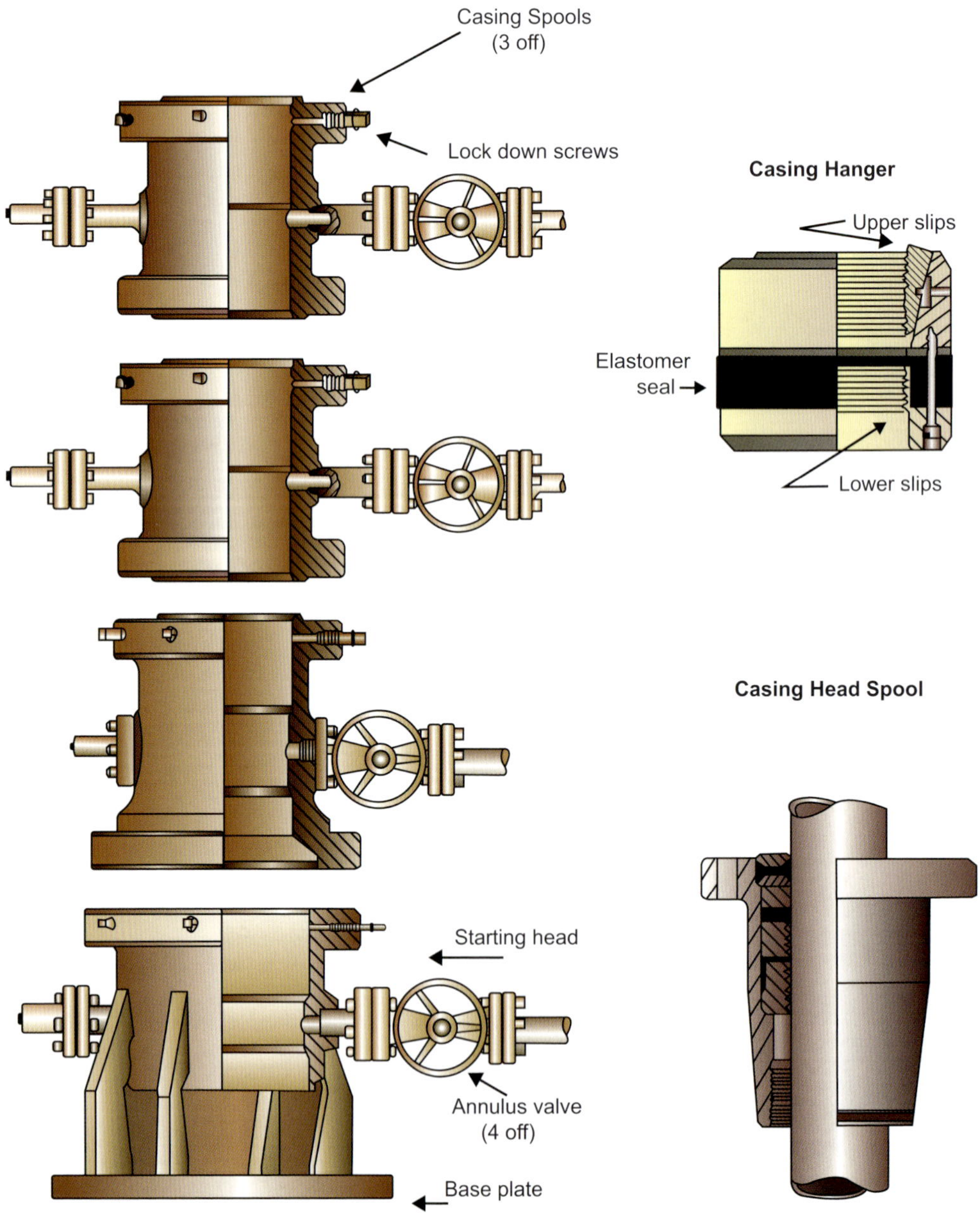

Figure 8.4 – The well component parts

to support and seal the upper section of casing against oil or gas pressure and provide connections through which the condition of the annulus spaces may be monitored. Generally speaking, the casing head spool is installed after the casing has been cemented, but before the cement has cured.

Examples of casing head support and seal arrangements are shown highlighted in Figure 8.4. The casing slips or hangers are manufactured with a serrated steel internal face designed to bite into the casing string and provide support within the casing head, and a metal to metal seal. A secondary rubber or neoprene 'pack off' seal may then be activated under the compression forces provided by external lock down screws or internal cap-head screws. This twofold sealing arrangement may vary slightly from one manufacturer to another, but the basic principles of operation remain the same.

8.2.4 Tubing Head Spool

The tubing head spool is designed to support and seal the production tubing within the final intermediate casing. The production tubing sits in a bushing or tubing hanger and is the only item within the well which receives direct exposure to reservoir pressure. The intermediate casings and casing head spools are designed to resist well pressure but will only be subjected to pressure should failure of the production tubing packing seals occur, or a gas lift enhanced oil recovery system is fitted.

8.2.5 Christmas Tree

The final operation consists of installing the christmas tree in preparation for bringing the well onstream. The christmas tree permits isolation of the reservoir products from the process equipment.

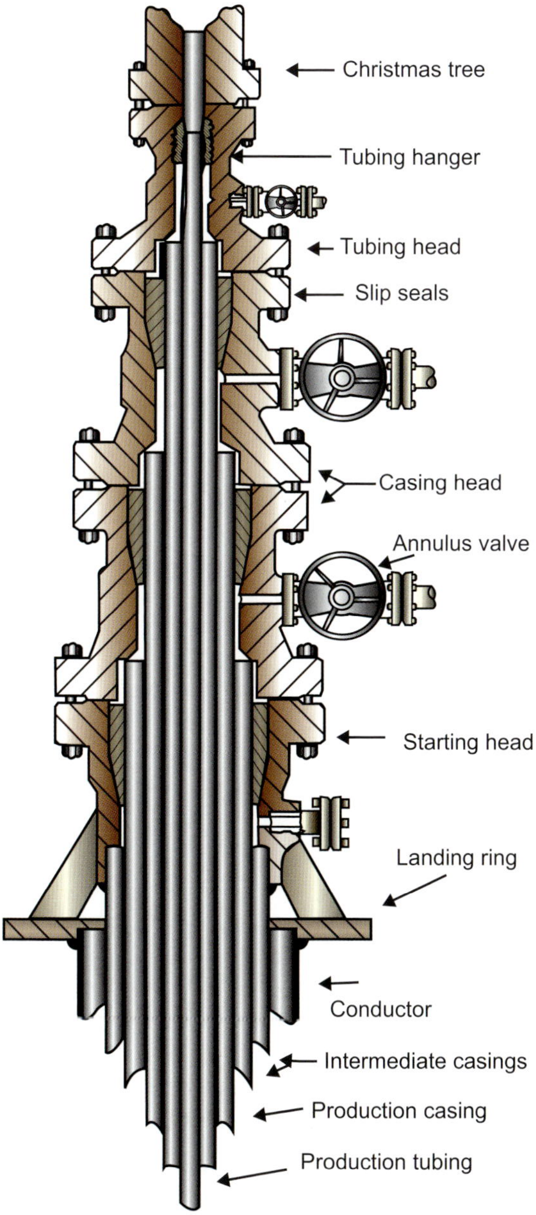

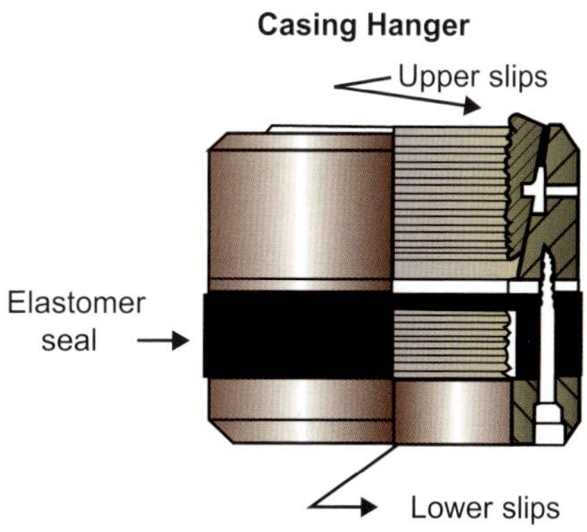

Figure 8.5 – Christmas tree assembly

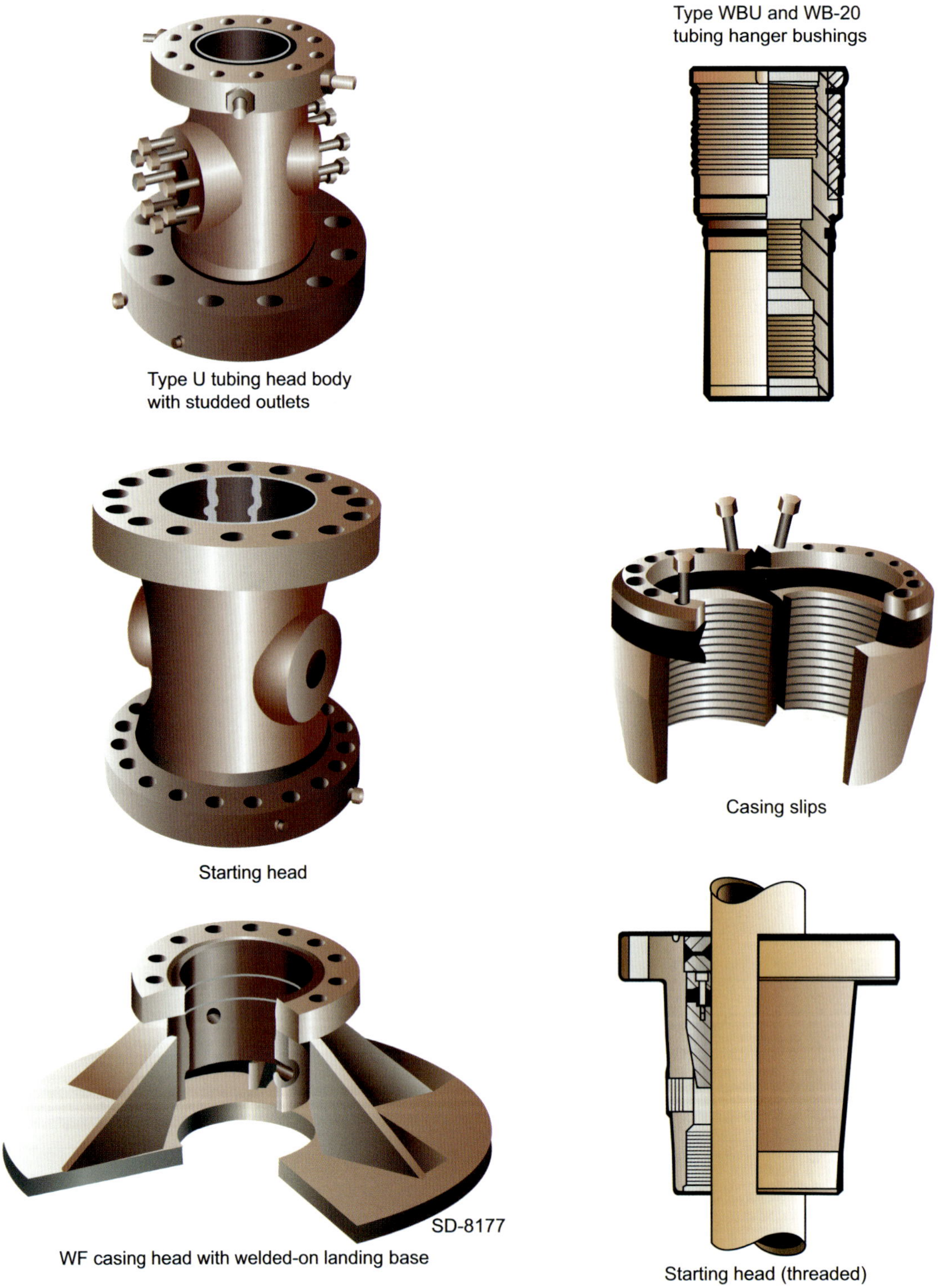

Figure 8.6 – Wellhead components

8.3 Mudline Suspension Systems – (MLSS)

A mudline suspension system provides an alternative to the surface wellhead for the support and sealing of intermediate casing strings. As the name suggests, the suspension system is located at the mudline or seabed and sits within the conductor pipe rather than on top of the conductor as is the case with the surface wellhead.

The illustration in Figure 8.6 shows a simplified layout of a MLSS of the type installed by a bottom mounted drilling rig or jack-up. Before continuing with the description of the various components, it would be beneficial to recap on how the conductor and intermediate casing strings are formed.

The conductor and casings are assembled from 9 metre (30 feet) lengths of pipe which have externally threaded ends. They are joined together and sealed against oil/gas pressure by internally threaded tubular couplings or joints. The MLSS employs special joints that incorporate a casing hanger that screws into the casing string as it is lowered into the hole.

The assembly process commences with the installation of the conductor. A joint containing a landing face is screwed into the conductor at a position that will coincide with the mud line. The first intermediate casing will locate or hang off this protrusion and each subsequent casing string will be supported in a similar fashion, ie hung off the preceding intermediate casing hanger joint.

Some designs of mudline suspension systems employ a spring loaded, expanding casing hanger which latches into the preceding casing to provide additional support.

The mudline suspension system has two main applications.

8.3.1 Pre-Drilling of Wells

The drilling of wells for a production installation is a particularly time consuming operation. A considerable saving in time and money can be achieved by drilling the wells prior to the arrival of the permanent installation and the design of the MLSS is ideally suited to this application.

On completion of drilling, plugs are set in the well and the conductor and intermediate casing strings which extend from the seabed to the drill floor are removed. They are simply unscrewed from the casing hanger and replaced with threaded abandonment caps. The process is reversed when the permanent installation arrives and the wells are required for production.

8.3.2 Subsea Wells

Subsea wells are drilled and completed in an identical manner to surface wells. However, the absence of a fixed structure to support a conventional wellhead necessitates the use of either a MLSS or subsea (marine) wellhead, both arrangements sharing the same basic installation concepts. As in the previous example, on completion of drilling, the casing strings extending from the casing hanger to the drill floor are removed. They are then replaced with a completion adaptor spool, a location device that provides a suitable profile for the attachment of the subsea christmas tree.

Figure 8.7 – The jack-up *'Neddrill 3'* testing an exploration well. The gas is being flared from a temporary overboard vent boom.

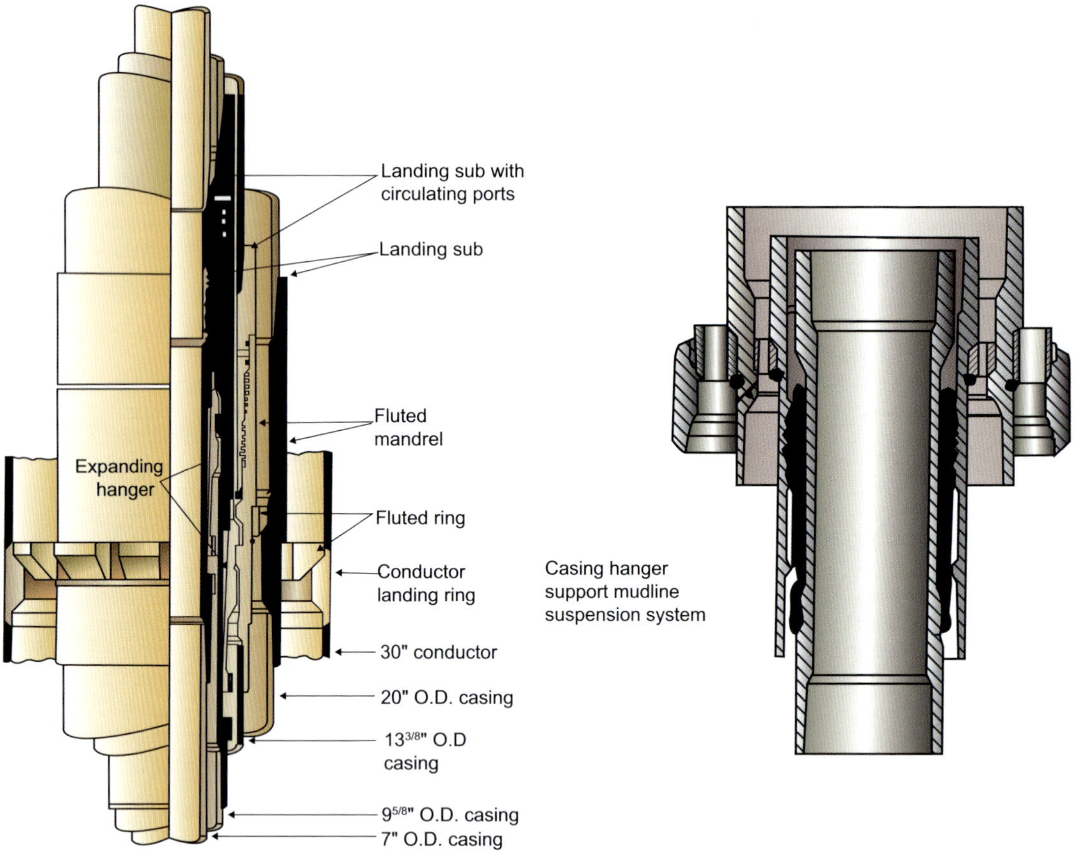

Figure 8.8 – Mudline suspension system

8.4 Mudline Safety Valve

The mudline safety valve (MLSV) or, to give it its correct title, the surface controlled subsurface safety valve (SSSV or SCSSV), is located in the production tubing at a depth of 60 to 90 metres (200 to 300 feet) below the surface of the seabed. It provides a last ditch barrier to isolate the well should the installation be destroyed by an explosion or collision with a ship. The valve is fail-safe in operation.

The most graphic example of SCSSV operation was provided when the '*Piper Alpha*' installation was destroyed by fire and explosion in July 1988. The mudline safety valves on all 36 wells closed satisfactorily, but unfortunately two of the wells provided fuel for the fire via the annulus spaces which were being used for gas lift purposes, the isolation of which ultimately required the attention of Red Adair's specialist fire-fighting team.

The valves should be checked for leakage on a regular basis, preferably monthly. This entails closing the valve and depressurising the production tubing up to the christmas tree. The production tubing should remain free of any significant pressure build-up for a period of 10 minutes. Should the valve fail or leak significantly, it should be replaced at the first available opportunity.

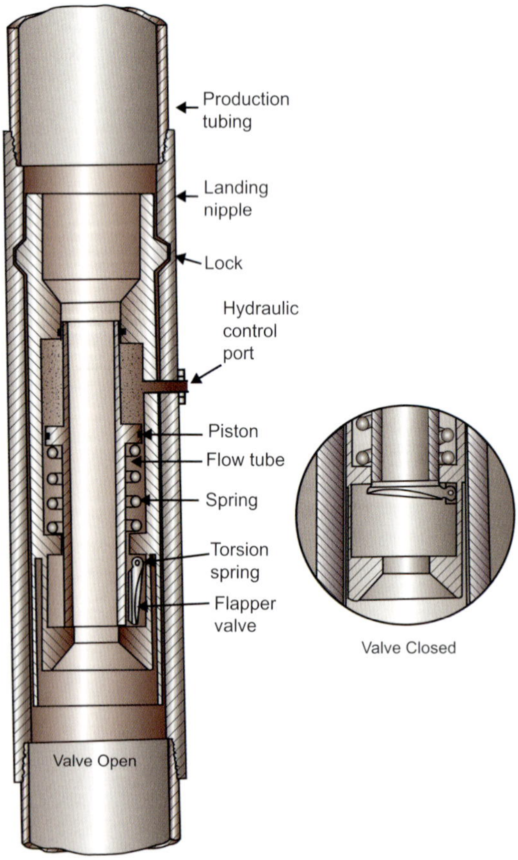

Figure 8.9 – Mudline safety valve

The actual valve, which may be of a ball, but more probably flapper type, is located in a tubular housing. The housing design determines the method by which it is installed in the production tubing, ie it may be tubing retrievable or wireline retrievable.

(i) Tubing retrievable
The tubing retrievable SCSSV is screwed into the production tubing as a joint when the tubing is run (installed). Replacement involves a workover and the pulling (removal) of the production tubing, a costly and time consuming exercise.

(ii) Wireline retrievable
The wireline retrievable SCSSV can be recovered, as the name suggests, by a wireline operation. These valves are often preferred for the ease in which replacement can be effected. The production tubing is fitted with a machined recess or landing nipple in which the valve is located and latched into position using a wireline tool string. The disadvantage of the wireline retrievable valve is that it restricts the throughput of the production tubing.

Figure 8.9 shows the operation of a flapper type wireline retrievable subsurface safety valve. Hydraulic oil pressure is required to keep the valve open against spring pressure. The hydraulic oil is normally supplied via stainless steel instrument tubing routed through the wellhead, although some of the older installations achieved the same result by pressurising the annulus space.

The hydraulic oil pressure is supplied to, or released from, the SCSSVs by the wellhead skid under the influence of electronic control signals initiated by emergency shutdown (ESD) system. The valves can also be operated manually from the wellhead skid, or by rupture of the fire loop.

Guidance Notes and Reference Standards

1. API Spec 14A Specification for Subsurface Safety Valve Equipment
2. API RP 14B Recommended Practice for Design, Installation, Repair and Operation of Subsurface Safety Systems. (Contains ISO 10417:1993).

8.5 The Production Packer

The function of the production packer is to seal the base of the production tubing into the intermediate casing string. Both permanent and temporary packers are available but the permanent type are generally preferred for their more reliable operation over prolonged periods.

Figure 8.10 explains the mode of operation of a permanent packer. It should be noted that the packer is installed prior to perforation of the casing or liner so the well is full of well preservation fluid and devoid of all hydrocarbon pressure.

A packer can be run (inserted) as part of the production tubing or installed on a wireline tool string. Once in position, the seals may be activated by a mechanical tool, hydraulic pressure, or an electrically activated gas charge, depending on design. Figure 8.10 shows a hydraulic packer and to activate the seals a steel blanking plug must first be set in the production tubing below the packer. The tubing is then pressurised to energise the setting piston, which moves upwards, snapping the shear pin and compressing the rubber sealing element. When the pressure is released, the steel slips bite into the intermediate casing and prevent re-expansion of the sealing element. The packer is permanently set and should removal be required, it must be milled out, a major operation.

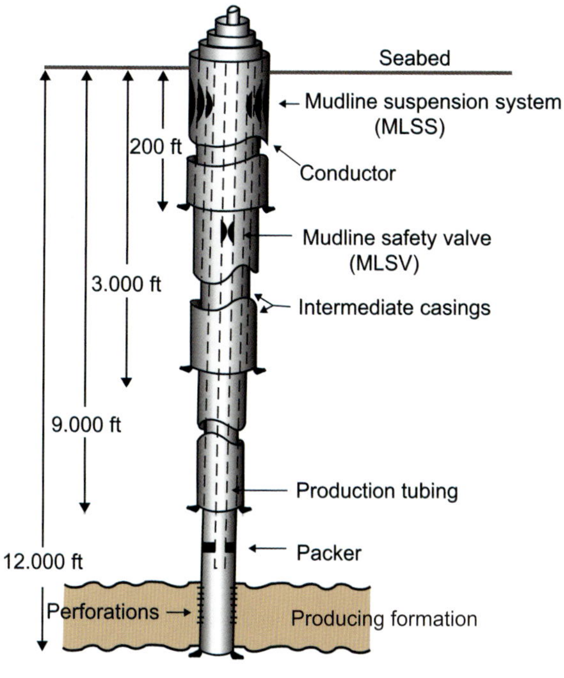

Typical Casing Arrangement

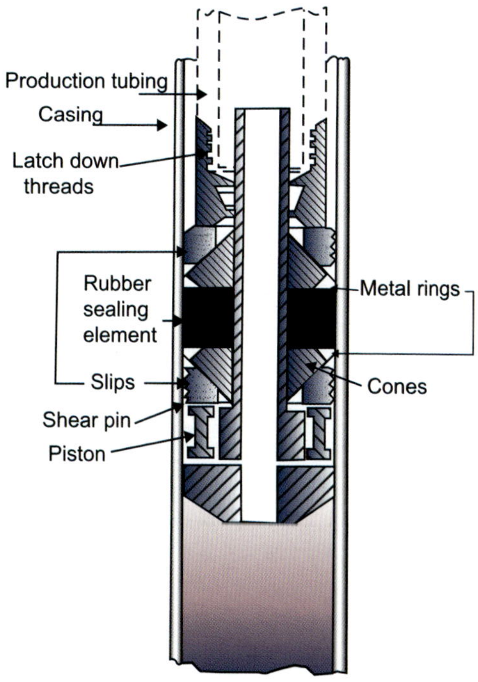

Production Packer

Figure 8.10 – Production packer

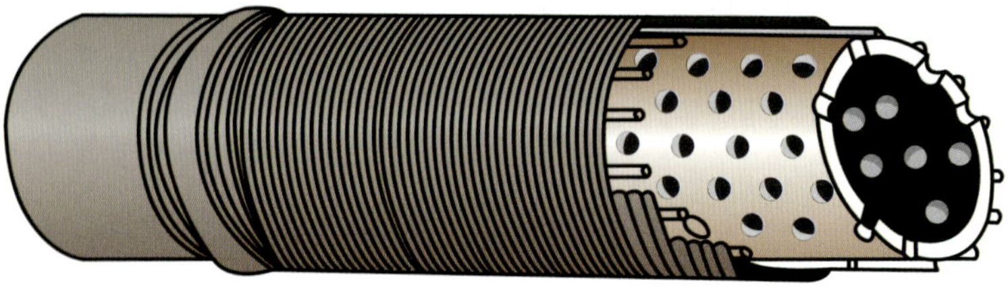

Figure 8.11 – All-welded wire-wrapped screens

Key shaped wire on ribs welded onto the base pipe decrease the chances of the screen becoming blocked.

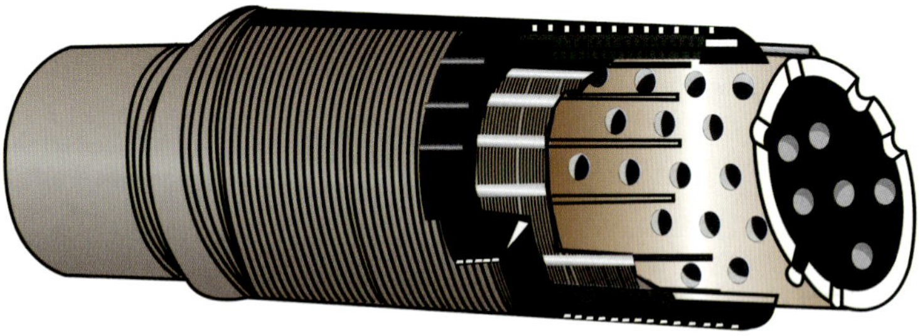

Figure 8.12 – Enhanced low-profile prepack screens

Suitable for installing in horizontal wells, these screens can also be installed in gravel packed wells. They incorporate a strong inner screen and a prepack sand annulus to optimise filtering capabilities.

9

Well Maintenance

9 Well Maintenance

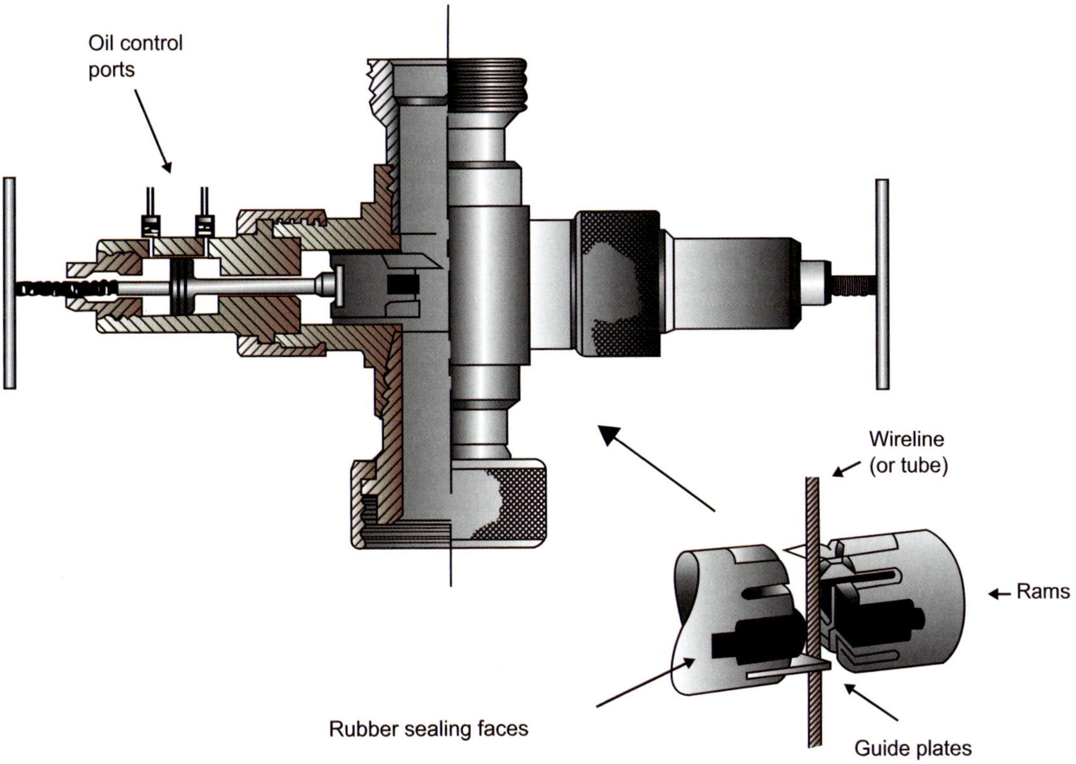

Figure 9.1 – Coil tubing/wireline valve

9.1 Wireline Operations

Wireline equipment has been in use since the early days of the oil and gas industry. It permits mechanical devices to be installed or removed in the production tubing with minimal interruption to the production process.

The equipment is attached (flanged or screwed) to the top of the christmas tree as shown in Figure 9.1. Various wireline tools can then be attached to the wire and lowered through the christmas tree swab valve and into the production tubing.

Most operations can be carried out using a single strand wire or slick line (also called piano or music wire) which is available in thicknesses ranging from 0.066 to 5/16 inch (1.65 mm to 8.0 mm) and supplied in lengths of 10,000 to 40,000 feet (3,030 to 12,120 metres). However, when a stronger line is required, a braided wire of 3/16 to 5/16 inch (4.75 mm to 8.0 mm) diameter is normally used (often called torpedo or well shooters line).

An electric line may be used as an alternative to the simple wireline. The electric line facilitates the passage of electrical signals and permits the use of more sophisticated equipment. The electric line can even be used to operate a video camera, which may be lowered into the well and used to visually identify leaks and blockages in the production tubing.

The wireline is stored on a winch drum that is powered by a diesel driven hydraulic pump. Hydraulic equipment is preferred on the grounds of safety for use in hazardous areas and a counter fitted to the winch drum provides the operator with an indication of the depth of the tool string in the well.

9.1.1 Wireline Equipment

The wireline assembly shown in Figure 9.2 contains the following items of equipment:

i) Wireline valve or blowout preventer (BOP)
The wireline valve lies at the heart of the wireline operation. It enables quick and safe isolation of well pressure from the lubricator sections mounted above it. The valve is often referred to as a BOP (blowout preventer), somewhat of a misnomer as the term blowout is only used in relation to drilling operations. However, the wireline valve is very similar in appearance to the drilling BOP if somewhat smaller, and they are both designed to contain reservoir pressure within the well bore so the confusion is understandable.

In its simplest form the wireline valve consists of a pair of rams fitted with rubber inserts which can be manually wound in to close and seal around the wire, sealing the production tubing.

A more sophisticated version is available that employs two pairs of hydraulically operated rams. The top rams are fitted with rubber inserts for sealing on the wire and the lower blind rams are designed to shut off well pressure should the wire become detached from the winch drum.

The wireline valve is attached to the christmas tree at the swab connection. A crossover assembly may be required between the two to compensate for diameter, flange or thread variations.

During wireline operations on a producing well, the christmas tree should remain connected to the emergency shutdown (ESD) system. This ensures that should an emergency situation develop, the christmas tree master valve will close and sever the wire line. The wire, complete with tools, will drop into the well allowing the mudline (subsurface) safety valve to close and effect sealing of the well bore. The wire and tools can be obtained by fishing at a later date.

The master valve on older christmas trees may not be designed to cut wires. Where this situation occurs, a fail safe wireline cutting gate valve should be installed between the wireline valve and the christmas tree and connected into the platform's ESD system.

ii) Lubricator or riser

The name lubricator is misleading as no lubrication actually takes place within this component. It consists of a length of pipe, approximately 8 to 10 feet (2.4 to 3.0 metres) in length and of a diameter to suit a particular tool string, normally 3½ to 5½ inches (90 to 140 mm). It is fitted with male (pin) and female (box) unions (screwed or welded) at each end to permit ease of assembly. Several sections of lubricator may be joined together to give the necessary height above the wireline valve to facilitate the entry of the tool string.

iii) Stuffing box and check valve union

The stuffing box is a simple device attached to the top section of the lubricator whose function is to seal the wire against gas pressure. The tubular body is filled with layers of packing and may incorporate a plunger or ball valve designed to shut off gas pressure should the packing become damaged or

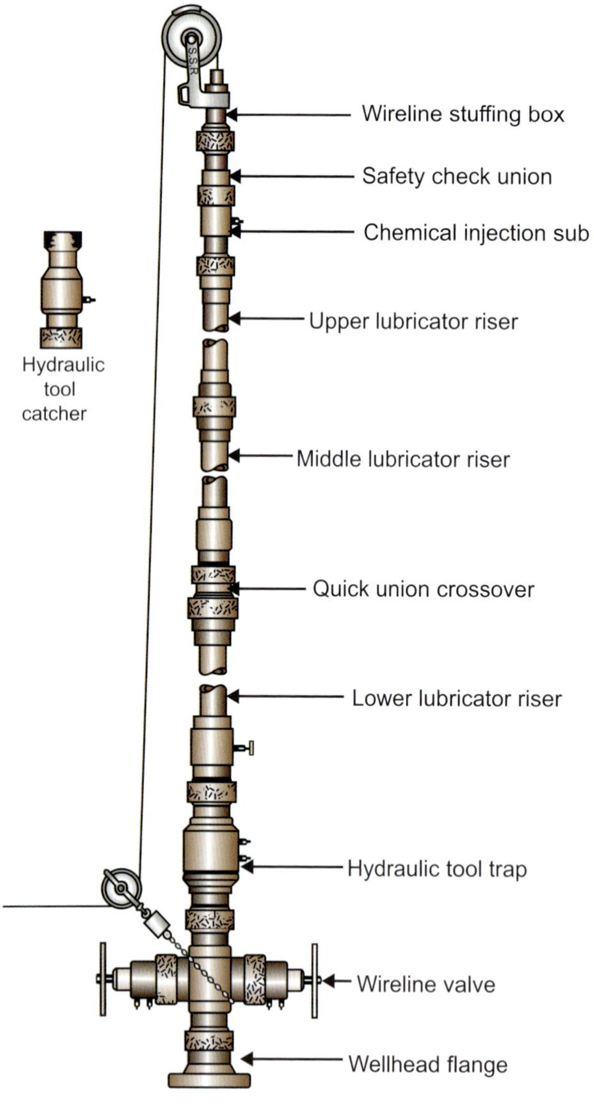

Figure 9.2 – (a) Slick line lubricator stack

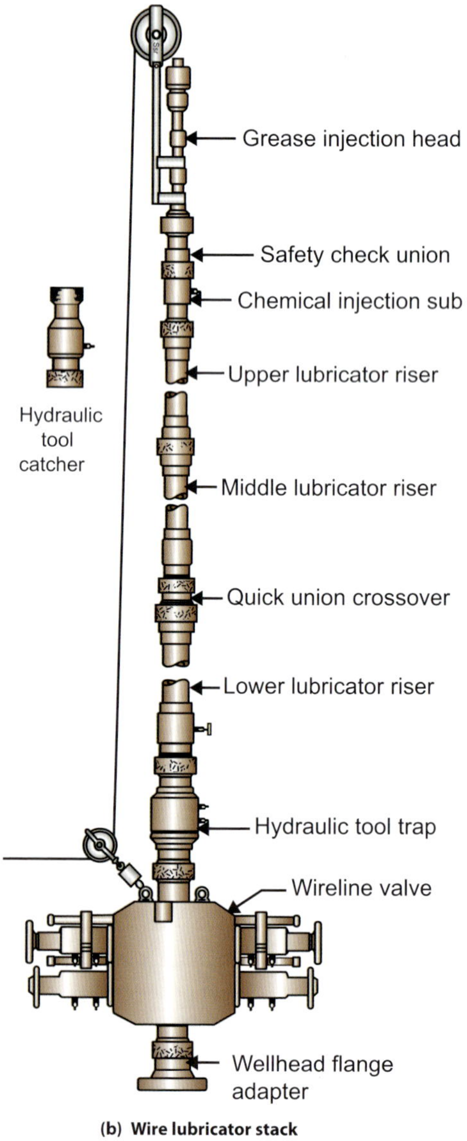

(b) Wire lubricator stack

the wire break. In the absence of a plunger or ball valve a separate check valve union will be mounted below the stuffing box.

iv) Hay pulley

The Hay pulley is attached to the top of the stuffing box, its function being to redirect the wire to the winch drum.

v) High pressure grease tube (HGT)

The high pressure grease tube or grease injection head is used in place of the stuffing box when braided wire is used in preference to a single strand wire. Grease is pumped through the injection head at a pressure approximately 500 psi (35 bar) above the gas pressure to seal and lubricate the wire.

vi) Tool trap

The wireline winch drum is fitted with a counter to enable the operator to determine the depth of the tool string in the well. However, there is always the risk of stripping the wire from the rope socket as it is winched into the stuffing box, so to prevent the tool string from dropping back into the well, a tool catcher may be fitted. Mechanical and hydraulic tool traps are available, some of which latch onto the rope socket under wireline tension and some which close under the toolstring as it is raised into the lubricator assembly. The tool trap will be located either at the top or the bottom of the lubricator stack, depending on the design.

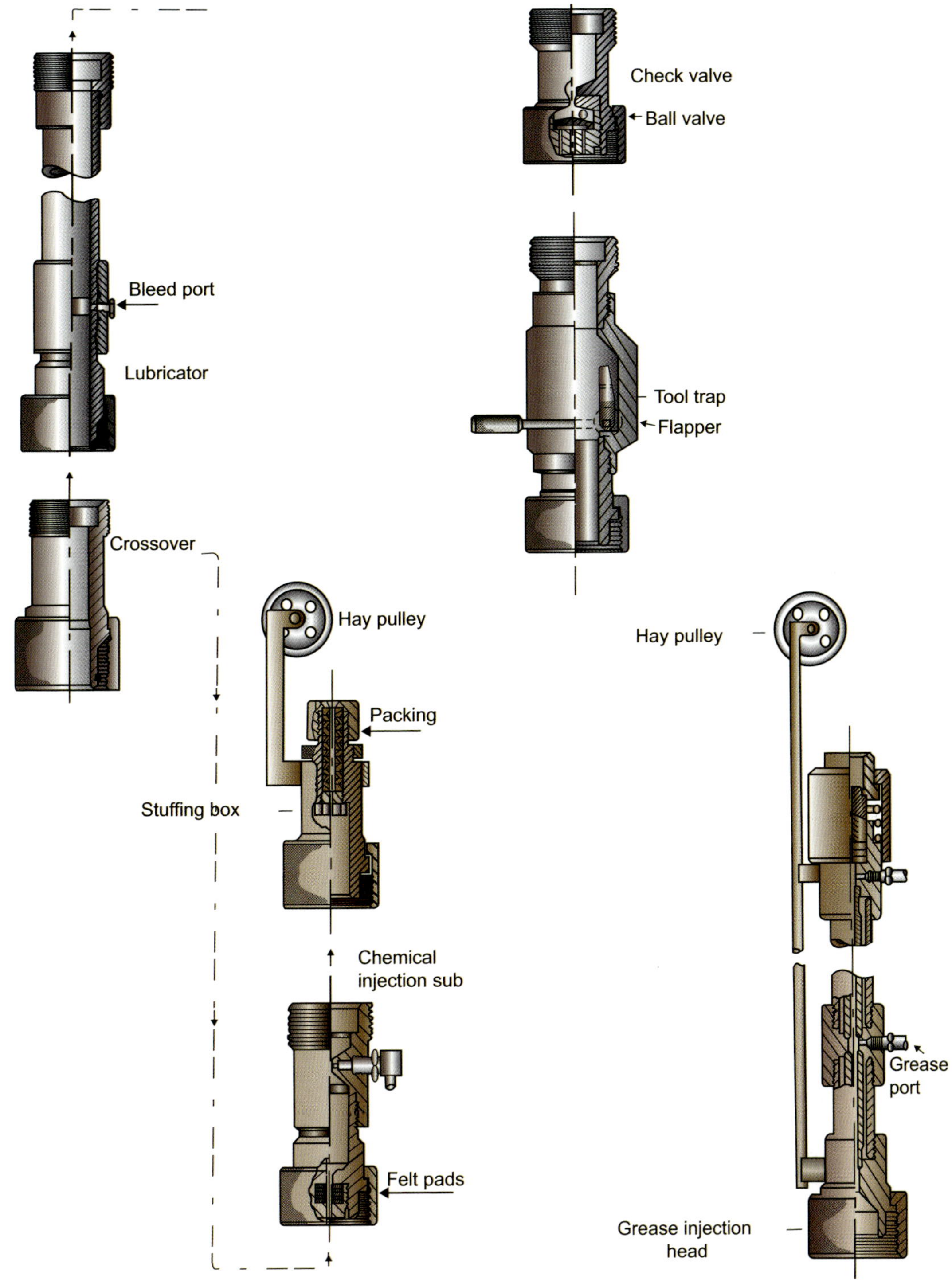

Figure 9.3 – Wireline stack components

vii) Chemical injection sub
The chemical injection sub provides the means by which chemicals may be applied to the wireline. The chemicals are introduced into the sub assembly at a side port and are absorbed by replaceable felt pads that wipe the wire as it enters and leaves the well. Methanol is frequently used to combat icing while corrosion inhibitors provide the wire with some protection against hydrocarbon products produced from sour reservoirs.

9.1.2 Wireline Tool String

Figure 9.4 shows a typical wireline tool string arrangement. The wire rope socket, jars and knuckle joint are present during all wireline operations and special tools are added to the basic string depending on the task in hand.

The wire is fed through the stuffing box and lubricator prior to addition of the tool string and the complete assembly is then mounted on top of the wireline valve. The wireline tool string is then lowered through the christmas tree swab valve, master valve or valves and into the production tubing.

The components of a tool string may be described as follows:

i) Wire rope socket
This device joins the tool string to the wire. The wire is threaded through the body and around a plug, which is then pulled into a taper. Slip rope sockets are a variation on this basic theme and are designed to break at 50%, 60%, 70%, 80% or 90% of the wire's tensile strength should the tool string become jammed. Failure of the wire simplifies tool string retrieval, the rope socket being designed to facilitate the attachment of a latching recovery or fishing tool.

ii) Jars
The simplest form of jar is a heavy length of tube and various weights are available. They are used to jar free tools stuck in the tubing (as in the previous case describing the rope socket) and to provide extra weight to assist latching operations.

Mechanical jars can only be used to jar in a downward direction. Hydraulic jars are required for jarring in an upward direction.

iii) Knuckle joint
The knuckle joint contains a ball and socket that provides flexibility within the tool string.

iv) Weight bars
Weights are used to increase the effectiveness of jars and weigh from 5 to 33 pounds (2.25 to 15 kilograms).

v) Pulling/running tools
Pulling refers to the removal of a device such as a mudline safety valve or pressure test plug from the production tubing. The installation process is known as 'running a tool' and relies on a latching mechanism for emplacement.

9.1.3 Wireline Operations

A multitude of operations can be carried out using wireline equipment and four of the basic tasks will now be described.

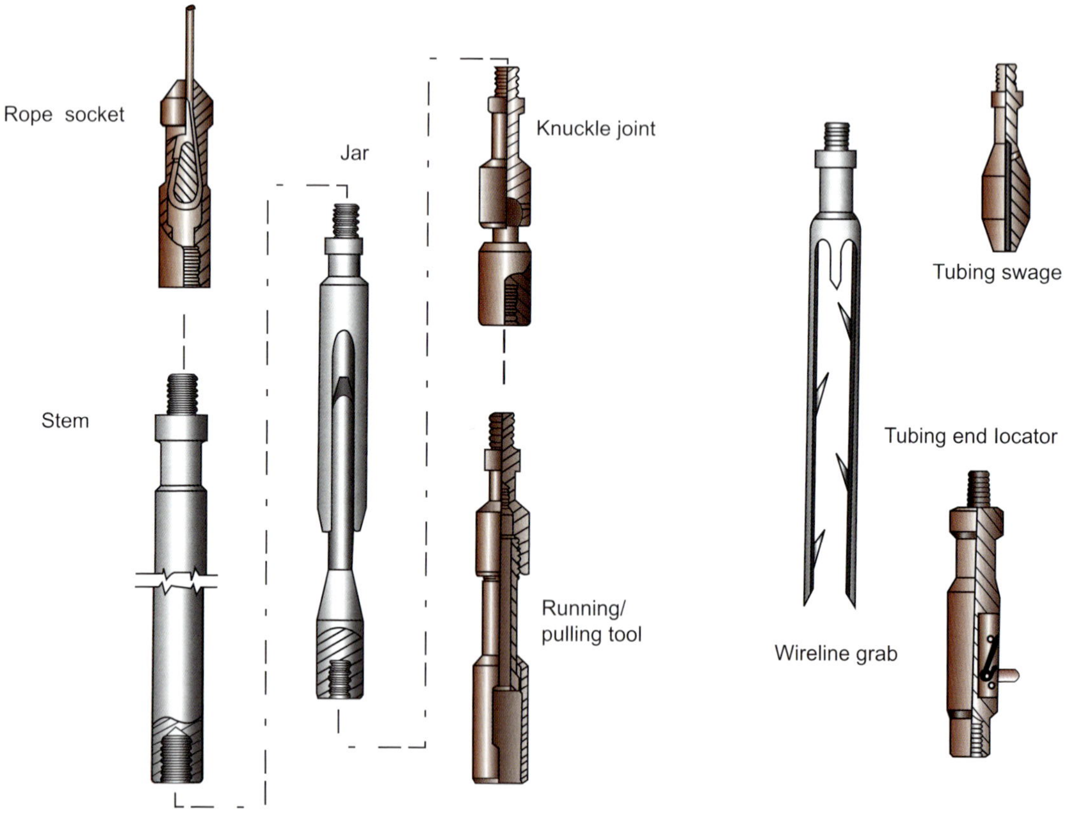

Figure 9.4 – (a) Wireline tool string **(b) Selection of tools**

i) Perforating

Perforating is carried out on a new well prior to bringing it into service and on existing wells that have been subjected to modifications in the vicinity of the hydrocarbon bearing formation or pay zone. It is a wireline process by which a string of explosive charges is lowered through the production tubing to the depth at which the gas or oil producing formation exists. When the explosive charges are detonated the concentrated blast punches holes through the casing and into the formation. The numerous small holes produced permit hydrocarbons to flow into the production tubing and up to the rig. The casing may be perforated over a distance of several hundred feet.

ii) Location plugs and valves

Two of the most frequent operations carried out using wireline equipment are the change of wireline retrievable mudline or subsurface safety valves (MLSV or SCSSV) and, to a lesser extent, the fitting of blanking plugs in the production tubing.

The production tubing is the innermost tubing string in the well through which hydrocarbon products are conveyed from the reservoir to the christmas tree. As the tubing is installed it is fitted with a number of landing nipples (shouldered joints) set at different depths, which are designed to accommodate the subsurface safety valve and, on occasions, steel blanking plugs. The blanking plugs permit pressure tests to be carried out on the production tubing when searching for leaks, and provide the means by which the well can be isolated prior to changing a christmas tree.

The installation of a subsurface safety valve is a straightforward wireline operation. They are attached to a suitable running tool and latched into the landing nipple, a pulling tool being used to separate the latch prior to removal.

iii) Formation surveys

(a) Bottom hole surveys

An accurate assessment of the condition of a production well can be made by lowering self-contained pressure and temperature sensing devices into the hydrocarbon bearing formation on the end of a wireline toolstring. The data obtained enables the reservoir engineers to determine whether or not the well is producing its maximum potential.

(b) Logging

During drilling operations the hydrocarbon bearing potential of the formation can be determined by logging the hole using electric wireline operated tools. A log is a record of readings generated by a logging tool which transmits electric, sonic or radioactive signals into the formation. The return signals are computer analysed and a profile of the logged zone developed. Subsurface formations possess certain measurable characteristics that help to identify the type of rock structure and the hydrocarbon content.

iv) Well maintenance

Typical maintenance operations carried out with wireline equipment are the cleaning of the production tubing of wax, clearing blockages with a swage and removing sand. Special tools have been developed for each of these operations.

There is also a fishing tool for recovering broken wires. The term 'fishing' may also be used to describe the process of latching on to the subsurface safety valves or plugs prior to their removal.

v) Wireline tractor

Deploying wireline equipment in highly deviated or horizontal wells presents problems because the equipment relies on gravity to take the tools to their destination. Wireline tools can be deployed with coiled tubing equipment, which can push the tools downhole, but more recently a downhole tractor has been developed for pulling wireline into horizontal wells.

The tractor system is controlled from a communications and motor control system located at the surface. An electrical AC motor integrated into the wireline tractor drives a hydraulic pump which forces the tractor wheels onto the side of the production tubing or casing. Once a pre-set pressure is reached, the wheels start to rotate and drive the tractor forward. When the oil pressure is released, the wheels retract into the tractor body and an electrical relay shifts the unit between the tractor and logging mode. Conventional wireline tools are used.

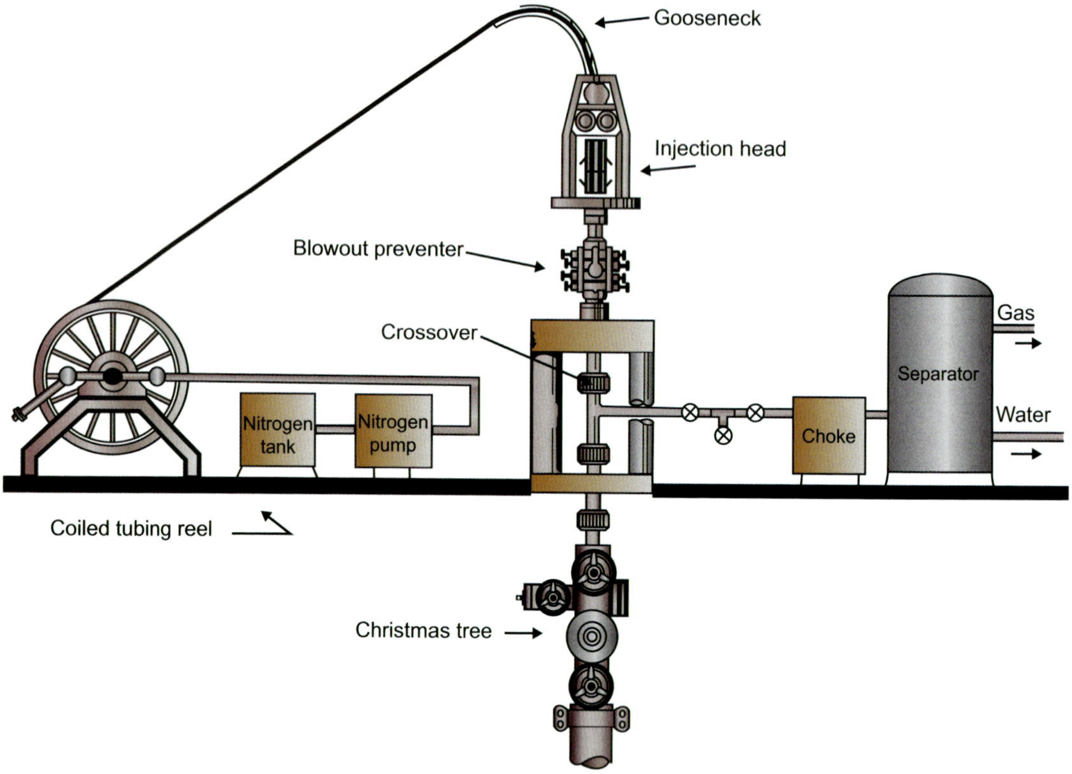

Figure 9.5 – Coil tubing equipment schematic

9.2 Workover Operations

Figure 9.6 – The *'Big Orange XVIII'* well service vessel alongside a gas production installation.
Inset – Sphere launcher, an interlocking door prevents inadvertent opening while the chamber is under pressure.

During the life of a well, which maybe 25 years or more, situations inevitably arise that call for the modification or repair of equipment associated with the production tubing. Mechanical problems such as leaks in packers, tubing joints and cement seals develop and the depletion of hydrocarbons may result in changes in reservoir conditions which can restrict output, typically caused by the build-up of sand. These problems can be corrected by suspending production on the affected well and carrying out a workover job. Workovers involve the use of highly sophisticated equipment, normally supplied by contractors who specialise in a particular field of operations. They may be loosely divided into two categories, ie workovers that deal with the rectification of defects in the production tubing and workovers designed to improve or stimulate the hydrocarbon bearing formation in the vicinity of the well bore.

9.2.1 Well Stimulation

Three of the more common well stimulation processes will now be discussed. They may be carried out using containerised equipment temporarily located on the platform or from purpose built well stimulation ships, depending on the size of the task in hand, the operations involving the injection of gases (nitrogen) or liquids (acids) into the well. Coil tubing equipment may be used for some of the smaller tubing cleaning operations, the equipment outwardly resembling the wireline valve and lubricator stack used for wireline operations. However, it is a small steel tube (¾–3½ inch diameter) that passes through a coil tubing BOP into the christmas tree and on into the well and not a wireline tool string. The inherent flexibility of the tubing enables it to be coiled around a large drum in continuous lengths of up to 20,000 feet when not in use.

i) Fracturing (Fracing)

Fracturing is used to improve the flow of hydrocarbons into the well bore. It involves the injection of oil or water based fluids at very high pressures into the rock formation. This induces cracks and opens fissures, enlarging and extending the drainage area through the rocks.

Sand, or artificially manufactured proppants may be injected with the fracture fluid so that when the hydraulic pressure is released, the solid particles will remain in place holding the fissures apart as the liquid drains away.

ii) Acidizing

Acid can be used to remove damage near the well bore in sandstone and carbonate formations and to produce long linear flow channels away from the well.

Matrix acidizing involves injecting acid below fracture pressure to dissolve deposits restricting the perforations.

Fracture acidizing involves injecting acid at a pressure greater than fracture pressure. The acid forces the fracture line apart and dissolves a flow path which remains when the fluid pressure subsides and the fracture closes once more.

Additives are mixed with the acid solution to prevent the corrosion of steel and the formation of sludges which could block the tubing perforations.

iii) Sand control

Sand can congregate at the well base and block the casing perforations. Small quantities can be scooped out using wireline equipment but where large quantities exist, more drastic action is required.

A popular solution is to install a slotted screen in the open hole, or in the casing. Gravel packing the outside of the screen will permit the passage of hydrocarbons whilst hindering the passage of sand.

Another solution involves the precipitation of resin between the sand grains in order to bond them together and prevent the flow and further accumulation of sand.

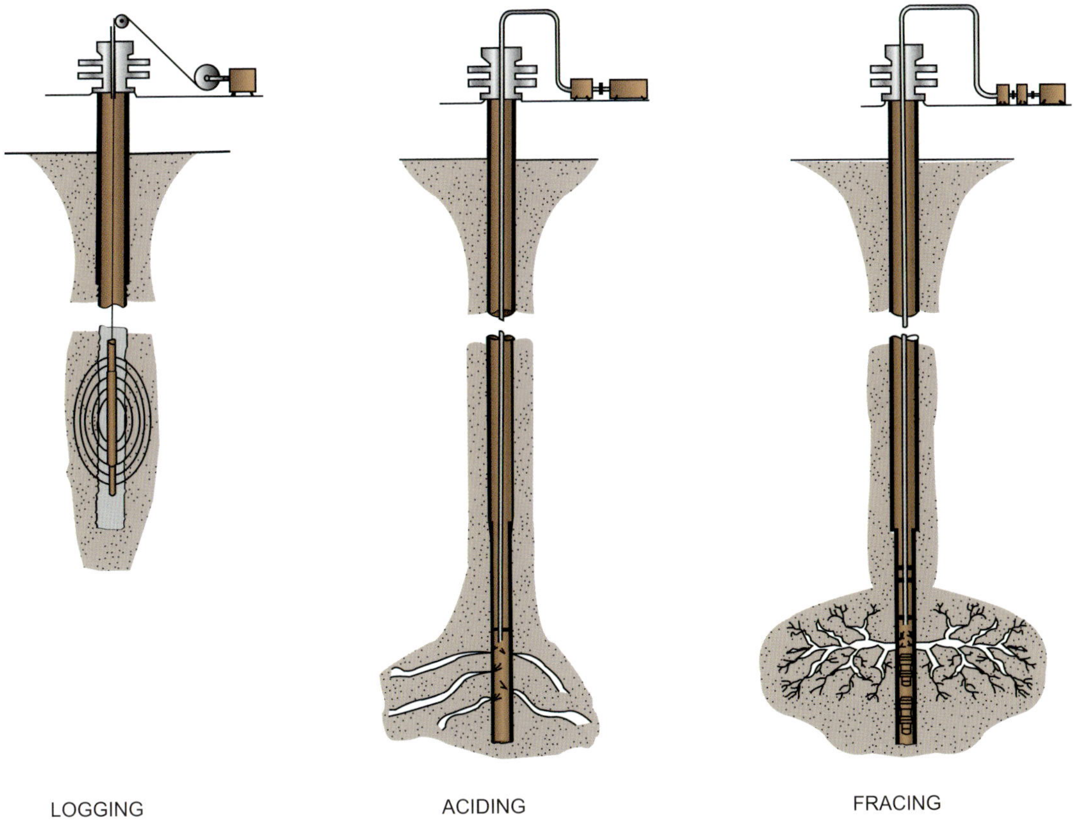

Figure 9.7 – Logging

9.2.2 Well Modifications

The removal of the production tubing is a major operation requiring the use of the drilling derrick, or a mobile workover rig. Problems which necessitate the removal of the production tubing tend to be fairly serious in nature and for safety reasons the well must be killed before work commences. This can be done most effectively by displacing the contents of the production tubing with brine in order to overbalance the hydrostatic head of the formation. With the well suitably killed, the christmas tree can be removed and replaced with a BOP stack prior to work commencing.

Some of the more basic well workover operations will now be described.

i) Production tubing damage
The production tubing is exposed to the full force of reservoir pressure and all the associated contaminants which accompany the hydrocarbons, such as formation acids and abrasive sands. Consequently, it is not unusual for problems to develop as the well ages and leaks occasionally occur which require the replacement of all, or at least part of the production tubing.

ii) Subsurface (Mudline) safety valve replacement
Mudline safety valves may be of the wireline retrievable or tubing retrievable type. As the name suggests, the tubing retrievable valves require the removal of the production tubing in order to facilitate their replacement. Whilst the mudline safety valves tend to be very reliable in operation, it is unlikely that they will operate reliably throughout the life of a well which could be 25 to 30 years. As the valve is located only 30 metres (100 feet) below the seabed, replacement does not necessitate the removal of the entire production tubing but it is still quite a considerable task.

iii) Production tubing replacement
As the reservoir ages it is common practice to replace the production tubing with one of a smaller diameter in order to increase the velocity of well fluids. This ensures that impurities such as water and sand are brought to the surface where they can be separated and disposed of by the process plant rather than risk their accumulation in the production tubing which will eventually kill the well.

iv) Packer replacement
Quite a common problem that afflicts older wells is leakage of the packer which seals the production tubing into the innermost casing string. The remedy involves the removal of the production tubing so that the defective packer can be machined out. This is achieved using a milling cutter on the end of a drill string. With the casing suitably cleaned and all drilling debris removed the production tubing can be replaced and a new packer activated.

v) Cement squeeze
Occasionally, leaks develop in the intermediate casing strings at tubing joints, or where corrosion has taken place. This may manifest itself as a reduction in well output as reservoir products are lost from the well into the formation, or it may be indicated by increased water production when water seeps from the formation into the well. Again, the solution necessitates the removal of the production tubing in order to isolate the damaged casing with blanking plugs. Cement is then pumped, under pressure, between the plugs from where it is squeezed out through the leak zone and will set to form a permanent repair. A similar procedure is used for blanking off redundant perforations.

vi) EOR equipment
Some enhanced oil recovery aids such as deepwell pumps and gas lift equipment are installed or run with the production tubing. Therefore, any repairs or modifications will involve a workover in order to remove the production tubing.

Appendices

Appendix I – Structural Steel

Introduction
It is beyond the scope of this book to enter into a discussion on the design of an offshore installation. Historically, the only readily available publication on the subject is produced by the American Petroleum Institute, API RP 2A – *Recommended Practice for Planning, Design, and Constructing Fixed Offshore Platforms*. The Code, which is now in its 20th Edition, dates back to 1969 and was developed largely as a response to the destruction of 28 platforms by hurricanes in the Gulf of Mexico in 1964 and 1965.

Having established a suitable design, the structure must then be constructed in accordance with the requirements of a recognised construction specification. Traditionally structures have been constructed in accordance with the requirements of the American Welding Society's AWS D1.1 – *Structural Welding Code*. In common with the majority of engineering reference books which originate in the USA, AWS D.1.1 provides a wealth of information on all aspects of welding (including weld procedures and welder qualification tests) and includes guidance on the selection of materials, workmanship, inspection and non-destructive examinations. More recently in Europe, the Engineering Equipment and Material Users Association have developed a standard aimed specifically at the North Sea, namely EEMUA 158, *Construction of Fixed Offshore Installations in the North Sea*. The EEMUA 158 defines the basic requirements for fabrication work on primary structures whilst recommending that steel work other than primary be fabricated in accordance with AWS D1.1.

If not referenced specifically, EEMUA 158 and AWS D1.1 will almost certainly provide the basis from which Company specific structural construction specifications are prepared and as such will be encountered repeatedly on new building, repair and modification projects.

1. Structural Steelwork

The component parts of the jacket and topside support framework of an offshore installation are categorised as either primary or secondary structures and the steels used for construction are similarly identified as primary and secondary steels.

i) Primary Structure
Primary structural members are those which contribute to the overall integrity of the installation and include such items as jacket legs, piles, braces, columns and beams. Where these components are subjected to significant tensile loads they must be constructed from steels with guaranteed through thickness properties and are often classed as Special Structural primary steels, examples of which are nodes and pad eyes/ears.

ii) Secondary Structure
Secondary structural members can be described as those components whose contribution to the integrity of the structure as a whole is not significant and includes items such as deck stringers, crane pedestal supports and stairways. Failure of a secondary structural member is unlikely to affect the overall integrity of the installation.

Some of the more readily identifiable structural components are:

Can – Large cylinder formed from rolled plate and longitudinally welded. Welded end to end, they are used in the construction of jacket legs.

Bracing – Tubular support member arranged to provide stiffening to the jacket legs. May be manufactured from cans, seamless or welded pipe depending on the diameter required.

Node – Intersection of bracings. May be welded or cast with cast steel components preferred for modern deepwater jackets.

Plate girder – "I" section beam manufactured from steel plate and used as a main deck support, normally referred to as a PG.

Pad eye/ear – Fabricated or cast steel lifting attachment through which (eye) or around which (ear) wire strops are attached for the lifting of the module during installation.

2. Material Specifications

Structural steels used for the construction of offshore installations will normally comply with one of four specifications.

i) British Standard 4360 – Specification for Weldable Structural Steels
For many years BS 4360 was used extensively both in the UK and abroad for the manufacture of structural steels with properties suitable for use offshore. Before being withdrawn, it was used to provide the basis of a European Standard with applications specific to offshore structures, EEMUA No. 150.

ii) EEMUA No. 150 – Steel Specification for Offshore Structures (now discontinued and replaced by BS EN 10225:2009 - Weldable structural steels for fixed offshore structures)
The EEMUA No. 150 specified parameters for the selection of structural steels and provided comparison tables for British, European and American material grades. It has now been superceded by EN 10025.

iii) EN 10025 – Weldable Structural Steels for Fixed Offshore Structures
This standard is primarily applicable to the North Sea but may also be used in other geographic locations provided that due consideration is given to local conditions, eg, temperature. It covers steels with a minimum yield strengths up to 460 MPa with low temperature impact properties at temperatures down to –40°C.

iv) ASTM – American Society for Testing and Materials
The ASTM organisation provides a range of specifications covering low strength carbon steels through to high strength low alloy steels, most of which are suitable for offshore structures.

v) API Specification 5L – Line Pipe
Whilst Spec. 5L pipes are manufactured primarily for pipeline service, they are also used extensively as jacket bracings. The project fabrication specification must be studied to confirm the suitability of the

materials and establish if any additional mechanical tests are required prior to use.

3. Material Properties

The specifications listed above contain numerous grades of steel which may be compared firstly by tensile strength, secondly by yield strength and thirdly by impact (charpy) properties. The materials are generally referenced using a number to indicate tensile or yield strength and a letter to give an indication as to the resistance to brittle fracture. For example, EEMUA 355D has a minimum yield strength of 355 N/mm^2 with impact values recorded at −20°C, this material is equivalent to the old BS 4360 50D material where the 50D designation refers to the minimum tensile strength of 500 N/mm^2 (actually 490 N/mm^2) and the D again refers to the achievement of acceptable impact values at −20°. The numbering system used in EN 10025 is very similar to that used in EEMUA although somewhat more comprehensive. For example, S355 G2+N is a steel with a minimum specified yield strength of 355 MPa with impact testing undertaken at −20°C. The +N indicates that the steel is supplied in the normalized condition whilst +M would indicate a steel supplied in the thermomechanically rolled condition. Similarly, +QT indicates a steel supplied in the quenched and tempered condition. The G2 refers to a steel group and a number of groups are available for which reference should be made to the standard.

In order to simplify the selection of steels for a particular project, the fabricator or client may tabulate materials into groups or types depending on their application. The numbering system employed has not been standardised so the previously mentioned S355 G2+N may be listed as a type 2 material by one fabricator whilst a competitor lists it as a type 4. To avoid confusion, all materials should be selected in accordance with the requirements listed by the latest revision of the project specific fabrication specification.

Appendix II – Welding

Welding is an extremely complex subject and the preserve of the Welding Engineer. However, an appreciation of the more significant specifications and terms used by the industry will greatly enhance the reader's understanding of construction activities.

1. Terminology

QA/QC – Quality assurance and quality control are not welding terms in themselves but wherever fabrication and welding are carried out, there will almost certainly be a Quality System in evidence. A Quality Assurance System in accordance with the British Standard 5750 (or an equivalent such as ISO 9000) is required to ensure that the end product is manufactured in accordance with the specification.

WP – The purpose of the Weld Procedure test is to prove the compatibility of the welding consumables with the construction materials, to confirm the suitability of the welding process and equipment and to establish the parameters for items such as the geometry of the weld preparation, preheat/interpass temperatures and heat inputs. On completion of welding the test piece is subjected to both non-destructive and destructive examinations in accordance with the specification to which approval is sought. Whilst some specifications such as AWS D1.1 *Structural Welding Code* permit the use of pre-qualified procedures, this practice tends to be frowned on by the offshore industry. The weld procedures are invariably qualified prior to commencement of construction using samples of materials and consumables that will eventually be used in production.

The American Petroleum Institute publication API RP 1104 – *Standard for Welding of Pipelines and Related Facilities,* is a self-contained reference document which deals specifically with welding procedures. It provides a step-by-step guide on how the tests should be conducted, tested and documented and is highly recommended reading. The AWS D1.1 – *Structural Welding Code* is also very informative.

WPQR – The Welding Procedure Qualification Record exists as a permanent record of the welding procedure test and contains all the relevant information categorised as either essential or non-essential variables, and includes the non-destructive and destructive test records.

WPS – The Weld Procedure Specification is prepared from the WPQR and provides the welder with specific guidance on how to approach a particular weld. It includes information pertaining to the size and type of electrode, the number of runs, the voltage, amperage and polarity to be used. A number of WPSs may be prepared within the limits of the essential variables permitted by the WPQR to cover such items as variations in component size, thickness and material type.

SWIS – The Site Welding Instruction Sheet is associated with EEMUA 158 *Construction Specification for Fixed Offshore Structures in the North Sea* and fulfils an identical function to the WPS.

Welder Qualification Test – To ensure that a welder possesses the necessary skills to produce a sound weld he must complete a test piece which is representative of conditions that will be encountered during production welding. Whilst the test weld must comply with a particular WPS it will normally permit the welder to weld a considerable number of similar WPSs.

Preheat – Preheat is used primarily to slow the cooling rate of a welded component in order to reduce the shrinkage stresses, prevent excess hardening and loss in ductility. It also assists in the diffusion of hydrogen and reduces the likelihood of underbead cracking. Preheat may be applied by propane gas torch, or by the attachment of electric resistance mats. The degree of preheat is governed largely by the thickness of the material and can be calculated from the Carbon Equivalent formula.

$$\text{Carbon Equivalent, Ceq} = \%C + \frac{\%Mn}{6} + \frac{\%Ni}{15} + \frac{\%Mo}{4} + \frac{\%Cr}{4} + \frac{\%Cu}{13}$$

The approximate preheat temperatures based on carbon equivalent values are listed below but in all cases the fabrication specification should be consulted to establish the exact requirements.

up to 0.45%	preheat optional
0.45% to 0.60%	95°C to 205°C (200°F to 400°F)
0.60% and above	205°C to 370°C (400°F to 700°F)

Interpass temperature – In a multi-pass weld, the temperature of the weld metal prior to commencement of the next pass is referred to as the interpass temperature and it must be controlled within pre-determined limits if weld defects are to be avoided.

Post-Weld Heat Treatment (PWHT) – Post-weld heat treatment or stress relieving is carried out to reduce the stresses caused by welding. For general fabrication work it is only required when material thickness exceeds 30–40 mm but for some process pipework it may be required for material thickness as thin as ¾ inch (19 mm).

Typically it involves heating the component to a temperature of 590°C to 650°C (1100°F to 1200°F), holding or soaking for a period of one hour per inch of material thickness, followed by controlled cooling to 400°C before cooling slowly under insulation to ambient temperature. PWHT may be effected by the attachment of electric resistance heating mats by insertion into a furnace.

As with preheat requirements, the fabrication specification should be consulted to establish the material thickness which requires stress relieving and the means by which the duration of heat application and the temperature are to be calculated.

SMAW – Submerged Manual Arc Welding using conventional stick electrodes is the most widely used of the various arc welding processes. The stick or welding rod consists of a steel wire covered with a flux designed to decompose under the heat of the arc and generate a shielding gas which will prevent the formation of oxides and nitrides in the molten weld pool.

SAW – Submerged Arc Welding, or sub-arc, is the term used to describe an automatic or semi-automatic welding process. The welding wire is fed from a reel whilst powdered flux is poured over the weld preparation to ensure that the arc remains submerged. Metal deposition may be up to 10 times that achieved with SMAW welding. The process is only suitable for downhand (flat) welding and whilst circular components can be welded, they must be rotated with the initial pass or root run being manually welded.

FCAW – Flux Cored Arc Welding is a development of SMAW welding, can be used in all positions and exists in both automatic and semi-automatic forms. The flux is contained within the core of the welding wire which is in effect a stick electrode turned inside out. The welding wire is supplied in reels and is automatically fed through a gun so that the operator can weld continuously without the inconvenience of the frequent interruptions required to change electrodes.

Figure II-1 – Manual flux cored arc welding of subsea template pile guides

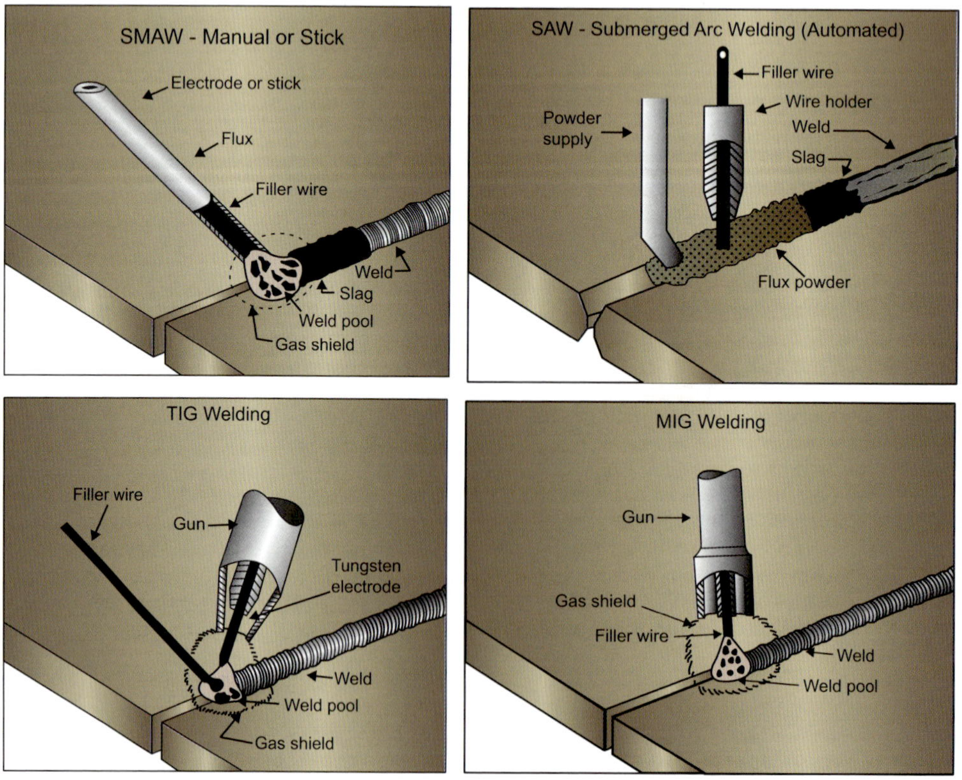

Figure II-2 – Welding processes

GMAW – Gas Metal Arc Welding is a semi-automatic welding process normally referred to as MIG (metal inert gas) and utilises either carbon dioxide, or an argon and helium gas mixture as a shielding agent in preference to a traditional flux. The gas is circulated around the welding wire which is fed from a reel. MIG welding permits high deposition rates and is suitable for use on steels, alloy steels, aluminium, copper and magnesium.

GTAW – Gas Tungsten Arc Welding is normally referred to as TIG (tungsten inert gas) welding and employs a non-combustible tungsten electrode to create an arc against the work piece whilst the filler wire is fed either manually or automatically into the molten weld pool. Protection against oxidation is provided by an argon, or argon/helium shielding gas mixture. Steels, aluminium and copper alloys can be TIG welded and, whilst a relatively slow process, it is particularly well suited to the welding of small diameter pipes where root penetration must be kept to a minimum.

Carbon Air Arc Gouging – Air arc gouging is frequently employed in preference to grinding for the removal of large quantities of weld metal prior to repairs or modifications to the weld preparation. The gouging process employs a carbon electrode to create an arc against the work piece and the subsequent pool of molten metal is removed by a jet of compressed air.

Plasma Arc Cutting – This is a relatively new process which is used primarily for the cutting out of steel components from plate in the workshop. An electric arc is generated against the work piece by a tungsten electrode, the arc subsequently being fuelled by a high velocity, super heated plasma gas stream which may be a mixture of argon and hydrogen, or nitrogen and hydrogen. The process is both quicker and more accurate than oxy-acetylene cutting and may be used on materials such as aluminium alloys in addition to ferrous materials. The finished article can normally be welded without further surface preparation.

Consumable – The term 'consumable' refers to the items consumed during the welding process, namely electrodes, filler wires, fluxes and shielding gases. With the exception of the gases, consumables are generally manufactured in accordance with AWS (American Welding Society) specifications.

2. Welding Specifications

The standards listed below represent those which will be most frequently encountered during the construction of an offshore installation and its equipment.

i) **British Standards**
BS EN 287 – Approval testing of welders for fusion welding (formerly BS 4871) (withdrawn)
BS EN 288 – Specification and approval of welding procedures for metallic materials (formerly BS 4870)
BS EN 1011-1: 1998, BS EN 1011-2: 2001 – Welding. Recommendations for welding of matallic materials. General guidance for arc welding/Arc welding of ferritic steels
EEMUA 158 – Construction specification for fixed offshore structures in the North Sea

ii) **American Standards**
AWS D1.1 – Structural welding code - steel
ANSI/ASME B31.3 – Process Piping
ASME IX – Boiler and pressure vessel code.

3. Supervision

The welding of all critical components such as pressure vessels, pressure piping and structural steelwork should be supervised by suitably qualified welding inspectors and subsequently examined by qualified NDT (non-destructive testing) technicians. The subject of supervision is dealt with in more detail in Appendix III.

For those who wish to expand on their knowledge of welding *The Procedure Book of Arc Welding* should be obtained. This moderately priced publication is produced by the Lincoln Electric Company, Cleveland Ohio, USA and is readily available in Europe. Over the years it has become a standard work of reference for welding both on and offshore and will be found sharing shelf space with the more formal specification wherever welding is carried out.

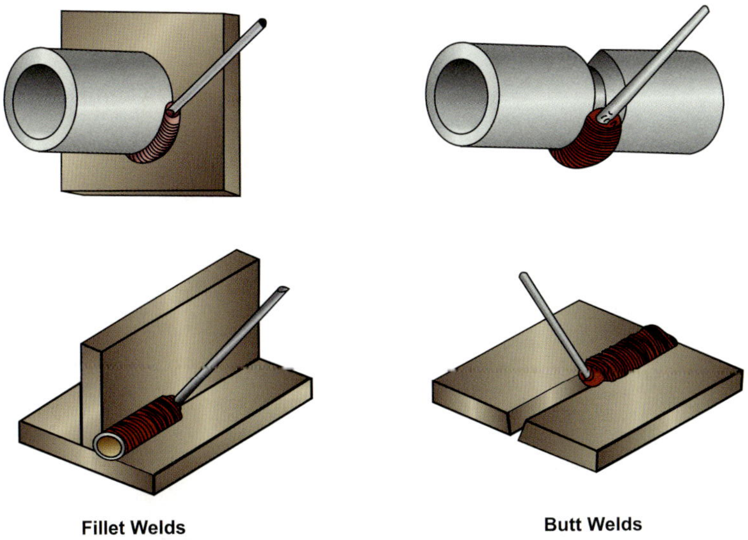

Figure II-3 – Types of welds

Appendix III – Non-Destructive Examination

Non-destructive examination (NDE), or non-destructive testing (NDT) provides the means by which components may be examined for defects during the construction of an offshore installation. The type and extent of NDE required will be determined by the fabrication specification and EEMUA 158 *Construction Specification for Fixed Offshore Structures in the North Sea* provides particularly informative reference tables on the subject.

There are a number of NDE processes, the more important of which will now be described. They can be divided into two categories, the first being used for the detection of defects on the surface of the component whilst the processes which fall into the second category are used for the detection of internal defects.

1. Surface Inspection Techniques

i) Close Visual Inspection – CVI

Close, or thorough visual inspection with the naked eye is the most basic and still the most important of all the non-destructive examination techniques and one which should never be overlooked.

ii) Magnetic Particle Inspections – MPI

MPI is the most sensitive and reliable of techniques for the detection of surface and near surface defects in ferro-magnetic materials (iron and steel). To enhance defect definition the object under examination is sprayed with white background paint prior to magnetisation with either a permanent magnet, or an electromagnet. The component is then liberally coated with particles of iron oxide in either liquid or dry powder form. The particles align themselves with any discontinuities in the magnetic field and the defects appear as a clear black line which is emphasised by the white background paint.

Note: Permanent magnets should be restricted to the examination of materials of thickness less than 6 mm (¼ inch). An AC electromagnetic yoke is preferred for thicker sections due to the more intense nature of the field it produces.

iii) Eddy Current Inspection

Like MPI, eddy current test equipment relies on the disturbing effect that defects have on a magnetic field. A test probe containing an electromagnetic coil is traversed over the article under inspection and the response signal is displayed on an oscilloscope. One of the main advantages of eddy current testing is that it can be used on painted components, although some of the offshore paint systems are particularly thick and may require removal prior to testing.

iv) Dye Penetrant Examination

Whilst not as effective as MPI or eddy current inspections, dye pen is used extensively on both ferrous and non-ferrous components. It involves the application of a red dye which is absorbed by capillary action into surface breaking defects. Surplus dye is removed prior to the application of a developer which highlights the dye retained within the defect.

2. Volumetric Inspection Techniques

i) Ultrasonic Examination – U/T

An ultrasonic examination can be carried out on most homogeneous materials and when performed by a skilled operator, it represents the most sensitive of volumetric NDT techniques. It employs an ultrasound signal to search for defects, the results generating a signal which can be interpreted on an oscilloscope. Ultrasonic testing is particularly well suited to the location of cracks and laminations in fusion welded components.

For many years radiography has been preferred to ultrasonic testing primarily because a permanent record of the examination is produced in the form of a radiograph. However, with the advances in computer power, automated ultrasonic testing equipment is now available which can produce both a pictorial report and an electronic record. Ultrasonic testing is therefore challenging the domain of radiography, being both quicker and requiring fewer safety precautions due to the absence of radiation.

ii) Radiography – R/T

Industrial radiography involves the passing of ionising radiation through an object in order to record the results on radiographic film. It is most frequently used for the examination of circumferential welds and often referred to as 'bombing'. The radiation may be generated by an X-ray machine or by a radioactive isotope such as Iridium or Cobalt. Whilst x-ray equipment generally produces superior results, particularly on thin components, isotopes are preferred for site work because they are more compact and do not require an external power source or a supply of coolant.

3. NDE Operator Qualifications

All welding should be examined by suitably qualified welding inspectors and when required, further examined by suitably qualified NDE technicians. All inspection work should be carried out in accordance with written procedures which are generally based on specifications prepared by National Standards Organisations (NSO) such as the British Standards Institute, ASME, AWS and EEMUA.

There are two universally recognised organisations which operate schemes for the assessment and certification of welding inspectors and NDT technicians, the Welding Institute in the UK and the American Society of Non-Destructive Testing (ASNT) in the USA.

The Welding Institute examines inspectors and technicians on behalf of an independent management board and issues CSWIP/PCN certificates to successful candidates. The certificates are valid for five years and are transferable from one company to another.

CSWIP – Certification scheme for weldment inspection personnel.

PCN – Personnel certification in non-destructive testing.

The ASNT scheme differs from the Welding Institute's because the qualification certificates are issued by the company that employs the technician. The qualifications are specific to the type of work carried by the issuing company and are not transferable to other companies.

SNT TC 1A – ASNT certificate categorised into three levels of qualifications.

4. Non-Destructive Examination Specifications

The British Standards (BS) listed below have formed the basis of non-destructive examinations in the UK since the offshore industry started in the 1960's. Whilst the majority of the standards have been withdrawn and replaced by the BS EN standards in recent years, the old standards are likely to be referenced for many years to come, particularly where installations are being modified and upgraded. They are also still referenced in other documents which have yet to be updated such as the British Standards relating to pressure vessels and pipeline welding.

i) British Standards

BS 2910 Methods for radiographic examination of fusion welded circumferential butt joints in steel pipes. (Withdrawn. Replaced by BS EN 1435: 1997)
BS 2600 Radiographic examination of fusion welded butt joints in steel. (withdrawn)
BS 3923 Methods for ultrasonic examination of welds. (withdrawn)
BS 5289 Code of Practice. Visual inspection of fusion welded joints. (withdrawn)
BS 6443 Method for penetrant flaw detection. (withdrawn)
BS 6072 Method for magnetic particle flaw detection. (Replaced by BS EN ISO 9934-1: 2001 but remains current)
BS EN 1435:1997 Non-destructive examination of welds. Radiographic examination of welded joints.
BS EN 1714:1998 Non-destructive testing of welded joints. Ultrasonic testing of welded joints. (withdrawn)
BS EN 970:1997 Non-destructive examination of fusion welds. Visual examination.
BS EN 571-1:1997 Non-destructive testing. Penetrant testing. General Principles.
BS EN ISO 17638:2009 Non-destructive examination of welds. Magnetic particle testing.

iii) American Standards

ASME V – Boiler and Pressure Vessel Code
Article 2 – Radiography
Article 4 – Ultrasonic inspection
Article 6 – Liquid penetrant flaw detection
Article 7 – Magnetic particle inspection
Article 8 – Eddy current testing
Article 9 – Visual inspection

AWS D1.1 – Structural Welding Code
(See Chapter 6 – Radiography/Ultrasonic.)

Appendix IV – Units of Measurements

1 tonne	= 7.5 barrels (crude oil)
1 long tonne	= 1.0165 tonnes
1 barrel	= 42 US gallons (35 imperial gallons)
	= 0.1589 cu metres

1 cubic metre	= 35.31 cubic feet
1 billion cubic metres	= 0.83 million tonnes of oil equivalent
1 cubic metre of gas	= 0.36 Therms
1 tonne of fuel oil	= 406 Therms
1 Therm	= 100 cubic feet = 1000,000 British Thermal Units

METRIC CONVERSIONS

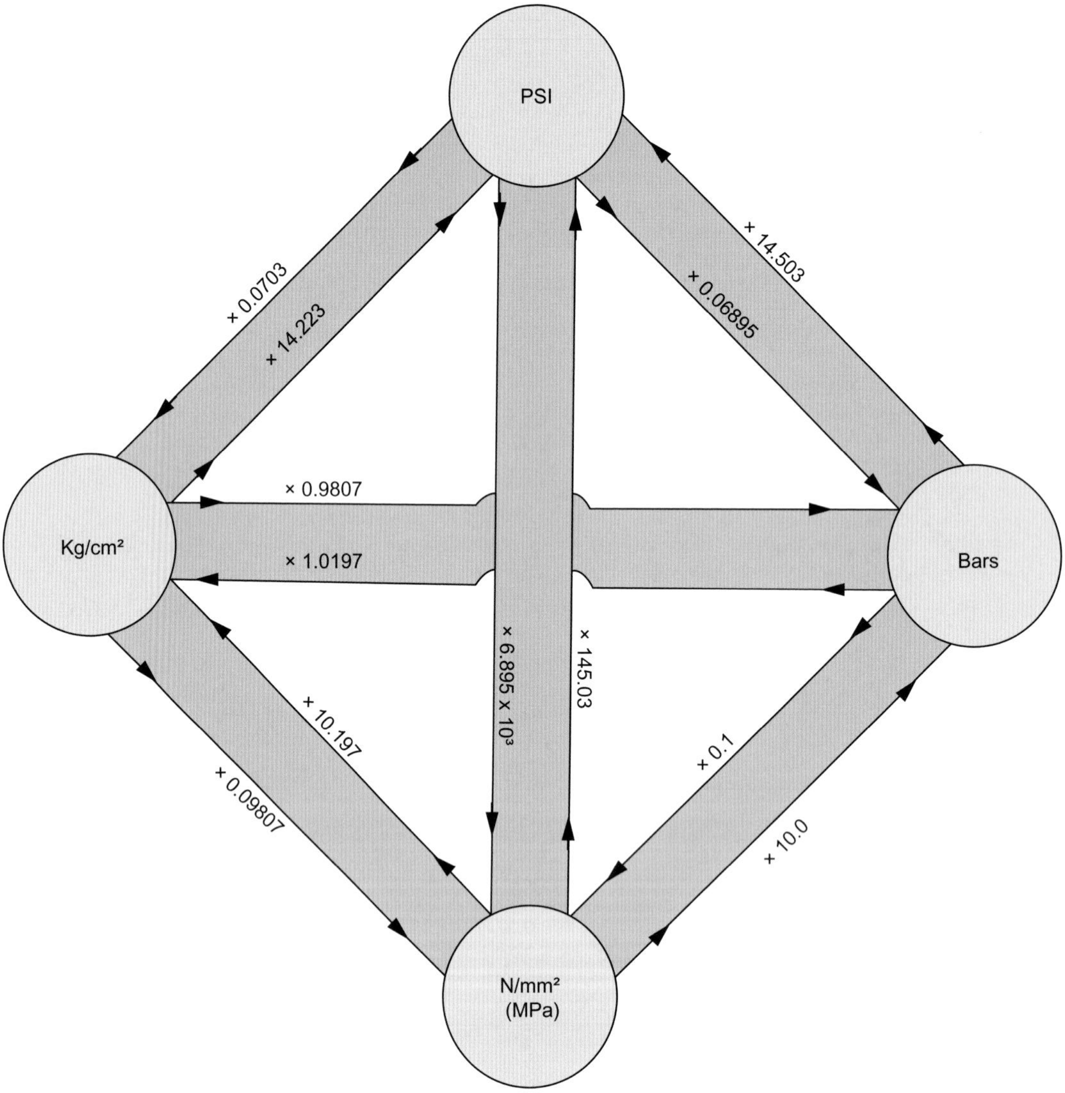

Figure IV-1 – Conversions diagram

Appendix V – Table of Line Pipe Dimensions

Platform Piping – Dimensions and Maximum Working Pressures

ASTM A106 Grade B Pipe
[Design temp. –29/204°C(–20/400°F)]

Nominal Size Inches	Outside Diameter Ins/mm	Nominal Wall Thickness Inches	Nominal Wall Thickness mm	Weight Class	Schedule No.	Working Pressure psig
2	2.375	0.218	5.34	XS	80	2489
	60.3	0.344	8.74	—	160	4618
		0.436	11.70	XXS	—	6285
2.5	2.875	0.276	7.01	XS	80	2814
	73.0	0.375	9.52	—	160	4194
		0.552	14.02	XXS	—	6850
3	3.500	0.300	7.62	XS	80	2552
	88.9	0.438	11.13	—	160	4123
		0.600	15.24	XXS	—	6090
4	4.500	0.237	6.72	STD	40	1140
	114.3	0.337	8.56	XS	80	2276
		0.438	11.13	—	120	3149
		0.531	13.49	—	160	3979
		0.674	17.12	XXS	—	5307
6	6.625	0.280	7.11	STD	40	1206
	168.3	0.432	10.97	XS	80	2062
		0.562	14.28	—	120	2817
		0.719	18.26	—	160	3760
		0.864	21.95	XXS	—	4660
8	8.625	0.322	8.18	STD	—	1098
	219.1	0.406	10.21	—	60	1457
		0.500	12.70	XS	80	1864
		0.594	15.08	—	100	2278
		0.719	18.26	—	120	2838
		0.812	20.63	—	140	3263
		0.875	22.22	XXS	—	3555
		0.906	23.00	—	160	3700
10	10.750	0.365	9.27	STD	40	1023
	273.0	0.500	12.70	XS	60	1485
		0.594	15.08	—	80	1811
		0.719	18.20	—	100	2252
		0.844	21.43	—	120	2700
		1.000	25.41	XXS	140	3271
		1.125	28.58	—	160	3737
12	12.750	0.375	9.52	STD	—	888
	323.9	0.406	10.31	—	40	976
		0.500	12.70	XS	—	1246
		0.562	14.28	—	60	1425
		0.688	17.48	—	80	1794
		0.844	21.43	—	100	2258
		1.000	25.40	XXS	120	2730
		1.125	28.58	—	140	3114
		1.312	32.00	—	160	3700

Nominal Size Inches	Outside Diameter Ins/mm	Nominal Wall Thickness Inches	mm	Weight Class	Schedule No.	Working Pressure psig
14	14.000	0.375	9.52	STD	30	807
	355.6	0.438	11.07	—	40	971
		0.500	12.70	XS	—	1132
		0.594	15.08	—	60	1379
		0.750	19.10	—	80	1794
		0.938	23.80	—	100	2304
		1.094	27.00	—	120	2734
		1.250	31.75	—	140	3171
		1.406	35.71	—	160	3616
16	16.00	0.500	12.70	XS	40	988
	406.4	0.656	16.66	—	60	1345
		0.843	21.41	—	80	1780
		1.031	26.19	—	100	2225
		1.218	30.95	—	120	2675
		1.437	36.51	—	140	3212
18	18.000	0.500	12.70	XS	—	876
	457.2	0.562	14.25	—	40	1001
		0.718	19.10	—	60	1319
		0.937	23.80	—	80	1771
		1.156	29.30	—	100	2232
		1.343	34.92	—	120	2632
20	20.000	0.325	9.52	STD	—	499
	508.8	0.500	12.27	XS	—	669
		0.812	20.62	—	60	1098
		1.031	26.19	—	80	1404
		1.280	32.52	—	100	1760
		1.500	38.10	—	120	2079
		1.750	44.50	—	140	2446
		1.968	50.00	—	160	2776
24	24.000	0.375	9.52	STD	—	415
	609.6	0.500	12.70	XS	—	556
		0.562	14.25	—	30	625
		0.687	17.48	—	40	768
		0.968	24.60	—	60	1091
		1.218	30.95	—	80	1383
		1.531	38.89	—	100	1753
		1.812	46.03	—	120	2093
		2.062	52.38	—	140	2399
		2.343	59.38	—	160	2752
30	30.000	0.375	9.52	STD		
	762.0	0.500	12.70	XS		
36	36.000	0.375	9.52	STD		
	914.4	0.500	12.70	XS		

Appendix VI – UK Legislation

Part 1. History

In terms of legislation, the offshore industry started with a clean sheet of paper and a golden opportunity to develop regulations specific to the requirements of what is a very specialised industry. In practice, the industry produced disjointed, conflicting regulations, mainly in response to disasters involving considerable loss of life. The '*Piper Alpha*' disaster in 1988 proved to be the catalyst for a radical change in the way the industry was both certified and regulated.

To fully appreciate the significance of the post *Piper* legislation and the move towards Verification, a basic understanding of the previous regulatory regime under Certification is essential to understand the shift in responsibilities that lie at the heart of the regulatory changes. An appreciation of Certification will also prove of value to readers involved with offshore installations at geographic locations, other than the UK Continental Shelf (UKCS), where governments have adopted and still operate schemes of certification similar to those originally instigated by the United Kingdom.

1.0 Initial Legislation

The first significant offshore industry legislation was enacted as a result of the 1958 Geneva Convention on the Continental Shelf, which recognised the right of countries to exploit the natural resources within their coastal waters.

The UK Government ratified the 1958 Convention agreements and incorporated them into domestic law under the Continental Shelf Act (CSA) of 1964. That same year the first gas was discovered offshore in the UK following a massive onshore discovery in Groningen in Holland. The first oil in the North Sea was discovered five years later, in 1969.

The Geneva Convention and the Continental Shelf Act may be regarded as legislation about property. Legislation pertaining to the safety of offshore installations and their personnel followed with the introduction of the Mineral Workings (Offshore Installations) Act in 1971 and the Health and Safety at Work Act (HSWA) in 1974.

1.1 Mineral Workings (Offshore Installations) Act 1971 (MWA)

The Mineral Workings Act was enacted as a consequence of the loss of 13 lives when the drilling jack-up, the '*Sea Gem*', capsized and sank in 1964. The Act empowered the Secretary of State to ensure that offshore installations were certified and remained fit for purpose in compliance with the Regulations. To enforce the requirements, the Secretary of State for Energy passed the Offshore Installations (Construction and Survey) Regulations, Statutory Instrument Number 289 in 1974, and this marked the start of certification regulations for offshore structures in the UK.

The Construction and Survey Regulations decreed that all offshore installations located in the territorial waters of the UK, be in possession of a Certificate of Fitness before producing hydrocarbons and, by this means, the requirements for certification referred to in the MWA would be met.

The Secretary of State decreed that Certifying Authorities would be established who would issue the Certificates of Fitness for offshore installations. The operational effectiveness of the Certifying Authorities was to be monitored by the Petroleum Engineering Division (PED) of the Department of Energy (DoE).

1.2 Health and Safety at Work Act (HSWA) 1974

The Heath and Safety at Work Act of 1974 marked a transformation in the approach to safety at work as it made the individual more responsible for their own safety and the safety of their working environment. It attempted to remove the considerable burden created by excessive and ineffective legislation.

The HSWA empowered the Secretary of State for Employment to provide for the occupational health, safety and welfare of individuals at their place of work. The Secretary of State created the Health and Safety Committee (HSC), whose function was to formulate policies and regulations pertaining to the Heath and Safety at Work Act. Policing of the regulations would be carried out by inspectors of the newly formed Health and Safety Executive (HSE), a body responsible to the Health and Safety Commission.

Initially, the Heath and Safety at Work Act was not applicable to offshore workers. However, in 1977 the Act was extended to include the territorial waters of the UK, which included offshore installations. The Secretary of State for Energy was made responsible for offshore safety, a responsibility that was discharged through the Petroleum Engineering Division (PED), Safety Directorate.

While the Petroleum Engineering Division monitored the offshore installations to ensure that they complied with the Heath and Safety at Work Act, they would themselves be responsible to the Health and Safety Commission. This arrangement was further complicated by the fact that the Health and Safety Commission could not introduce regulations for offshore work without consulting the Petroleum Engineering Division, and vice versa.

While the Mineral Workings Act provided ample scope for the development of legislation which reflected the specialist needs of the offshore industry, it appeared to conflict with the Heath and Safety at Work Act and created an air of confusion that would persist for nearly 15 years. The situation was not resolved until 1991 and the outcome of the '*Piper Alpha*' inquiry when all the subsequent new regulations were introduced under the Health and Safety at Work Act, rather than the Mineral Workings Act.

2.0 Certification Offshore Installations (Construction and Survey) Regulations, Statutory Instrument No. 1974/289

2.1 Certifying Authorities

While a Government department could have fulfilled the functions required of a Certifying Authority, the Secretary of State elected to make use of the existing ship Classification Societies in the role of Certifying Authority because they operate similar certification schemes within the shipping industry and it was felt that the offshore industry would benefit from their established organisational framework and experienced personnel. They were adjudged to be both competent and independent and so ideal for the task.

Six classification societies were authorised by the Secretary of State for Energy to issue Certificates of Fitness:

- Lloyd's Register of Shipping – LRS.
- Bureau Veritas – BV.
- Det Norske Veritas – DNV.
- Offshore Certification Bureau – OCB.
- American Bureau of Shipping – ABS.
- Germanischer Lloyd – GL.

2.3 Certificate of Fitness (C of F)

The point at which a Certificate of Fitness was issued normally coincided with the location of the drilling template or the installation of the jacket. To issue a Certificate of Fitness, the Certifying Authority would carry out a design appraisal of the main structure and equipment and monitor the construction, installation and commissioning of the installation.

The issue of the Certificate of Fitness was dependent on the acceptance of the installation by the Certifying Authority, the final decision resting on the shoulders of the attending surveyor as the regulations stated that the installation and its equipment would be subjected to whatever tests were required in the opinion of the surveyor. This placed the Certifying Authorities in a position of great authority and responsibility.

A Letter of Limitation and a Letter of Qualification accompanied each Certificate of Fitness. These letters enabled the Certifying Authority to highlight operational and environmental constraints, and any deficiencies in operating procedures or equipment. Typically, a limitation amounted to a restriction on the use of a particular item of equipment or operating procedure, such as a defective crane, while a qualification related to items of a less critical nature, such as a requirement to update the Statutory Operations Manual or the Planned Preventative Maintenance system.

Once issued, a Certificate of Fitness remained valid for a period of five years, subject to annual mechanical, structural and electrical inspections being carried out by the Certifying Authority.

Annual Surveys
The Annual Survey consisted of a thorough visual inspection of the installation, a review of equipment certification and maintenance records, and selective function testing of the safety systems. Changes to any item of equipment on the installation required the prior approval of the Certifying Authority.

Major Surveys
The fifth annual inspection was referred to as the major survey and had to be completed prior to renewal of the Certificate of Fitness. In addition to items examined during the annual survey, a major survey included a more detailed investigation to ensure that no significant changes to approved designs had been made. The major survey included an underwater inspection of the structure of the installation. This was, and still is, normally spread over the five year period as part of a planned maintenance programme carried out by a specialist diving contractor.

Before July 1998, the Construction and Survey Regulations, or SI 289 as they were more commonly referred to, was the single, most important piece of offshore legislation having been in existence for more than 20 years. The revocation of SI 289 heralded a further attempt by the regulators to find a suitable means of controlling an extremely hazardous working environment.

With the introduction of the Design and Construction Regulations in 1996 and the amendments therein to the Safety Case Regulations, the Certificate of Fitness, Annual and Major Surveys, and the role of the Certifying Authority disappeared as Operators, or Duty Holders, as they are referred to, being tasked with certifying their own installations. However, while the use of an approved Certifying Authority is no longer a mandatory requirement of the legislation, all the Certifying Authorities previously listed continue to provide services to the offshore industry in the new role of Independent Competent Bodies or Independent Competent Persons, a position that is clearly identified within the new regulations.

3.0 Fire Fighting and Life Saving

In addition to complying with the requirements of SI 289, regulations were set in place to cover requirements for the provision of both fire fighting equipment and life saving appliances.

3.1 Offshore Installations (Fire-Fighting Equipment) Regulations, Statutory Instrument No. 1978/611 Offshore Installations (Life-Saving Appliances) Regulations, Statutory Instrument No. 1977/486

Up until 1995, the requirements for active fire fighting equipment on offshore installations located in the territorial waters of the United Kingdom were governed by SI 611 with SI 486 covering life saving appliances. For compliance with SI 611 and SI 486, each and every piece of fire fighting equipment and life saving appliance had to be of a type approved by the Department of Transport (DoT), a duty subsequently transferred to the Maritime and Coastguard Agency (MCA). The DoT also carried out a bi-annual examination of the fire-fighting equipment and life-saving appliances, a duty subsequently transferred to the Certifying Authorities.

In 1995, SI 611 and SI 486 were revoked and replaced by the *Offshore Installations (Prevention of Fire and Explosion, and Emergency Response) Regulations, SI 1995/743,* the Regulations being referred to as PFEER. The PFEER Regulations do not require any significant changes to be made to the type of equipment previously installed under SI 611 and SI 486, but rather than dictating exactly what must be provided and what precautions must be taken as was previously the case, a fire and explosion analysis (FEA) must be undertaken. It is the results of the analysis which will ultimately dictate what equipment should be installed to detect, control and mitigate major accident events.

4.0 The Cullen Inquiry

The explosion and resulting fires which engulfed the *'Piper Alpha'* installation on the 6th July 1988 claimed the lives of 165 of the 226 persons onboard the platform and two of the crew of a rescue boat. The death toll was the highest in any accident in the history of the offshore industry and lead to the biggest shake up of the offshore safety regime in the UK for 20 years.

The public inquiry delivered a report that contained 106 recommendations and proposed far reaching changes in the way that offshore installations would be both certified and regulated.

The central recommendation of the report was that every operator or owner of an offshore installation should prepare a Safety Case, a decision influenced by the HSE's experience of regulating major hazards onshore under the Control of Industrial Major Accident Hazards Regulations 1984 (CIMAH) (SI 1984/1902). There were to be changes to the regulatory body and the progressive replacement of the existing regulations.

As a result of the Inquiry, the Offshore Safety Division (OSD) was created in 1990 as a department within the Health and Safety Executive (HSE). By the Spring of 1991, all duties pertaining to the occupational health, safety and welfare of offshore installations were transferred from the Petroleum Engineering Division (PED) of the Department of Energy, to the Offshore Safety Division of the Health and Safety Executive. The following year the first of the legislative reforms took place with the introduction of the Safety Case Regulations.

Part 2. New Legislation

The current offshore legislation was introduced in 1996 so it is perhaps a little misleading referring to it as new legislation. However, in the author's opinion, the current legislation cannot be fully understood or appreciated without a basic knowledge of the prescriptive certification regime that preceded it, so the terms 'old' and 'new' legislation have been retained.

1.0 Summary of Applicable Legislation

Health and Safety at Work Etc. Act 1974 (HSWA)
The Health and Safety at Work Act places general duties on employers to ensure, so far as is reasonably practicable, the health and safety of their employees, and others may be affected by their undertaking. These general duties are supported by the specific requirements in regulation 3 of the Management of Health and Safety at Work Regulations (SI 1992/2051) (MHSWR) for employers to undertake risk assessments for the purpose of identifying the measures which need to be put in place to prevent accidents and protect people against accidents.

Safety Case Regulations (SCR) – SI 2005/3117 (originally SI 1992/2885)
The Safety Case Regulations are the key focal point of the new regime of offshore health and safety legislation and they are underpinned by the DCR, MAR, PFEER and PUWER. Key elements of the Safety Case Regulations (SCR) are the identification of hazards with the potential to cause a major accident and the consideration of measures that will reduce the risks associated with the hazards to levels that are as low as reasonably practicable. Also, the requirement to produce a verification scheme to identify, maintain and inspect those items which are considered to be critical to the safety of the installation, these being referred to as safety-critical elements.

Design and Construction Regulations (DCR) - SI 1996/913
The objectives of the Design and Construction Regulations (DCR) are to replace the certification regime established by the Offshore Installations (Construction and Survey) Regulations 1974 (SI 289). The Regulations contain requirements for ensuring that offshore oil and gas installations and oil and gas wells are designed, constructed and kept in a sound structural state for purposes of health and safety. The Regulations deal with the integrity of installations and the provisions in it impose duties on the duty holder relating to ensuring that integrity is achieved and maintained.

Prevention of Fire and Explosion, and Emergency Response Regulations (PFEER) - SI 1995/743
The PFEER Regulations focus on identifying and preventing fire and explosion hazards and as these hazards have the potential to affect the structure, they are also relevant to complying with Design and Construction Regulations (DCR). The intention PFEER is to provide a less prescriptive form of

legislation and to encourage a more integrated approach to the hazards associated with fires and explosions. The Regulations require the duty holder to carry out an assessment, essentially a fire and explosion analysis (FEA) of the installation. The object of the assessment is to identify events which could give rise to a major accident resulting in a fire or explosion and to evaluate the likelihood and consequences of such an incident occurring. It is the results of the analysis which will ultimately dictate what equipment should be installed and what precautions should be taken to prevent such an occurrence arising and should one occur, recommend steps which will detect, control and mitigate the effects.

Management and Administration Regulations (MAR) - SI 1995/738
One of the important aspects of the Safety Case Regulations is the requirement to demonstrate that management systems are adequate to ensure compliance with relevant statutory health and safety provisions. The verification arrangements introduced into the SCR by the DCR will be an important component of such management systems.

Provision and Use of Work Equipment Regulations (PUWER) - SI 1992/2932
PUWER seeks to ensure the safe provision and use of work equipment and covers most offshore plant. Since the hazards associated with plant will, to some degree, affect an installation's structure and integrity, compliance with PUWER is expected to contribute significantly to meeting the requirements of the Design and Construction Regulations, particularly at the design stage.

Pipeline Safety Regulations (PSR) - SI 1996/825
The PSR lays down duties on pipeline operators relating to the design, construction, operation, maintenance and decommissioning of pipelines and any associated apparatus or works.

Aviation
The duty holder should ensure that the design of the installation also meets the requirements of relevant aviation legislation. Guidance on this subject may be obtained from the Civil Aviation Authority (CAA) publication CAP 437, Offshore Helicopter Landing Areas.

2.0 The Safety Case

At first glance the general format of the new regulations appears to be very similar to the old. There are still regulations covering the design and construction of the installation, pipelines, fire-fighting and lifesaving equipment, and the provision of emergency procedures, and there is a good for reason for this. While some operators thought that the whole of the offshore industry could be controlled under one set of regulations, the Safety Case Regulations, incorporating a requirement for a Formal Safety Assessment (FSA), this idea was rejected.

3.0 Goal Setting

The intention of the regulations was to provide a less prescriptive form of legislation and to encourage a more integrated approach to dealing with potential hazards. Each of the regulations provides a stated objective in the form of a General Duty and the duty holder must specify exactly how compliance with these duties will be achieved, effectively setting out what are considered to be acceptable levels of risk, and establishing performance goals. Ensuring that stated objectives are met rather than prescribing the detailed measures to be taken is referred to as goal setting.

4.0 Verification

Having installed the necessary equipment and instigated procedures that define how the goals will be achieved, the regulations require the duty holder to set in place a verification scheme for items deemed to be safety-critical elements by the Safety Case Regulations, the object being to ensure that the efficiency of the equipment and the effectiveness of the procedures are being maintained in service. The preparation of the verification scheme and the subsequent examinations are to be carried out by, or in conjunction with, an independent and competent body selected by the duty holder.

Summary
To summarise, the Safety Case Regulations set the duty holder three basic tasks;

i) To set goals by which compliance with the regulations will be achieved,
ii) to prepare a verification scheme to ensure that these goals are achieved and maintained in practice,
ii) to select an independent and competent person(s) to carry out the ongoing inspections and examinations associated with the written schemes of examination.

The regulations completely change the philosophy of offshore legislation and the operator or duty holder is in a position where they effectively self-certify their installations, rather than being obliged to employ a certifying authority to fulfil this function.

Part 3. The Safety Case

The central recommendation of Lord Cullen's report on the *'Piper Alpha'* disaster was that every operator or owner of an offshore installation should prepare a Safety Case for each installation, both fixed and mobile, old and new. The new Regulations, cited as the Offshore Installations (Safety Case) Regulations, came into force on the 30th November 1993, with older installations having until the 30th November 1995 to comply.

The Safety Case is a written document prepared by the duty holder that sets out comprehensive information on arrangements for managing health and safety, and a means of demonstrating that the hazards on an installation have been identified and assessed and that they are under control. The Regulations are accompanied by a set of Guidance Notes to assist in the interpretation of the requirements and in the preparation of the Safety Case, which is a very involved and complex task.

The format of the Safety Case is similar to the Installation Operations Manual required under the SI 289 Regulations. Both are self-contained documents designed to provide factual information such as the geographic, meteorological and oceanographic conditions that affect the installation, and include information and drawings showing the design and layout of the structure and plant. However, in addition to providing factual information, the principal aims of the Safety Case are to ensure that:

(a) The management system is adequate to ensure compliance with statutory health and safety requirements
(b) adequate arrangements have been made for audit and preparation of audits reports
(c) all hazards with the potential to cause major accidents have been identified, evaluated and means taken to reduce the risks to personnel.

It is quite clear from a) and b) that the Health and Safety Executive (HSE) place considerable emphasis on the importance of employing sound management practices to minimise the risks to personnel from working activities and the working environment. Consequently, a large part of the guidance on Regulations is dedicated to an explanation of the extremely onerous requirements of Safety Management Systems (SMS).

Subsequent amendments to the Safety Case Regulations, made by regulation 26 of the Design and Construction Regulations in 1996, introduced the requirement to verify the effectiveness of items identified as being safety-critical.

1.0 The Safety Management System (SMS)

Objectives
The term 'Safety Management System' refers to the elements in the management system that are particularly concerned with health and safety aspects, and with loss control. The duty holder's policies and aims for achieving high standards of health and safety should be described and the plans and programmes which are aimed at delivering these objectives and safety objectives should be outlined. Safety goals must be set and performance standards established, in order to provide a level against which the systems can be measured.

Furthermore, the Safety Management System should specify how the assessment of risks has been carried out and contain a summary of the methods and criteria used. A clear statement must be made on the company's policy towards the use of Quantitative Risk Assessments (QRA).

Obligations
The Safety Management System should describe the duty holder's proposals to ensure that its aims for achieving high standards of health and safety performance are communicated to all concerned. There should be clear unbroken lines of command and accountability from top management down to the lowest operational level. Clear charts must be provided to demonstrate formal acceptance of responsibility for the safety policy at Board level for UK based companies and arrangements for holding individuals accountable for their health and safety performance should be included. Responsibility to coordinate and monitor the implementation should also be allocated at this level.

High standards of health and safety performance require a positive and informed commitment from the work force and the SMS should describe the methods for securing this commitment. Safety measures must be discussed with safety representatives to assist in the development of health and safety measures and representatives should have direct access to the responsible senior company officer at Board level. Statutory requirements for formal representation of the workforce in discussions with management on health and safety matters are contained in the Offshore Installations (Safety Representatives and Safety Committees) Regulations 1989.

Line managers and supervisors should be aware of their responsibilities and formally accept them and the SMS should clearly state the responsibilities placed on the offshore installation manager (OIM). The responsibility placed on managers may not extend beyond matters which are in practice within their control, the employer remaining responsible for ensuring that managers have clear guidelines.

Procedures
The SMS should indicate the form in which detailed procedures are to be documented and made available to managers and employees and of particular importance are:

(a) Written schemes providing for the systematic examination, maintenance and, where appropriate, testing by a competent person, of all of the parts of the installation and its equipment; for the intervals at which this should be done; and arrangements to ensure that defects are suitably acted upon;
(b) permit to work (PTW) systems, specifying activities covered, the duties of all responsible persons, records required to be kept, and other procedures.

Monitoring
Monitoring, as distinct from auditing, should be carried out at prescribed intervals by personnel responsible for the operations under scrutiny, assisted as necessary by health and safety advisors. Two forms of monitoring should be used because a low incident rate is not a guarantee that risks are being effectively controlled;

(a) Active Monitoring – provides feedback on performance, before an incident occurs. Examples of active monitoring are achievements of objectives assigned to individuals or groups, periodic inspection of plant and equipment, and the monitoring of operation of the permit to work system.

(b) Reactive Monitoring – provides information on accidents, ill health, dangerous occurrences and near misses and the system for internal reporting and recording of events should be described in the SMS.

Audit
Audit is described as the structured process of collecting independent information on the efficiency, effectiveness and reliability of the Safety

Management System and for the drawing up of plans for corrective action.

Operators or duty holders must satisfy themselves by means of regular audits that the Safety Management System is being adhered to. The independent audit constitutes the final stage in the management control cycle and should be carried out by competent persons who are outside of the line management chain for the areas or activities being audited. Staff from different parts of the organisation or external consultants may be used and there should be a facility for conveying constructive recommendations to senior management with a view to making improvements in the management systems.

2.0 Risk Assessments

Risk is a subjective concept and its quantification is universally accepted as being extremely difficult to define and open to subjective judgement. It is defined in British Standard 4778 as *"the process whereby decisions are made to accept a known or assessed risk and/or the implementation of actions to reduce the consequences or probability of occurrence."*

Risk assessments may be qualitative or quantitative studies. The term 'Formal Safety Assessment (FSA)' was used extensively by Lord Cullen during the '*Piper Alpha*' Inquiry and while an FSA could be an entirely qualitative study, it would normally be a combination of both qualitative and quantitative risk assessments. A qualitative assessment often identifies the more obvious problems which can be rectified immediately, prior to commencing with the more detailed quantitative assessment.

Qualitative assessments – These tend to be based on actual operational experience, engineering standards, and sound engineering judgement.

Quantitative risk assessments – These involve calculations to assist with the identification of the risks and to determine the frequency, magnitude and consequence of hazardous events.

Quantitative Risk Assessment (QRA)
QRA has two main functions:

- To assist in the comparison of design and operating philosophy options (with a view to effective and efficient safety management)
- to demonstrate to the Regulatory Authority that specified safety standards and risk targets are being achieved (or maintained).

It was Lord Cullen's wish that quantitative risk assessments would be used extensively as a means of establishing compliance with the new regulations. Whilst the use of QRA is considered to be highly desirable, its use is only mandatory under the PFEER regulations in relation to the assessment of the survivability of the temporary safe refuge and escape routes. However, it is expected that most duty holders will use QRA extensively and where they choose not to, an acceptable alternative form of engineering assessment must be provided.

QRA Analyses
Quantitative risk assessments essentially involve an analysis of the installation, equipment and operational activities to identify any hazards which have the potential to cause a major accident such as the escape of hydrocarbons, impact with an aircraft or ship, structural failure, fire or explosion. It normally involves the following processes:

- Hazard identification, essentially to determine what could go wrong
- initiating events analysis to determine where, and how often, things could go wrong
- event tree analysis to determine what contributes to an accident
- consequence analysis to determine how severe the accident is likely to be, and what the consequences are likely to be
- risk calculation and sensitivity analysis
- risk assessment - how do risk levels compare against company criteria?
- identify practices and measures to reduce risk.

The setting of an acceptance criteria for risk provides means for evaluating the relative significance of the results obtained from the risk analysis. The acceptance criteria are arrived at by discussion, a company effectively setting what it considers to be acceptable levels of risk, as part of the goal setting process. The risk levels should compare favourably with established industry standards and criteria to show that the measures taken to prevent such occurrences happening are adequate. The Health and Safety Executive use the concept of As Low as Reasonably Practicable (ALARP) in judging the level of acceptable risk.

ALARP
The guiding principles of ALARP are:

- There is an upper limit above which risk cannot be tolerated on any grounds
- there is a lower limit below which the risk is acceptable and no further risk reducing measures are necessary
- in between these two limits there lies a region in which the benefits justify the risks. This is the ALARP region.

The main safety goal is that risks should be as low as reasonably practicable (ALARP). This means that a situation can be achieved when further improvements in safety are not required if their costs are grossly disproportionate to their benefits.

HAZOP and HAZAN
In discussions involving an investigation of hazards, the terms 'HAZOP' and 'HAZAN' will be frequently encountered and a brief explanation of these terms will clarify the subject of hazard analyses. These operations afford a means of producing on paper in a systematic and thorough fashion, and in advance of plant start up, potential hazards to the plant, process and personnel, and ultimately making recommendations to eliminate these hazards.

HAZOP – A hazard operational study involves going through an installation system by system, drawing by drawing, operation by operation to identify what potential hazards may arise. Once identified, these potential hazards are subjected to a hazard analysis.

The ALARP Principle

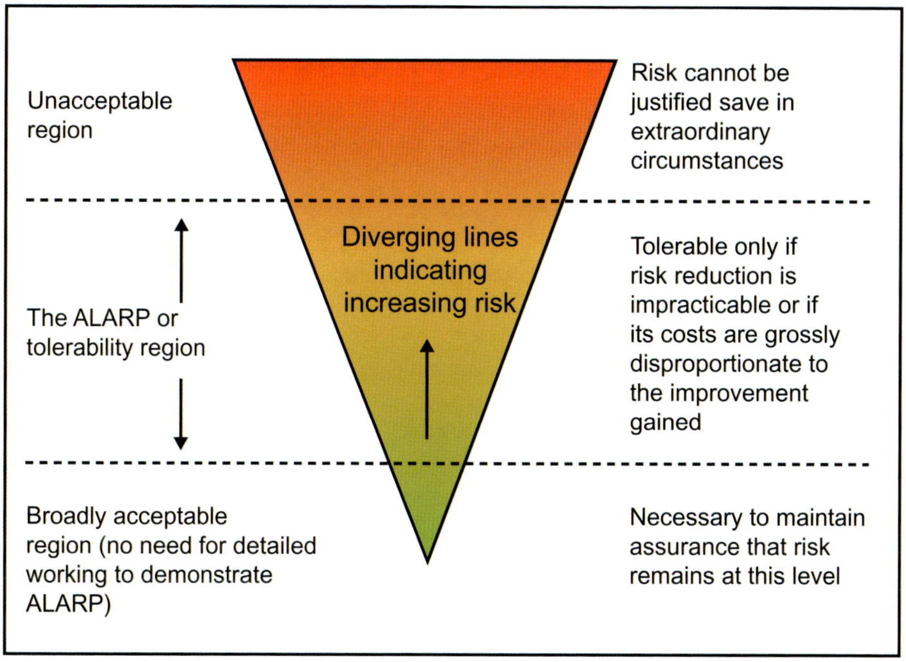

HAZAN – A hazard analysis is a technique for assessing the probability of a hazard occurring and determining the subsequent consequences for comparison with a set target, or known criteria.

3.0 Verification Schemes

Introduction
Regulation 2 of the Safety Case Regulations deals with the requirement to provide a verification scheme for safety-critical elements.

The requirements relating to verification are applicable to the following regulations:

- Safety Case Regulations (SCR) – SI 2885
- Prevention of Fire and Explosion and Emergency Response Regulations (PFEER) – SI 743
- Design and Construction Regulations 1996 – SI 913.

Objective
The overall objective of the verification scheme is to set in place independent and competent scrutiny of those parts of an installation which are critical to safety. The verification scheme should be written and provide independent checks to confirm the continuing suitability of safety-critical elements throughout the life cycle of the installation.

The verification scheme should address the items contained in Schedule 7 of the Safety Case Regulations and it should be drawn up by, or in consultation with, an independent and competent person(s) (ICP). It may be set out in a single document, or in a series of documents, and it may be recorded electronically, but information about the scheme should be readily accessible and a summary of the scheme is to be included in the Safety Case.

Matters To Be Provided For In A Verification Scheme

1. The duty holder is required to provide particulars of how the independent and competent person(s) will be selected and to specify the appropriate levels of independence.
2. An effective communication system must exist to convey information to those carrying out verification work. They must have a thorough understanding of the implications of all facets of the different stages of the installation's life cycle and how they fit together.
3. The nature and frequency of examination and testing.
4. Arrangements for review and revision of the scheme.
5. The arrangements for the making and preservation of records showing:

    - The examination and testing carried out
    - the findings
    - remedial action recommended; and
    - remedial action performed.

6. Arrangements for communicating the output of the verification process to an appropriate level in the organisation to ensure that the action required in the light of verification findings is taken.

3.1 The Verification Process

The first step, and an essential preliminary to developing a verification scheme, is the accurate identification and listing of the safety-critical elements that require verification. Performance standards must then be determined, followed by the preparation of a suitable written scheme, often referred to as a written scheme of examination, to ensure that all the verifiable items are examined to confirm that they remain suitable for their intended purpose.

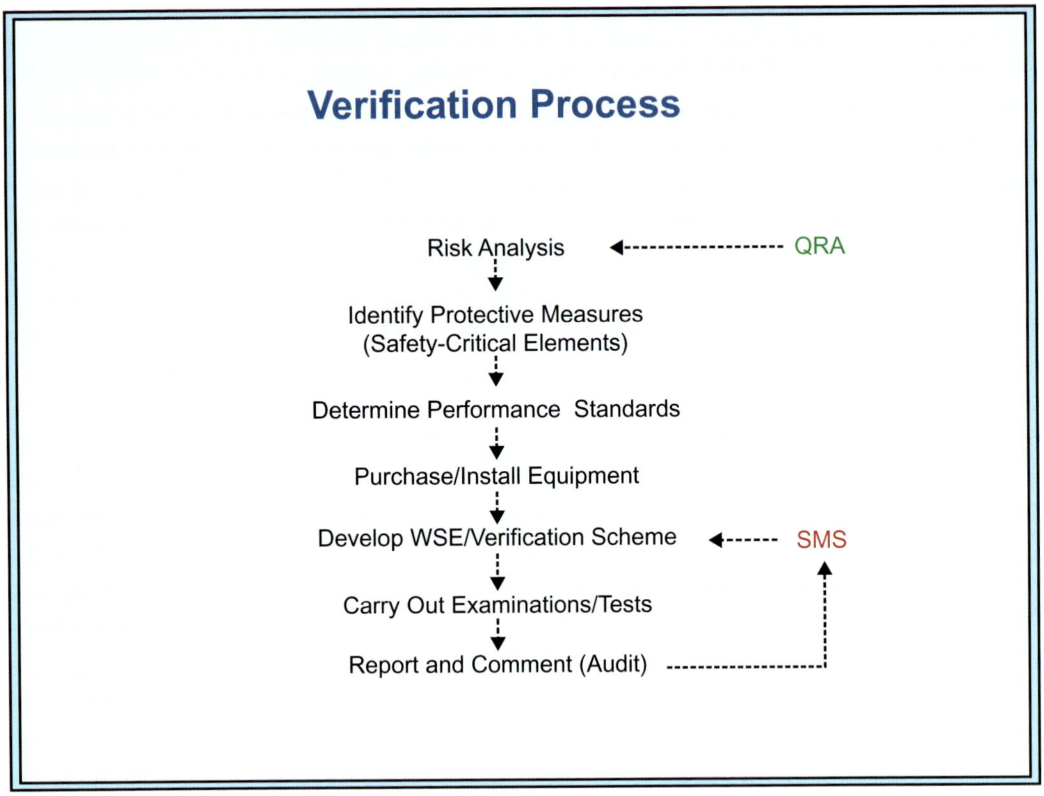

3.1.1 Safety-Critical Elements (SCE)

Safety-critical elements are defined as those structures, plant, equipment, systems (including computer software) or component parts whose failure could cause or contribute substantially to a major accident, and includes equipment which is intended to prevent or limit the effect of a major accident. The definition of a major accident event can be found in Regulation 2 of the Safety Case Regulations and in:

- A fire, explosion or the release of a dangerous substance involving death or serious personal injury
- any event involving major damage to the structure of the installation or plant
- the collision of a helicopter with the installation
- the failure of life support systems for diving operations in connection with the installation, or the detachment or trapping of a diving bell
- any other activity involving death or serious personal injury to five or more persons.

The list of safety-critical elements is likely to include most of the equipment on the platform. If we consider the definition, the jacket structure, pipeline risers and the hydrocarbon process equipment in their entirety, all are likely to cause or contribute to a major accident should they fail, so will be classed as safety-critical elements. The ESD system, fire and gas systems and fire-fighting equipment are all designed to prevent or limit the effect of a major accident, so they will also be classed as safety-critical elements. These are just examples to illustrate the potential scope of the scheme, there are many more safety-critical elements besides these.

The safety-critical elements may be established as part of the study carried out to identify hazards in the preparation of the Safety Case and in the Assessment of Major Accident Hazards, required by PFEER. The safety-critical elements may then be reviewed for completeness against a generic list of SCEs developed by specialist safety engineers for similar installations. This should ensure that there are no obvious anomalies.

Once the list of safety-critical elements is complete, it must be reviewed and commented on by the duty holder's nominated independent and competent persons. A detailed assessment by the independent and competent person of the methodology used to identify the safety-critical elements is not required. They are expected to use professional judgement, expertise and experience to reach a view on whether the list is appropriate for the installation in question. Any problems identified by the ICP which cannot be resolved with the duty holder must be listed as reservations and maintained on file.

Having identified and confirmed the completeness of the list of safety-critical elements, the next stage in the verification process is to establish levels of performance, or performance standards, with which the equipment must comply in order to fulfil its function.

Note: In addition to the items described as safety-critical elements (SCE) by Regulation 2 of the Safety Case Regulations, duty holders must now include the plant and equipment identified under Regulations 18 and 19 of the PFEER Regulations in the verification scheme.

3.1.2 Performance Standards

Performance Standards are clearly defined in PFEER:

> *A performance standard is a statement, which can be expressed in qualitative or quantitative terms of the performance required of a system, item of equipment, person or procedure, which is used as the basis for managing the hazard – e.g. planning, measuring, control or audit – through the life cycle of the installation.*

Setting performance standards is a crucial aspect of the verification process. They may be described in 'FARSI' format, ie functionality, availability, reliability survivability, and on how they interact with other systems. They should be measurable and auditable.

Using a fire pump as an example, the pump performance or functionality can clearly be specified, typically to supply a calculated volume of water at a particular pressure in order to meet the demands of the largest section of fire main and deluge. As regards the other aspects, it should have a 100% availability and operate reliably for a period of time as designated in the PFEER assessment. It should be capable of surviving in all accident situations, other than if it is destroyed by the accident itself and in order that it is not immobilised by the failure of other systems, it should be fully self-contained, ie it should not rely on interaction with other systems that could immobilise it if they fail.

Once the performance standards have been established, suitable equipment can be purchased and an examination scheme set in place to ensure that the performance of the equipment can be achieved and maintained in practice.

3.1.3 Examination Process

The arrangements in writing for systematic examination, or written scheme of examination as the examination process may be referred to, is the means by which it will be established that all verifiable items are being effectively maintained and remain suitable for their intended purpose. The scheme may be complimentary to, but not a substitute for, routine maintenance programmes.

The written scheme should detail exactly what items are to be examined and the frequency of the examinations. The examinations, which will be carried out by independent and competent persons, will involve careful and critical scrutiny of the safety-critical elements using appropriate techniques, which may include testing. The findings and any remedial action required should be clearly stated and recorded.

It should be noted that in addition to the examinations which are carried out during production, examinations are also to be carried out during design, fabrication, construction and repair. Information channels must be in place to follow design verification, fabrication and commissioning activities through into production and into ongoing verification.

3.1.4 Independent and Competent Persons

Duty holders are responsible for selecting persons to carry out verification work and they will need to justify their selection. This may be done in a number of ways, one of which would be to use a formally accredited organisation such as the National Certification Scheme for In-service Inspection Bodies.

The competence of a person should consider such matters as technical competence and experience and these persons should be available in adequate numbers. Professional judgement will be an important aspect of their work.

The independent persons must be sufficiently impartial and objective in their judgement to ensure that safety is not compromised. They should not verify their own work, and their management lines should be separate from those whose work they are checking. While it is acceptable in principle for a duty holder's in-house team to check work done elsewhere in the same organisation, it would influence objectivity if their management chain included the manager responsible for meeting production targets that might be unachievable if the plant were shut down on safety grounds.

3.1.5 Records and Review of the Scheme

The operator or owner of an installation must ensure that its verification scheme is kept at an address in Great Britain and that the records are kept for a period of six months after the scheme ceases to exist. The purpose of this requirement is to provide an auditable trail showing what work has been done, its findings, any recommendations made and any work carried out as a result.

The operator, in the case of a fixed installation, and the owner, in the case of a mobile installation, shall ensure that, as often as may be appropriate, the verification scheme is reviewed and, where necessary, revised or replaced by or in consultation with an independent and competent person. A note is made of any reservations expressed by such persons in the course of drawing it up.

3.1.6 Responsibility

Whilst the preparation of the list of safety-critical elements, the compilation of performance standards and indeed the writing of the verification scheme can be carried out by a safety engineering consultancy of the duty holder's choice, the responsibility for all the elements which constitute the scheme rests with the duty holder.

The Statutory Instruments that are directly relevant to offshore operations in the North Sea will be found in the appendices.

Part 4. Statutory Instruments

Introduction to Part Four – The more significant of the new regulations which are currently applicable to offshore installations in the UK sector of the North

Sea are covered in this Appendix. They have been dealt with on a regulation by regulation basis with the more salient points from the regulations and the various guidance notes being quoted in order to give the reader an overall view of the intent of the regulations.

1.0 Safety Case Regulations 2005 – SI 3117

Regulation 1 Citation and commencement
The Regulations may be cited as the Offshore Installations (Safety Case) Regulations 2005. They replace and revoke the previous Safety Case Regulations which originally came into force on the 31st May 1993.

Regulation 2 Interpretation
Each installation is to be provided with a Safety Case which is to be kept current throughout the life cycle of the installation.

Regulation 2 defines certain words and phrases used in the Regulations.

Some of the more relevant definitions are explained.

Major Accident means:

(a) A fire, explosion or the release of a dangerous substance involving death or serious personal injury
(b) any event involving major damage to the structure of the installation or plant
(c) the collision of a helicopter with the installation
(d) the failure of life support systems for diving operations in connection with the installation, or the detachment or trapping of a diving bell
(e) any other activity involving death or serious personal injury to five or more persons.

'Safety-critical elements' means such parts of an installation and such of its plant (including computer programmes), or any part thereof,

(a) The failure of which could cause or contribute substantially to; or
(b) a purpose of which is to prevent, or limit the effect of, a major accident.

Any reference to a verification scheme is a reference to a suitable written scheme for ensuring that the safety-critical elements:

(a) Are or, where they remain to be provided, will be suitable; and
(b) where they have been provided, remain in good repair and condition.

The verification scheme should provide independent checks to confirm the continuing suitability of safety-critical elements throughout the life-cycle of the installation. The scheme may be complimentary to, but not a substitute for, routine maintenance programmes.

The verification scheme should include means to cover:

(a) Examination, including testing where appropriate, of the safety-critical elements by independent and competent persons
(b) examination of any design, specification, certificate, CE marking or other document, marking or standard relating to those elements by such persons
(c) examination of such work in progress by such persons
(d) the taking of appropriate action following reports by such persons.

In choosing competent persons, the following need to be considered:

(a) Matters such as technical competence (suitably qualified staff in adequate numbers, relevant experience etc), and
(b) aspects such as the allocation of tasks to staff competent to undertake them and arrangements for the flow of information.

A person shall be regarded as independent only where:

(a) His function will not involve the consideration by him of an aspect, of a thing liable to be examined, for which he bears or has borne such responsibility and might compromise his objectivity; and
(b) he will be sufficiently independent of a management system, or of a part thereof, which bears or had borne any responsibility for an aspect, which he might consider, of a thing liable to be examined, to ensure that he will be objective in discharging his function.

Mobile Installation – means an installation (other than a floating production platform) which can move without major dismantling or modification, whether or not it has motive power.

Floating Production Facilities (FPS and FPSO) are considered to be fixed installations by virtue of their specific exclusion from the definitions of a mobile installation.

Quantitative Risk Assessment – means identification of hazards and the evaluation of the extent of risk arising therefrom incorporating calculations based upon the frequency and magnitude of hazardous events.

Installation – means a structure used for the exploitation or exploration of mineral resources, the storage for recovery of gas under the seabed, the conveyance of things by pipe and the provision of accommodation for people who work on or from an installation.

Construction Activity – construction activities are identified as being the installation or dismantling of the platform, or any activity which requires the use of a heavy lift vessel.

Regulation 3 Communication and storage of information by electronic means
The regulation permits the communication of information by electronic means subject to certain conditions being complied with. Firstly, the recipient must be in agreement that information can be exchanged in this manner and provide a suitable address. The recipient must also be able to read and print the information and store it in a manner where the sender can not subsequently interfere with it. All communications must be accurately dated and timed.

Regulation 4 Application
The Safety Case Regulations apply to all installations located in the tidal waters and parts of the sea in or adjacent to Great Britain up to the seaward limits of territorial waters and parts of the Continental Shelf designated by the Act of 1964.

Regulation 5 Duties of licensee
The licensee is responsible for appointing the installation operator and the operator in turn is one of the main duty holders as defined by the Safety Case Regulations. Selecting the right operator is regarded as being the key to complying with this regulation although it is still incumbent on the licensee to monitor the effectiveness of the operator. These duties fall jointly on any partners in any licence even if they are not active in pursuing the goals of a licence and they will need to agree and put in place a management system to ensure these duties are discharged.

Regulation 6 Design and relocation notifications for production installation
Duty holders are required to give appropriate consideration to matters of health and safety at the earliest stages of the design of an offshore installation so that risks can be eliminated at source. Operators are required to send to the HSE a design notification which contains a broad overview of the design of the installation and they cannot finalise the design without taking into account the HSE's comments.

Regulation 7 Safety Case for production installation
All production installations are required to be in possession of a Safety Case approved by the HSE prior to the commencement of production. The Safety Case must be submitted to the HSE for review at least 6 months prior to the proposed production commencement date.

Regulation 8 Safety Case for non-production installation
Non-production installations, such as accommodation support vessels, require a Safety Case and it must be submitted to the HSE for approval at least 3 months before arrival on station.

Regulation 9 Design notification and Safety Case for non-production installation to be converted
Where a non-production installation is to be converted into a production installation, it must comply with all the requirements of a production installation, including the submission of a design Safety Case.

Regulation 10 Notification of combined operations
A combined operations Safety Case is required where two or more installations (each, themselves required to have a Safety Case when operating remotely) interact. The combined Safety Case must be submitted at least 21 days prior to commencement of combined operations and be approved by the HSE prior to work commencing. Examples of combined operations are:

(a) A mobile installation providing accommodation to another installation
(b) a mobile drilling unit cantilevered over a fixed wellhead platform
(c) a mobile drilling unit working on a subsea well linked back to a fixed production platform.

A combined operational Safety Case is NOT required for the placing of jackets and the hook up phases of a new installation unless the installation is producing hydrocarbons. A combined operational Safety Case is also not required for vessels employed in activities connected with the installations although their activities should be covered in the Safety Case for the installation for which they are working.

The primary reason for a combined Safety Case is to demonstrate that all the risks have been considered, that the co-ordination of management activities has been addressed and that provisions are in hand for dealing with evacuation, escape and rescue in the event of an emergency. Detailed advice on the content of the combined operations Safety Case is provided in Schedule 4 of the Guidance on Regulations.

Regulation 11 Safety Case for dismantling fixed installations
The final dismantling of a fixed installation requires a specific revision of the Safety Case to take account of the special hazards and risks involved.

Regulation 12 Management of health and safety and control of major accident hazards
The principle matters to be demonstrated in Safety Cases are that:

(a) The management system is adequate to ensure compliance with statutory health and safety requirements, including the management of contractors and sub-contrators
(b) adequate arrangements have been made for audit, and the preparation of audit reports
(c) all hazards with the potential to cause a major accident have been identified, their risks evaluated, and measures taken to reduce risks to persons to as low as reasonably practicable.

The Safety Case must include details of the operator's safety management system (SMS) to ensure that the relevant statutory provisions are complied with and that all risks and hazards have been satisfactorily evaluated and measures taken to reduce the risks to persons affected by the hazards to the lowest level that is reasonably practicable.

For each identified major accident hazard, the Safety Case should contain sufficient details to show that a range of potential measures have been identified which will contribute to further risk reduction, that these measures have been fully evaluated and that the benefits of these measures have been established and their practicality.

The Safety Case must include details of the audit arrangements for the management system. The audit must be an independent and systematic assessment of the adequacy of the management system to achieve the requirements of the Regulations.

Regulation 13 Review of Safety Case
A duty holder must review the Safety Case when directed to do so by the HSE and in the absence of a specific direction to do so shall review the Safety Case after a period of five years. The purpose of the review is to ensure that the Safety Case as a whole continues to be fundamentally sound and is not simply a check to confirm that it is still up to-date. A review of the Safety Case at an interval of less than five years may be required if there has been a major change in circumstances such as a change in duty holder, a change in the main production contractor or a major change to the management system.

The review of the Safety Case should examine the basic assumptions that were made in the original submission and the installation safety representatives must participate in the review process as required by the Offshore Installations (Safety Representatives and Safety Committees) Regulations 1989. The duty holder must send a summary of the review to the HSE within a period of 28 days of undertaking the review and this must include details of the consultation process with the safety representatives. A copy of the summary is to be maintained as required by Regulation 18. The summary should be short and factual and include details of the time the review took, the identities of the persons who took part in the review, the format of the review and include a summary of all revisions made to the Safety Case as a consequence of the review.

Regulation 14 Revision of Safety Case
This regulation requires the duty holder to keep the Safety Case up to-date so that it reflects the current status of the installation and its operations. Even apparently minor changes are to be assessed and logged and all the relevant documentation updated. Careful logging of alterations is considered to be essential to demonstrate that the duty holder is operating the installation in accordance with the requirements of the Safety Case. The need for revision may be triggered by an accident, an incident, an audit or the change of duty holder or emergency procedures. Following a revision of a Safety Case it need not be submitted to the HSE for review unless the changes are significant such as a major material change on which the original Safety Case was based. It is the responsibility of the duty holder to decide what a major material change is but they can discuss this with the HSE if they are unsure.

Regulation 15 Power of Executive in relation to the Safety Case and related documents
The HSE may direct a duty holder to revise a current Safety Case. Once accepted the Safety Case remains valid until the installation ceases to exist.

Regulation 16 Duty to conform with the Safety Case
A duty holder must ensure that the installation is operated in conformity with the Safety Case and it is an offence not to do so.

Regulation 17 Notification of well operations
At least 21 days before well operations are commenced, the operator must send notification of their intentions to the HSE. Detailed advice on the notification of well operations is provided in Schedule 6 of the Guidance on Regulations.

Regulation 18 Keeping of documents
This regulation requires copies of several documents prepared for the purpose of the Safety Case to be kept on the installation, where they are accessible to safety representatives and at a nominated company address in the UK. These documents include copies of any audit reports of the Safety Case, including the findings, recommendations, actions and any agreed time scales for the implementation of such actions. It is quite acceptable to keep the documents in electronic form.

Regulation 19 Verification schemes
The duty holder of an installation must ensure that a record is made of the safety-critical elements and that this record is commented on by an independent and competent person. A verification scheme must then be drawn up based on the requirements provided in Schedule 7 of the Regulations. These requirements are discussed more fully elsewhere in this chapter.

The essential preliminary to developing an effective verification scheme is the accurate identification of the safety-critical elements. This work may be done in-house or it may be done by a suitable contractor, but the responsibility for the list rests with the duty holder. As well as the safety-critical elements, the plant specified in PFEER must be included in this scheme. The verification scheme itself should be subjected to a review by an independent competent person. The responsibility for the verification scheme rests with the duty holder.

Regulation 20 Review and revision of verification schemes
The duty holder is required to review and where necessary revise the verification scheme and this review should take place in conjunction with their independent competent person. They are to make note of any reservations raised by the ICP during the course of any review. A review of the scheme with the ICP should take place annually where consideration should be given to any changes to safety-critical elements.

Regulation 21 Continuing effect of verification schemes
The duty holder is to ensure that the verification scheme continues to be effective whilst the installation remains in being.

Regulations 22 Defence
In any proceedings taken against a person for failure to comply with Regulations 19 to 21, it shall be a defence for the person charged to prove that the commission of the offence was due to the action or default of another person not being one of his employees and that he took all reasonable precautions, and exercised all due diligence, to avoid the commission of the offence.

Regulations 23 Exemptions
The HSE may exempt by a certificate in writing any person or installation from the provisions of the Safety Case Regulations, provided that they are satisfied that the health and safety of persons who are likely to be affected by the exemption will not be prejudiced in consequence of it and that such an exemption would not conflict with EU health and safety laws. The granting of an exemption, whilst possible, is considered to be extremely unlikely.

Regulation 24 Appeals

There may be occasions when the duty holder does not agree with a decision made by the HSE and wishes to challenge it, for example if HSE does not accept a Safety Case or a revision to it. This regulation allows the duty holder to appeal directly to the Secretary of State.

Regulation 25 Amendments

These regulations amend regulation 19 of the PFEER Regulations. The requirement to provide written schemes for the systematic examination of safety equipment was originally introduced by the PFEER Regulations. This revision to the Safety Case Regulations removes the requirement for written schemes from the PFEER Regulations, the written schemes now falling under the heading of the Verification Scheme, Regulation 19, although the intent remains the same, that is, to ensure that equipment is examined by a competent and independent person(s).

Explanatory Note

(This note is not part of the Regulations)

1. These Regulations replace the Offshore Installations (Safety Case) Regulations 1992 (S.I. 1992/2885) ("the 1992 Regulations") to provide for the preparation of safety cases for offshore installations and for the notification of specified activities to the Health and Safety Executive ("the Executive").

2. The 1992 Regulations implemented Article 3(2) of Council Directive 92/91/EEC (O.J. No. L348, 28.11.92, p.9) concerning the minimum requirements for improving the safety and health protection of workers in the mineral-extracting industries through drilling. These Regulations continue to provide for the implementation of this provision in Great Britain.

3. A safety case is defined in the Regulations as a document containing specified information relating to the management of health and safety and the control of major accident hazards and containing the particulars specified in the Schedule referred to in the provision of the Regulations under which it is prepared *(regulations 2(2) and 12)*.

4. The Regulations—

(a) require a licensee to ensure that any operator he appoints is capable of carrying out his functions and discharging his duties satisfactorily *(regulation 5)*;
(b) require an operator to prepare and send to the Executive a design notification for a production installation which is to be established *(regulation 6(1))* and a relocation notification for a production installation that is to be moved to a new location *(regulation 6(2))*;
(c) prohibit the operation of a production installation unless a safety case has been sent to and accepted by the Executive *(regulation 7 and Schedule 2)*;
(d) prohibit the movement of a non-production installation in relevant waters (as defined in regulation 2(1)) with a view to its being operated there unless a safety case has been sent to and accepted by the Executive *(regulation 8 and Schedule 3)*;
(e) require a design notification to be sent to the Executive in respect of the conversion of a non-production installation to a production installation *(regulation 9(1))* and prohibit the operation of a converted installation unless a safety case has been sent to and accepted by the Executive *(regulation 9(5))*;
(f) prohibit the engagement of an installation in a combined operation with another unless a notification has been sent to the Executive *(regulation 10 and Schedule 4)*;
(g) prohibit the dismantling of a fixed installation unless a revised safety case has been sent to and accepted by the Executive *(regulation 11 and Schedule 5)*;
(h) require a safety case to be reviewed when directed by the Executive and at intervals of 5 years *(regulation 13)*;
(i) require a safety case to be revised when appropriate and when directed by the Executive *(regulation 14)*;
(j) grant to the Executive powers in respect of safety cases and related documents *(regulation 15)*;
(k) require any procedures or arrangements in safety cases to be followed and provide for specified defences for contravention of the requirement *(regulation 16)*;
(l) prohibit the commencement of a well operation unless a notification has been sent to the Executive *(regulation 17)*;
(m) impose requirements with respect to the making and keeping of documents *(regulation 18)*;
(n) impose requirements with respect to the creation, revision and continuing effect of a verification scheme in respect of an installation and provide a defence for contravention of the requirements *(regulations 19 to 22)*;
(o) provide for the granting of exemptions from the Regulations by the Executive *(regulation 23)*;
(p) provide for an appeal to the Secretary of State against certain decisions of the Executive *(regulation 24)*;
(q) contain transitional provisions in respect of activities in existence or commenced within a specified time of the coming into force of the Regulations *(regulation 27)*.

5. A copy of the regulatory impact assessment prepared in respect of these Regulations can be obtained from the Health and Safety Executive, Rose Court, 2 Southwark Bridge, London SE1 9HS. A copy has been placed in the Library of each House of Parliament.

2.0 Prevention of Fire and Explosion, and Emergency Response Regulations 1995 – SI 743

Introduction – The PFEER Regulations apply to fixed and floating production installations and to accommodation vessels in support of such operations. They also apply to drill ships and jack-ups involved in exploration, production or accommodation support. The PFEER Regulations do not apply to diving support vessels. The definition of installations is taken from the Management and Administration Regulations (MAR), SI 1995/738.

Regulation 1 Citation and Commencement
These Regulation may be cited as the Offshore Installations (Prevention of Fire and Explosion, and Emergency Response) Regulation 1995 and shall come into force on the 20th June 1995.

Regulation 2 Interpretation
Major accident means:

(a) A fire, explosion or other release of a dangerous substance involving death or serious personal injury to persons on the installation or engaged in an activity on or in connection with it
(b) any event involving major damage to the structure of the installation or plant affixed thereto or any loss in the stability of the installation
(c) the collision of a helicopter with the installation
(d) the failure of life support systems for diving operations in connection with the installation, the detachment of a diving bell used for such operations or the trapping of a diver in a diving bell or other subsea chamber used for such operation; or
(e) any event arising form a work activity involving death or serious personal injury to five or more persons on the installation or engaged in an activity in connection with it.

Regulation 3 Application
This regulation applies to installations located in the territorial waters of Great Britain and the UK Continental Shelf.

Regulation 4 General Duty
(1) The duty holder shall take appropriate measures with a view to:

(a) Protecting persons on the installation from fire and explosion; and
(b) securing effective emergency response.

The purpose of the regulations overall is to promote a risk-based, systematic approach to managing fire and explosion hazards and emergency response. The general principles of risk control and health and safety management set out in the Approved Code of Practice on MHSWR are relevant to this process and should be followed. They are summarised as follows:

Principles for preventative and protective measures
- It is always best if possible to avoid a risk altogether
- combat risks at source, rather than by palliative measures
- wherever possible adapt work to the individual
- take advantage of technological and technical progress to improve work methods and make them safer
- risk prevention measures should form part of a coherent policy and approach to reduce risks progressively that cannot be prevented or avoided altogether
- give priority to measures which protect the whole workplace
- make sure workers understand what they need to do
- create an active health and safety culture which affects the organisation as a whole.

Principles for health and safety management
- **Planning.** Adopting a systematic approach which identifies priorities and sets objectives. Whenever possible, risks are eliminated by the careful selection and design of facilities, equipment and processes or minimised by the use of physical control measures.

- **Organisation.** Putting in place the necessary structure with the aim of ensuring that there is a progressive improvement in health and safety performance.

- **Control.** Ensuring that the decisions for ensuring and promoting health and safety are being implemented as planned.

- **Monitoring and Review.** Like quality, progressive improvement in health and safety can only be achieved through the constant development of policies, approaches to implementation and techniques or risk control.

These principles should be applied for fire and explosion hazard management and emergency response by:

- Developing an approach to fire and explosion hazard management based on hazard identification and risk assessment, involving a balanced and integrated approach to designing out hazards, preventing hazardous events occurring, detecting that they have occurred, are occurring, controlling their escalation, and mitigating their consequences
- a systematic approach to every response by considering the various stages of an emergency and how each of those stages can be most effectively managed, to reduce risks to health and safety.

Regulation 5 Assessment
The duty holder must perform an assessment to identify which events could give rise to a major accident involving fire or explosion, or the need to evacuate the installation. The assessment must establish appropriate performance standards to be met by the protective measures introduced. All requirements needed to comply with the rest of the regulations will be determined by the results of the assessment.

Regulation 6 Preparation for Emergencies
This regulation reflects the importance of adequate preparation for emergencies, including training and deals mainly with the organisation of personnel. One person, the offshore installation manager (OIM), should be given responsibility for taking overall charge in an emergency and the roles and responsibilities of those in the command structure should be clearly defined and displayed on an emergency duties plan.

Appropriate instruction should be given to all persons on the installation on or before arrival at the installation.

Regulation 7 Equipment for Helicopter Emergencies
The duty holder must ensure that equipment is provided that can be used to deal with an accident involving a helicopter, including crash rescue equipment. These arrangements are in addition to those provided for dealing with a fire or explosion which are covered under Regulations 9, 12 and 13.

Regulation 8 Emergency Response Plan
This regulation requires the duty holder to prepare an emergency response plan, which documents the organisation and arrangements for dealing with an emergency on the installation. The plan should contain procedures which cover all reasonably foreseeable emergencies, ie it should set out who does what, when, where, how and to what effect.

The plan should cover both the offshore and onshore arrangements and take account of the views of those who are likely to become involved, such as HM Coastguard and the pipeline owners and operators. The plan should be exercised and tested with sufficient frequency and depth so that it can be relied upon to work effectively in an emergency, and should be monitored and reviewed in the light of these exercises and tests.

Regulation 9 Prevention of Fire and Explosion
This regulation requires the duty holder to take appropriate measures to prevent fire and explosion. Good design is the most effective means of preventing fire and explosion and this involves addressing issues such as the choice of process and production methods to minimise the risk of fire and explosion by optimising plant layout.

Measures to control activities which might lead to a release or ignition hazard should include adequate procedures and arrangements for the control of hot work and the use of electrical equipment not designed for use in areas of flammable and explosive atmospheres, ie a permit to work system.

Regulation 10 Detection of Incidents
Detection measures need to be provided for the full range of reasonably foreseeable events which require emergency response with a view to enabling information regarding such incidents to be conveyed forthwith to places from which control action can be instigated. These measures should cover:

(i) Fire detection
(ii) Flammable or toxic gas detection
(iii) Detection of leakages of flammable liquids
(iv) Man overboard
(v) Monitoring of ships and supply boat operations
(vi) Monitoring of helicopter flights
(vii) Monitoring of diving operations.

Detection measures for major accident hazards should be based on the findings of the assessment and should be automatic in operation, where this is reasonably practicable, as should the transmission of information from detection systems to the point at which control action is instigated. Where detection and the relay of information cannot be done automatically, adequate arrangements should be in place to detect incidents and to instigate control action.

Regulation 11 Communication
The duty holder shall make appropriate arrangements to give an audible warning (and visual where necessary) of an emergency, to all persons on the installation and for the purpose of emergency response, for communication between persons not on the installation but engaged in activities in connection with it and beyond it such as diving operations. These arrangements are to remain effective in an emergency.

All arrangements should be based on the findings of the assessment required by Regulation 5 for major accident hazards and should take into account the findings of the risk assessment required by Regulation 3 of MHSWR for non-major accident hazards.

The alarm and communications systems provided should be capable of transmitting clear information to personnel wherever they are likely to be on the installation. Where the method of raising the alarm does not automatically initiate a general visual or acoustic alarm, there should be clear procedures for instituting the appropriate alarm and conveying information to the relevant personnel. This might include HM Coastguard, other installations, shore-based personnel and rescue and recovery services.

Duty holders should ensure that there are arrangements in place to alert personnel to an incident on the installation and that everyone is aware of the meaning of the different alarm signals. This should be included in installation induction and refresher training. There should be:

(i) Red flashing lights to provide warning against a release of toxic gas.
(ii) Yellow flashing lights for all other warning signals.
(iii) Acoustic signals provided for evacuation:
  a) continuous signal of variable frequency;
  b) continuous signal of constant frequency for toxic gas;
  c) intermittent signal constant frequency for all other cases.

Regulation 12 Control of Emergencies
Duty holders must have appropriate control measures to limit escalation of an emergency and this includes measures to monitor the extent of an emergency for the purpose of exercising managerial command and control.

Measures should be provided for the timely and effective shutdown of systems which could exacerbate an emergency and for the initiation of emergency shutdown systems and plant used to limit the spread of fire. These systems should be operable from a safe location such as the control point or TR which should be staffed at all times. These measures will include plant, equipment and workplace procedures and examples are:

(i) Emergency response procedures.
(ii) ESD systems.
(iii) Vents and drains.
(iv) Automatic isolation valves.
(v) Blow-down and flare systems.
(vi) Portable fire-fighting equipment.

There also specific requirements in respect of emergency shutdown valves in the Pipeline Safety Regulations (PSR), SI 1996/825.

Regulation 13 Mitigation of Fire and Explosion
Based on the findings of the assessment, the duty holder must put in place measures that will mitigate the effects of fire and explosion and the measures are likely to include:

(i) Deluge systems.
(ii) Fixed extinguishing systems.
(iii) Fire-resistant coatings.
(iv) Manual response equipment and procedures.
(v) Ventilation control systems.
(vi) Fire and blast walls.

The measures taken should provide for automatic systems unless manual operation can be justified in the assessment.

In determining what portable equipment is to be provided, duty holders should take into account the numbers and location of personnel on the installation, the practicality of its effective use, the hazards and the availability of other systems.

Regulation 14 Muster Areas etc.
The regulation requires the duty holder to make provision for clearly defined muster areas where personnel on the installation can assemble safely while the emergency is assessed and control action taken. It also requires procedures to be in place for accounting for people and provision for personnel to access safely means for leaving the installation if necessary.

Muster points should remain useable and egress routes passable based on the scenarios identified in the assessment and the emergency response plan and appropriate communications facilities should be provided.

Adequate emergency lighting should include, where appropriate, floor level illumination and direction of escape indicators.

Personnel should be provided with appropriate information about the location of their muster station.

Regulation 15 Arrangements for Evacuation
The duty holder shall ensure that such arrangements are made as will ensure, so far as is reasonably practicable, the safe evacuation of all persons to a place of safety. Evacuation in this context means by helicopter, direct sea transfer or by lifeboat, ie without getting wet. It does not include liferafts or methods which involve escape directly into the sea. These are considered to be a last resort and are covered in Regulation 16.

There are a number of means of evacuation and the preferred means should be the normal means of getting persons to and from the installation, that is by helicopter. Alternative means must also be provided for when the normal means fail and in most cases, lifeboats (TEMPSC – totally enclosed motor propelled safety craft) provide the alternative means. There should be sufficient places provided by TEMPSC for 150% of the persons onboard unless an alternative standard is justified.

Regulation 16 Means of Escape
Escape means leaving an installation in an emergency when the evacuation arrangements have failed due to a catastrophic incident when a planned and orderly evacuation cannot be achieved. Means are to be provided for descent to the sea such as ladders and stairways, davit launched liferafts, chute systems, and individually controlled descent devices; and items in which personnel can float on reaching the sea such as throw-over liferafts. Preference is to be given to devices which provide some protection from the elements and avoid the need to enter the sea directly. Entering the sea directly is described as a last resort and should be avoided if at all possible.

Regulation 17 Arrangements for Recovery and Rescue
This regulation deals with arrangements which must be set in place for the recovery of persons following their evacuation or escape from the installation and for their subsequent safety. It involves the recovery of individuals who have fallen into the sea or who have ditched into the sea following a helicopter incident.

There are many circumstances for which only a suitable vessel standing by will provide effective arrangements and in these circumstances such a vessel will need to be provided.

Regulation 18 Suitability of Personal Protective Equipment for Use in an Emergency
This regulation deals with the requirement to supply personal protective equipment and to ensure that there is, prepared and operated, a written scheme for the systematic examination and testing by a competent person of the equipment.

Personal protective equipment must be designed to increase the chances of survival of an individual in an emergency by enabling them to reach muster areas, the temporary refuge, evacuation or escape points. The equipment should provide protection against the effects of fire, heat, smoke, fumes or toxic gas and in the event of immersion in the sea. A set of protective equipment should be located in the accommodation for every individual and should include a smoke hood, portable light source and heat proof gloves, survival suit and lifejacket. Additional equipment should be provided at appropriate locations on the installation.

There should also be appropriate personal protective equipment for use by those with specific emergency duties such as for helicopter duties and fire-fighting. The provision of personal protective equipment for use in major accident hazards should be based on the findings of the assessment required by Regulation 5, and for non-major accident hazards on the assessment required by Regulation 3 of MHSWR.

The regulations do not cover clothes or equipment required for normal working activities.

Regulation 19 Suitability and Condition of Plant
All plant and equipment installed to comply with the PFEER Regulations 7 to 19 must be suitable by design, construction or adaptation for the actual work it is provided to do. Performance standards determined in the assessment required by Regulation 5 should be the basis for assessing that

plant required to deal with major accident hazards is suitable for its purpose. They may be described in terms of (FARSI) functionality, availability, reliability survivability and on how they interact with other systems. They should be measurable and auditable.

The requirement to provide written schemes for the systematic examination of safety equipment was originally introduced in the PFEER Regulations but in the revision to the Safety Case Regulations in 2005, the requirement for written schemes was transferred from the PFEER Regulations to the Safety Case Regulations. The written schemes now fall under the heading of the Verification Scheme, Regulation 19, although the intent remains the same, that is, to ensure that equipment is examined by a competent and independent person(s).

The written schemes originally replaced the examination of lifesaving appliances and fire-fighting equipment required under previous law; the scope of the scheme is similar, but now the duty holder is responsible for preparing the scheme. The written schemes are in addition to any schemes required as part of the routine maintenance and testing programmes.

The written scheme is designed to provide a careful and critical scrutiny to access the suitability of the plant for its purpose. It may be set out in a single document, or a series of documents or recorded electronically, but information about the scheme should be readily accessible.

Regulation 20 Lifesaving Appliances
All survival craft, liferafts, lifebuoys, lifejackets and plant are to be suitably marked and equipped.

Regulation 21 Information Regarding Plant
Personnel are to be advised of the location of manual fire-fighting equipment, lifesaving appliances and personal protective equipment, and hazardous areas. This may be by plans, signs, etc.

3.0 Design and Construction, etc Regulations 1996 – SI 913

Introduction – The objectives of the Design and Construction Regulations (DCR) are to replace the certification regime established by the Offshore Installations (Construction and Survey) Regulations 1974 (SI 289). They came into force on the 30th June 1996 and apply to both fixed and mobile installations. In doing so, the Regulations also implement the relevant health and safety aspects of the Extractive Industries (Borehole) Directive (EID)(92/91/EEC), and support and complement the Offshore Installations (Safety Case) Regulations 1992 (SI 1992/2885)(SCR), particularly the requirement to provide a verification scheme for the safety-critical elements of an installation.

The Regulations contain requirements for ensuring that offshore oil and gas installations, and oil and gas wells are designed, constructed and kept in a sound structural state, and other requirements affecting them, for purposes of health and safety. Part II of the Regulations deals with the integrity of installations and the provisions in it impose duties on the duty holder relating to:

- The maintenance of its integrity at all times
- its design
- work to it
- the way it is operated
- arrangements for maintaining its integrity
- the reporting of danger to it
- decommissioning and dismantlement.

Part III contains requirements relating to the helicopter landing area and Part IV imposes requirements affecting oil and gas wells and covers such areas as well design, well control, well examination and information, instruction, training and supervision of operatives.

Part V deals mainly with extremely important changes to the Safety Case Regulations, in particular the requirement to provide a verification scheme for safety-critical elements of an installation.

Part I – Interpretation and General

Regulation 1 Citation and Commencement
These Regulations may be cited as the Offshore Installations and Wells (Design and Construction etc,) Regulations 1996 and shall come into force on the 30 June 1996.

Regulation 2 Interpretation
The definition of an installation in DCR differs from the definition given in the MAR and PSR regulations. Any well connected to the installation, and any pipeline or equipment connected to the pipeline (within or without 500 metres of the main structure) of the installation, is excluded from the DCR. This is to avoid duplication with Part IV of DCR – Wells, which covers wells separately and explicitly, and PSR which covers pipelines and associated equipment up to and including the 500 metres adjacent to the installation.

Regulation 3 Application
The Regulations apply to all installations (fixed and mobile) located in the relevant waters of the UK and to all installations covered by inter-governmental treaties.

Part II – Integrity of Installations

Regulation 4 General Duty
The duty holder shall ensure that an installation possesses such integrity as is reasonably practical. Integrity means structural soundness and strength, stability and, in the case of a floating installation, buoyancy in so far as they are relevant to the health and safety of persons.

Regulation 5 Design of an Installation
The duty holder shall ensure that the designs to which an installation are constructed are such that:

- It can with stand such forces acting on it as is reasonably practical
- its layout and configuration, including those of its plant, will not prejudice its integrity
- fabrication, transportation, construction, commissioning, operation, modification, maintenance and repair of the installation may proceed without prejudicing its integrity
- it may be decommissioned and dismantled safely; and
- in the event of reasonably foreseeable damage to the installation, it will retain sufficient integrity to enable action to be taken to safeguard the health and safety of persons on or near it.

The phrase "*capable of withstanding reasonably foreseeable forces*" has been used to ensure that the design takes appropriate account of forces imposed by environmental conditions, the weight of equipment that may be placed on it, the activities in connection with it, such as vessel or aircraft impact and accidents, fire and explosion.

The greatest scope for risk reduction is at the design stage and the layout and configuration of the plant should be optimised to assist in reducing the risks to the installation's integrity.

Regulation 6 Work to an Installation
The duty holder shall ensure that the work of fabrication, construction, commissioning, modification, maintenance and repair of an installation and activity in preparation for its positioning are carried out in such a way that, so far as is reasonably practicable, its integrity is secured. The duty holder may contribute to meeting the requirements of this regulation through contractual agreements with various parties involved in the processes.

Regulation 7 Operation of an Installation
The duty holder must ensure that the installation is operated in a manner which does not prejudice its integrity. This regulation refers primarily to the meteorological and environmental conditions and is most relevant to mobile installations.

Regulation 8 Maintenance of Integrity
This regulation requires the duty holder to make suitable arrangements to ensure that the integrity of an installation is maintained throughout its lifecycle. Such arrangements include planned maintenance, inspection of structures and periodic assessment of the installation and the frequency, scope and method of inspection should be sufficient to provide assurance that the integrity of the installation is being maintained.

Regulation 9 Reporting of Danger to an Installation
The duty holder shall ensure that within 10 days after the appearance of evidence of a significant threat to the integrity of an installation, a report is made to the HSE. Examples of what may be reportable are the identification of defects in the structure which could be a threat to the integrity of the installation.

Regulation 10 Decommissioning and Dismantlement
The duty holder shall ensure that an installation is decommissioned and dismantled safely.

Part III – Further Requirements Relating to Installations

Regulation 11 Helicopter Landing Area
The duty holder shall ensure that the helicopter landing area is adequate for its purpose and meets the requirements of relevant aviation legislation. Further information on offshore helicopter landing areas may be obtained from the Civil Aviation Authority (CAA) guidance notes CAP 437.

Regulation 12 Additional Requirements
The duty holder shall ensure that the additional requirements set out in the Extractive Industries (Boreholes) Directive (EID) 92/91/EEC are complied with by 3rd November 1999. The Schedule deals primarily with the health and safety aspects likely to be encountered on an installation such as the temperature requirements for work and rest places, ventilation, lighting, sanitation arrangements and accommodation. The requirements are broadly equivalent to the onshore Workplace (Health, Safety and Welfare) Regulations 1992 (SI 1992/3004). Sleeping quarters should contain no more than two persons.

Part IV – Wells

Introduction – The well provisions of the DCR Regulations apply to wells drilled onshore and offshore and the Regulations replace SI 1980/1759, the Offshore Installations (Well Control) Regulations 1980, and SI 1991/308, the Offshore Installations (Well Control)(Amendment) Regulations 1991, and Regulation 4 and Schedule 15 of SI 1976/1019, the Offshore Installations (Operational Safety, Health and Welfare) Regulations 1976.

It is likely that wells on, or attached to, an installation will be safety-critical to the safety of that installation and so should be included in the verification scheme.

Regulation 1 Citation and Commencement
These Regulations may be cited as the Offshore Installations and Wells (Design and Construction, etc.,) Regulations 1996 and shall come into force on the 30 June 1996.

Regulation 2 Interpretation
The Regulations apply to all wells in the offshore sector, including wells used for injection and monitoring, whether they are connected to a fixed installation, mobile installation or ship, or stand-alone on the seabed.

The well is defined in terms of its pressure containment boundary and any equipment that is vital to controlling the pressure within the well is therefore covered. This equipment includes blowout preventers and christmas trees. The well may be considered to end at:

- Above the top blowout preventer in the BOP stack and outside the choke and kill valves
- downstream of the swab and production wing valves of a christmas tree
- at the top of the stuffing box of a wireline BOP.

Regulation 13 General Duty
The well operator shall ensure that a well is designed, modified, commissioned, constructed, equipped, operated, maintained, suspended and abandoned so that, as far is reasonably practicable, there can be no unplanned escape of fluids from the well and that, the risks to the health and safety of persons from it or anything in it are as low as is reasonably practicable. The focus overall is therefore on the safe physical condition of the well rather than the actual operation being carried out in the well.

Regulation 14 Assessment of Conditions Below Ground
Before the design of a well is commenced, the well operator shall assess the hazards associated with the geological strata and formations, and fluids within them.

Regulation 15 Design with a view to Suspension and Abandonment
The well operator shall ensure that a well is so designed and constructed that, so far as is reasonably practicable, it can be suspended or abandoned in a safe manner.

Regulation 16 Materials
The well operator must ensure that every part of a well is composed of materials which are suitable. This requirement applies not only to such items as cement, casing or other well tubulars, but also the well-head equipment, eg drilling spools, casing heads, tubing heads and the well control equipment.

Regulation 17 Well Control
The well operator shall ensure that suitable well control equipment is provided. Well control equipment includes equipment whose primary purpose is to prevent, control or divert the flow of fluids from the well. As such, well control equipment includes blowout preventers, downhole preventers, christmas trees, wireline lubricators and stuffing boxes, rotating heads, tubing injection heads, circulating heads, internal blowout preventers and kelly cocks, choke and kill lines, choke manifolds and diverters. Plugs or other isolating devices installed in a borehole to prevent the well from flowing are also included.

In terms of a well operator discharging his duty for ensuring the provision of well control equipment, this will normally be achieved by review of the contractor's arrangements. If necessary, the well operator should check that the contractor has suitable policies, procedures and management controls to ensure that suitable equipment is supplied.

Regulation 18 Arrangements for Examination
Before the design of a well is commenced, the well operator shall put into effect arrangements in writing for an examination scheme for the well and its equipment, which will be carried out by independent and competent persons.

The objective of the examination is to obtain assurance that the well is designed and constructed properly, and maintained adequately. The examination must demonstrate that the pressure boundary of the well and the associated pressure containment equipment is suitable for this purpose. However, it is not anticipated that examination schemes will necessarily rely on the physical examination of wells but more on documentary evidence of well safety.

When a well is connected to a fixed or a mobile installation, there may be an overlap between the well examination scheme and the verification scheme for the installation. Where equipment is deemed to be safety-critical, it would need to be included in the installation verification scheme. Where a well is not attached to an installation, a separate well examination scheme will be required.

Regulation 19 Provision of Drilling Information
The well operator must report to the HSE on a regular basis when carrying out drilling, workovers or any other operation involving substantial risk of the unplanned escape of fluids from a well. The report is required to include a significant amount of operational data such as the setting of casings and the depths achieved.

Regulation 20 Cooperation
Every person to which the Regulations apply shall cooperate with the operator and owner of an installation.

Regulation 21 Information, Instruction, Training and Supervision
The well operator must ensure that persons involved in the drilling, well intervention or workover operations are suitably trained.

Part V – Miscellaneous

Regulation 22 Defence

Regulation 23 Certificates of Exemption
The HSE may exempt by a certificate in writing any person or installation from the provisions of the Regulations, provided that they are satisfied that the health and safety of persons who are likely to be affected by the exemption will not be prejudiced in consequence of it.

4.0 Management and Administration Regulations 1995 – SI 738

Introduction – The MAR regulations primarily deal with the more routine health and safety arrangements which relate to offshore installations. The Regulations apply to fixed and mobile offshore installations, but not to crane barges or subsea installations.

These Regulations cover:

(a) The notification to the HSE of changes of owner of an installation
(b) installation managers powers
(c) the keeping of records of persons onboard;
(d) permit to work systems
(e) various operational matters such as providing information to workers on the location of the relevant HSE offices
(f) the provision of health surveillance
(g) the provision of food and water supplies
(h) cooperation
(i) arrangements for helideck operations
(j) amendments to the Safety Representatives and Safety Committees Regulations SI 1989/971
(k) consequential amendment to other regulations such as the Pipeline Works (First Aid) Regulations, SI 1989/1671.

The Regulations add more detailed requirements necessary for the management of offshore operations.

Regulation 1 Citation and Commencement
These Regulations may be cited as the Offshore Installations and Pipeline Works (Management and Administration) Regulation 1995 (SI 1995/738) and came into force on the 20th June 1995, except Regulation 23(2) which came into force on 20th June 1997.

Regulation 2 Interpretation
A list of relevant definitions are provided.

Regulation 3 Meaning of Offshore Installation
An offshore installation is an installation standing in, or stationed in, relevant waters and used for the exploitation or exploration and storage of mineral reserves, or is used mainly for the provision of accommodation for persons who work on or from one of the previously defined structures.

The definition of an offshore installation in these Regulations replaces the definition previously used in the Mineral Workings Act and the Safety Case Regulations. It is worth noting that vessels providing accommodation for workers on offshore installations such as Hotels are classed as installations whilst heavy lift vessels are not, even if they are providing accommodation as a secondary service to a heavy lift capability. Well service vessels, survey vessels and shuttle tankers are not considered to be installations.

Things which are part of offshore installations
Subsea wells are classed as part of an installation if they are connected by a pipeline or cable, and any pipeline, apparatus or work within 500 metres of the main structure of the offshore installation to which it is attached. Subsea installations are not subject to these Regulations, or the PFEER and SCR if they are free standing and not connected to a fixed or floating installation.

Regulation 4 Application
The Regulations apply to all installations (fixed and mobile) located in the relevant waters and to all installations covered by inter-governmental treaties. Relevant waters are defined as tidal waters and parts of the sea in or adjacent to Great Britain up to the seaward limits of territorial waters and parts of the Continental Shelf designated by the Act of 1964.

Regulation 5 Notification Concerning Offshore Installations
The duty holder shall notify the HSE prior to the arrival or departure of a mobile offshore installation to or from relevant waters and also prior to the change of duty holder of an offshore installation.

Regulation 6 Managers
The duty holder shall appoint an installation manager (OIM) to ensure that the installation is at all times under the charge of a competent person.

The installation manager will be in overall charge of the installation and responsible for the day-to-day management, and in charge of the health, safety and welfare of persons on or about the installation, including ensuring the maintenance of order and discipline. The OIM will also take charge of the installation in an emergency.

The OIM will be competent within the meaning of Regulation 6 of the Health and Safety at Work Act, competent meaning one who has sufficient training and experience or knowledge and other qualities. Competence assessment involves taking account of the persons, skills, judgement, experience, qualifications and relevant knowledge and training in combination, all in relation to the tasks to be undertaken. To ensure that an individual remains competent, his/her performance should be monitored and refresher training and demonstrations of sustained competence be provided.

Regulation 7 Restraint and Putting Ashore
The OIM has legal powers to put people under restraint and send them ashore, when necessary, to ensure the safety of the offshore installation and the health and safety of those onboard.

Regulation 8 Cooperation
Safety requires cooperation between everyone who has a contribution to make in ensuring health and safety on the installation and this includes operators, owners, concession owners, employers, managers and people in charge of visiting vessels or aircraft.

Duty holders cannot evade their responsibilities by seeking to pass their responsibilities to others.

Regulation 9 Records
To ensure that in an emergency everyone can be accounted for, a record must be maintained on the installation of all the individuals onboard, a record often referred to as Personnel on Board (POB). This record will identify each individual's name, and the name and address of his employer, a more complete record of each individual's particulars is to be maintained at an onshore address. The records must be amended whenever anyone arrives or departs from the installation.

Regulation 10 Permits to Work
The duty holder must establish a system of written permits to work to be used in circumstances where the nature and scale of the risk arising from the work to be carried out demands such a stringent system of control. These permits must be issued by a competent person who has sufficient knowledge to decide on conditions and precautions necessary to ensure that safety is not jeopardised.

The HSE provide a guidance booklet on permit to work systems.

Regulation 11 Instructions
Written instructions should be provided and all relevant individuals made aware of these instructions for activities considered to have an element of risk in their execution. Examples of such activities are isolation procedures, over-side working, transfer procedures, helideck operations and inspection/maintenance procedures.

Regulation 12 Communication
Effective communications must be provided which are appropriate for health and safety purposes. Communications must be provided between the installation and the shore, vessels, aircraft (including helicopters) and other installations.

Regulation 13 Helicopters
A competent person, the helicopter landing officer (HLO), must be appointed who will be responsible for the day-to-day management of the offshore installation helideck, in control of the associated helideck operations and the helideck crew. Equipment should be provided for safe helideck operations, including chocks and tie-down straps/ropes, refuelling equipment and a suitable power source for starting helicopters may also be required in some instances.

Procedures must be established for safe helideck operations and an HLO should accompany any flight to an unmanned offshore installation to carry out the HLO's function immediately after the helicopter has landed and up to the point of its subsequent take-off.

Regulation 14 Operational Information
Arrangements should be provided to measure and record such meteorological and oceanographic information as is necessary for securing the safe operation of the installation and the persons on or near it. Environmental conditions may affect the safety of activities such as the loading or unloading of ships and helicopter operations. Operational parameters will have been established as part of the Safety Case.

Regulation 15 Information to Persons
The address and telephone number of the office of the HSE for the onshore installation must be posted where it can be seen by all individuals on the installation, and on vessels supplying a service to the installation such as pipelaying or well servicing operations. This is to ensure that the workforce can contact the HSE inspectors if required.

Regulation 16 Health Surveillance
Health and safety surveillance must be provided which is appropriate to the health and safety risks likely to be encountered onboard the installation. This surveillance should commence before workers are exposed to any hazardous substances. The regulations are not concerned with screening workers to identify any pre-existing health conditions or to meet fitness requirements for offshore work.

Regulation 17 Drinking Water
An adequate supply of clean drinking water must be provided and marked as such on each installation. Supplies should be tested regularly.

Regulation 18 Provisions
The duty holder shall ensure that all provisions for the consumption of persons on the installation are fit for human consumption, palatable and of good quality.

Regulation 19 Identification of the Offshore Installation
Each offshore installation, where physically possible, must display a name or other marking which will allow any vessel or helicopter approaching the installation to be able to identify it without confusion or ambiguity. The HSE publish guidance on the subject under the cover of Standard Marking Schedule for Offshore Installations.

Regulation 20 Certificates of Exemption
The HSE may issue a certificate exempting any person or offshore installation from these regulations provided that the health and safety of people who are likely to be affected are not prejudiced if an exemption is issued. The HSE may not grant exemptions from any of the provisions of relevant EC health and safety directives.

Regulation 21 Application of the Employer's Liability (Compulsory Insurance) Act 1969

Regulation 22 Repeals and Modifications of the 1971 Act

Regulation 23 Revocation and Modification of Instruments.

5.0 The Pipelines Safety Regulations 1996 – SI 825

Introduction – The Pipeline Safety Regulations replace earlier prescriptive legislation and introduce a goal-setting, risk-based approach to managing pipeline safety. The statutory instrument is accompanied by a publication, Guidance on Regulations.

Relationship With Other Regulations
The Pipeline Safety Regulations cover pipelines and associated equipment up to, and including, the 500 metres adjacent to the installation. Pipelines are not included in the DCR regulations but fall partly within the scope of the offshore safety case regime, protection of people from risks associated with the pipeline, and certain parts such as the risers and ESDVs are likely to be classed as safety-critical and will require verification.

Regulations 1 Citation and Commencement
These regulations may be cited as the The Pipeline Safety Regulations and revoke the Emergency Pipelines Regulations, SI 1029 in their entirety, and a number of the regulations within the other pipeline statutory instruments.

Regulation 2 Interpretation
Emergency shutdown valve means a valve which is capable of adequately blocking the flow of fluid within the pipeline. Minor leakage may be tolerated providing it does not represent a threat to safety. The rate of leakage should be based on the installation's ability to control safely the hazards produced by such a leak.

Regulation 3 Meaning of Pipeline
The Regulations cover the subsea pipeline and where pig traps are included, they are classed as part of the pipeline. The limit of the pipeline on the offshore installation is the inboard flange of the emergency shutdown valve (ESDV) or the inboard flanges of the primary shut off valves on the pig receiver, where fitted, which connect the pipeline to the installation. The limit of the pipeline at the onshore reception facility is similarly defined.

Regulation 4 Application
This regulation defines the scope of the requirements as onshore pipelines in Great Britain, and all pipelines in territorial waters and the UK Continental Shelf.

Regulation 5 Design of a Pipeline
The operator shall ensure that no fluid is conveyed in a pipeline unless it has been so designed that, so far as is reasonably practicable, it can withstand:

(a) Forces arising from its operation
(b) the fluids that may be conveyed in it; and
(c) the external forces and the chemical processes to which it may be subjected.

The external forces and the chemical processes to which the pipeline will be subjected will need to be identified and evaluated and this will include terrain, sub-terrain or seabed conditions. Account should also be taken of foreseeable mechanical and thermal stresses and strains to which the pipeline may be subjected during its operation.

In general, British Standards provide a sound basis for the design of pipelines. Other national or international standards are likely to be acceptable provided that they provide an equivalent level of safety.

Regulation 6 Safety Systems
Safety systems cover a means of protection such as emergency shutdown valves and shut-off valves. Safety systems are not meant to cover all control or measuring devices. However, safety systems do include control or monitoring equipment, such as flow detectors and pressure monitors, which have to function properly in order to protect the pipeline or to secure its safe operation.

Regulation 7 Access for Examination and Maintenance
The design of the pipeline should take due account of the need to facilitate examination and maintenance. Consideration should be given at the design stage to any requirement to provide suitable and safe access and operation for in-service inspections, such as pigging.

Regulation 8 Materials
The operator shall ensure that no fluid is conveyed in a pipeline unless it is composed of materials which are suitable. Any changes to the fluid conveyed or the operating conditions of the pipeline, including an extension of the pipeline design life, will warrant a reassessment of the pipeline material to ensure it is capable of conveying the fluid safely.

Regulation 9 Construction and Installation
The operator shall ensure that no fluid is conveyed in a pipeline (save for the purpose of testing it) unless it has been so constructed and installed that, so far as is reasonably practical, it is sound and fit for the purpose for which it has been designed.

Regulation 10 Work on a Pipeline
The operator shall ensure that modification, maintenance or other work on a pipeline is carried out in such a way that its soundness and fitness for the purpose for which it has been designed will not be prejudiced.

Regulation 11 Operation of a Pipeline
The operator shall ensure that:

(a) No fluid is conveyed in a pipeline unless the safe operating limits of the pipeline have been established; and
(b) a pipeline is not operated beyond its safe operating limits, save for the purpose of testing it.

Regulation 12 Arrangements for Incidents and Emergencies
Adequate arrangements must be in place to cover the eventuality of an accident occurring or damage being discovered which requires immediate action to be taken, particularly where pipelines are interconnected.

The Prevention of Fire, Explosion and Emergency Response Regulations (PFEER) requires the installation owner to draw up an emergency response plan to cover emergencies which may affect the pipeline.

Regulation 13 Maintenance
The operator shall ensure that a pipeline, and all attachments such as the pig trap and any safety systems, is maintained in an efficient state, and in good repair. This includes both how and when the pipeline should be surveyed and examined.

Regulation 14 Decommissioning
The operator shall ensure that a pipeline which has ceased to be used for the conveyance of any fluid is left in a safe condition. Where decommissioning is considered, advice must be sought from the Department of Trade and Industry whose formal approval must be sought prior to decommissioning. The pipelines are subject to the Petroleum Act of 1998.

Regulation 15 Damage to Pipelines
No person shall cause such damage to a pipeline as may give rise to a danger to persons. This regulation applies to the operator of the pipeline and to third parties, for example when anchoring vessels within the field.

Regulation 16 Prevention of Damage to Pipelines
Consideration should be given to reducing the potential for damage to the pipeline by use of concrete coatings, trenching, burial, protection structures or mattresses, etc.

The operator shall inform persons of its existence and whereabouts. As part of the offshore Pipeline Works Authorisation issued by the Department of Trade and Industry under the Petroleum and Submarine Pipelines Act 1975, information regarding the location of offshore pipelines is normally passed to the Hydrographer of the Royal Navy for inclusion on Admiralty charts.

Regulation 17 Cooperation
Where there are different operators for different parts of a pipeline, each operator shall cooperate with the other so far as is necessary to enable the operators to comply with the requirements of these regulations.

Regulation 18 Dangerous Fluids
These are identified in some detail in Schedule 2 but are fluids which are flammable, toxic, react violently with water, are oxidants or acrylonitrile.

Regulation 19 Emergency Shutdown Valves
Emergency shutdown valves (ESDVs) are required to be fitted to all risers of major accident hazard pipelines of 40 mm or more in diameter at offshore installations. (Schedule 3 sets out the requirements more fully and they are described in greater detail in the explanation of ESD systems in Chapter 3 of this book.)

Regulation 20 Notification Before Construction
The operator must notify the HSE of their intentions at least six months prior to the construction work on a pipeline commencing. This would normally be at the end of the conceptual stage of design where the proposals are sufficiently detailed to assess arrangements concerning the safe operation of the pipeline and its construction.

The HSE will be sympathetic to requests for shorter notification periods where good reasons can be demonstrated. This notification does not form part of the role the HSE undertakes as consultee on the route of the pipeline for planning purposes. Onshore pipeline routing will be dealt with by the Department of Trade and Industry.

Regulation 21 Notification in Other Cases
The HSE requires a notification period of 14 days prior to a pipeline being brought into use for the first time, or after major modifications have been carried out. In exceptional circumstances a shorter notification period will be approved.

Regulation 23 Major Accident Prevention Document
This regulation deals with the operator's methods for controlling the design, construction and installation, operation, maintenance and final decommissioning. Before the design of a pipeline is completed, a document relating to the pipeline must be prepared which contains sufficient particulars to demonstrate that:

(a) All hazards with the potential to cause a major accident have been identified
(b) the risks arising from those hazards have been evaluated
(c) the Safety Management System is adequate; and
(d) adequate arrangements of audit and the making of reports.

Major Accident Prevention Document – MAPD
Major accident means death or serious injury involving a dangerous fluid. The major accident prevention document (MAPD) is a management tool, the intention of which is to ensure that the operator has assessed the risks from major accidents and has introduced an appropriate safety management system to control those risks. The MAPD can be made up of a number of documents and a single MAPD can cover a number of pipelines.

Safety Management System – SMS
The SMS should cover the organisation and arrangements for preventing, controlling and mitigating the consequences of major accidents. Extensive guidance is given on the requirements of the SMS and the requirements for audit in the Guidance on Regulations.

The remaining regulations cover administration requirements and onshore requirements:

Regulation 24	Emergency Procedures
Regulation 25	Emergency Plans in Case of Major Accidents
Regulation 26	Charge by a Local Authority for a Plan
Regulation 27	Transitional Provision
Regulation 28	Defence
Regulation 29	Certificates of Exemption
Regulation 30	Repeal of Provisions of the Pipelines Act 1962
Regulation 31	Revocation and Modifications of Instruments

6.0 The Provision and Use of Work Equipment Regulations 1992 – SI 2932

Introduction – The Regulations apply to onshore and offshore activities covered by the 1989 Order on, or associated with, oil and gas installations, including mobile installations, diving support vessels, heavy lift barges and pipe-lay barges. The regulations cover an extremely wide range of equipment from overhead projectors to the process plant in its entirety.

Regulation 1 Citation
These Regulations may be cited as the Provision and Use of Work Equipment Regulations (PUWER) 1992. The Regulations came into force on 1st January 1993 for new equipment with existing equipment having until the 1st January 1997 to comply.

Regulation 2 Interpretation
Use in relation to work equipment means any activity involving work equipment and includes starting, stopping, programming, setting, transporting, repairing, modifying, maintaining, servicing and cleaning, and related expressions shall be construed accordingly.

Work equipment means any machinery, appliance, apparatus or tool and any assembly of components which, in order to achieve a common end, are arranged and controlled so that they function as a whole.

Regulation 3 Disapplication of These Regulations
These Regulations shall not apply to or in relation to the Master or crew of a sea-going ship.

Regulation 4 Applications of Requirements Under These Regulations
Employers have a general duty under Section 2 of the Health and Safety at Work etc. Act 1974 to provide and maintain, so far as is reasonably practicable, machinery, equipment and other plant that is safe. The Regulations place duties on all employers providing work equipment to ensure that it is suitable, properly maintained, etc.

Application to the Offshore Industry
These matters will also be considered in the context of the Offshore Installation (Safety Case) Regulations 1992 which will place a duty on owners or operators to demonstrate in their Safety Case that their management system is adequate to ensure that relevant statutory provisions in respect of the installation and connected activities will be complied with.

Regulation 5 Suitability of Work Equipment
Every employer shall ensure that work equipment is so constructed or adapted as to be suitable for the purpose for which it is used or provided.

This Regulation lies at the heart of this set of Regulations. It addresses the safety of work equipment from its initial integrity, the place where, and the purpose for which, it will be used. For example, is the equipment suitable for use in a wet environment, or in a flammable atmosphere?

Regulation 6 Maintenance
Every employer shall ensure that all equipment is maintained in an efficient state, in efficient working order and in good repair. The extent of maintenance required will vary tremendously depending on the equipment and a substantial, integrated programme will be required for a complex process plant.

Equipment may need to be checked frequently to ensure that safety-related features are functioning correctly. A fault in a safety-critical system could remain undetected unless maintenance procedures provide adequate inspection or testing. The frequency at which equipment needs to be checked is dependent on the equipment itself and the risk involved.

Routine maintenance – This includes periodic lubrication, inspection and testing, based on the recommendations of the equipment manufacturer, and should take into account any legal requirements.

Planned preventative maintenance – When inadequate maintenance could cause the equipment, guards or other protection devices to fail in a dangerous way, a formal system of planned preventative maintenance may be necessary.

There is no requirement for a maintenance log to be kept but if there is a log, it must be kept up-to-date.

Regulation 9 Training

Every employer shall ensure that all persons who use work equipment have received adequate training for the purposes of health and safety. An employer's obligation to train extends not only to those who use work equipment but also to those supervising or managing them. Statements of competence may be embodied in qualifications accredited by the National Council for Vocational Qualifications (NCVQ) and the Scottish Vocational Education Council (SCOTVEC).

Regulation 10 Conformity with Community Requirements

Every employer shall ensure that any item of work equipment provided for use in the premises or undertaking of the employer complies with any enactment (whether in an Act or Instrument) which implements in Great Britain any of the relevant Community Directives which are applicable to that item of work equipment.

There are legal requirements covering all those involved in the manufacturing and supply chain of work equipment which are designed to ensure that new work equipment is safe and complies with current legislation.

Article 100A Directives

The aim of this group of Directives is to achieve the free movement of goods in the Community Single Market by eliminating differing national controls and harmonising essential technical requirements.

The Directives set out essential safety requirements that must be met before products may be sold in the Community and products which comply with the Directives must be given free circulation within the Community.

Suppliers must ensure that their products when placed on the market comply with the legal requirements implementing the Directives applicable to their product. Compliance can be claimed by the manufacturer by affixing a mark – the 'CE Mark' to the equipment.

The situation is complicated by the fact that not all work equipment is covered by a product Directive, nor are product Directives retrospective. If an employer provides second-hand equipment for the first time in the workplace, it does not need to be modified to meet the essential safety requirements of the relevant product Directives but it must comply immediately with Regulations 11 to 24 of PUWER. (Second-hand equipment imported from outside the European Community has to comply immediately with essential safety requirements of the relevant product Directives.)

There is a movement within the EC to harmonize standards and this is being lead by the CEN and CENELEC. When developed, these standards will be transposed in the UK by the British Standards Institution and will bear a common number (ie EN XXXX will be BS EN XXXX here). Once these Directives are fully in force, only products which conform and bear the 'CE Mark' may be placed on the market in the UK.

Regulation 11 Dangerous Parts of Machinery

Employers must take effective measures to prevent contact with dangerous parts of machinery.
The term 'dangerous part of machinery' is well established in health and safety law.

A risk assessment carried out under Regulation 3 of the MHSWR should identify hazards presented by machinery. The risk assessment should not just deal with the machine in its normal operating mode, but must also cover activities such as setting, maintenance, cleaning or repair.

The Regulation sets out various requirements for guards and protection devices. These are largely common sense, and in large part are detailed in relevant national and international standards.

Regulation 12 Protection Against Specified Hazards

Every employer shall take measures to ensure that the exposure of a person using work equipment to any risk is either prevented, or, where that is not reasonably practicable, adequately controlled.

This Regulation covers risks arising from certain listed hazards during the use of equipment such as:

(a) Material falling from equipment such as scaffold boards
(b) overheating or fire due to friction in electric motors, ignition by welding torch, thermostat failing, cooling system failure or similar
(c) explosion of the equipment due to pressure build-up, perhaps due to the failure of a pressure-relief device or the unexpected blockage or sealing off of pipework.

A risk assessment carried out under Regulation 3 of the Management of Health and Safety at Work Regulations 1992 should identify these hazards, and assess the risks associated with them.

Regulation 13 High or Very Low Temperature

All parts of work equipment and substances produced which are at a high or very low temperature shall have protection where appropriate so as to prevent injury to any person by burn, scald or sear. The risk from contact with hot surfaces should be reduced by engineering methods, ie reduction of surface temperature, insulation, shielding, barricading and guarding.

Regulation 14 Controls for Starting or Making a Significant Change in Operating Conditions.
The controls provided should be designed and positioned so as to prevent, so far as possible, inadvertent or accidental operation. Buttons and levers should be of appropriate design, for example including a shrouding or locking facility.

Regulation 15 Stop Controls
Every employer shall ensure that, where appropriate, such equipment is provided with one or more readily accessible control, the operation of which will bring the work equipment to a safe condition in a safe manner.

Regulation 16 Emergency Stop Controls
An emergency stop control may be required in addition to the other safeguards to prevent risk when some irregular event occurs. The location of emergency stop controls should be determined by the risk assessment.

Regulation 17 Controls
Every employer shall ensure that all controls for work equipment shall be clearly visible and identifiable, including appropriate marking where necessary.

Regulation 18 Control Systems
Every employer shall ensure, so far as is reasonably practicable, that all control systems of work equipment are safe. Failure of any part of the control system or its power supply should lead to a fail-safe condition and not impede the operation of the stop or emergency stop controls.

Regulation 19 Isolation From Sources of Energy
Every employer shall ensure that, where appropriate, work equipment is provided with suitable means to isolate it from all its sources of energy. Isolation means establishing a break in the energy supply in a secure manner, ie by ensuring that inadvertent reconnection is not possible. The possibilities and risks of reconnection should be identified as part of the risk assessment, which should then establish how security can be achieved. For some equipment, an isolating switch or valve may have to be locked in the off or closed position to avoid unsafe reconnection.

Regulation 20 Stability
All work equipment should be stabilised by clamping or otherwise where necessary for safety.

Regulation 21 Lighting
Every employer shall ensure that suitable and sufficient lighting, which takes account of the operations to be carried out, is provided at any place where a person uses work equipment.

Regulation 22 Maintenance Operations
Regulation 6 requires that equipment is maintained. Regulation 22 requires that equipment is constructed or adapted in a way that takes into account the risk associated with carrying out maintenance work, such as routine and planned preventative maintenance, as described in the guidance to Regulation 6. Compliance with this Regulation will help to ensure that when maintenance work is carried out, it is possible to do it safely without risk to health, as required by Section 2 of the HSW Act.

Regulation 23 Markings
Every employer shall ensure that work equipment is marked in a clearly visible manner with any marking appropriate for reasons of health and safety.

Regulation 24 Warnings
Warnings or warning devices should be provided where risks to health or safety remain after other hardware measures have been taken. They may be incorporated into the permit-to-work systems and can reinforce measures of information, instruction and training. A warning is normally in the form of a notice or similar. Examples are positive instruction such as 'Hard Hats Must Be Worn'. A warning device is an active unit giving a signal; the signal may typically be visible or audible, and is often connected into equipment so that it is active only when a hazard exists.

Warnings must be easily perceived and understood, and unambiguous.

Regulation 25 Exemption Certificates
The Secretary of State for Defence may, in the interests of national security, exempt any of the home forces, or any visiting force, from these Regulations.

Regulation 26 Extension Outside Great Britain
The Regulations apply to offshore activities covered by the 1989 Order on, or associated with, oil and gas installations, including mobile installations, diving support vessels, heavy lift barges and pipe-lay barges.

Regulations 27 Repeals, Saving and Revocations
Refer to regulations.

Part 5. Associated Information

1.0 Health and Safety Executive Offshore Safety Division (OSD)

The Offshore Safety Division (OSD) was created in 1990 as a department within the Health and Safety Executive (HSE). In 1991, all duties pertaining to the occupational health, safety and welfare of offshore installations were transferred from the Petroleum Engineering Division (PED) of the Department of Energy to the Offshore Safety Division of the Health and Safety Executive.

The Offshore Safety Division is first and foremost a regulator whose core purpose is to ensure that risks to people from work activities in the offshore industry are properly controlled. Their objectives are influenced by political, economic, social and technical issues and their duties range from direct enforcement to contributing to research.

Two of the main tasks performed by the OSD are the assessment of safety cases and the inspection of offshore installations. The inspections are aimed primarily at a review of operating practices to ensure that the duty holder is complying with the procedures outlined in their safety case. Other duties undertaken are as follows:

- Enforcement of regulations
- accident investigation
- development of policies

- production of guidance
- research
- influence technical standards
- advise, guide and inform.

The OSD are represented at the three main geographic locations of offshore activity in the UK, Aberdeen, Bootle, and Norwich, and are organised into the following five groups:

OSD 1 Aberdeen
- Inspection Teams
- Inspection Management Support

OSD 2 Aberdeen
- Inspection Teams
- Well Engineering and Operations

OSD 3 Bootle
- Process integrity
- Fire, explosion and risk assessment
- Emergency Response
- Maritime and Aviation
- Mechanical Engineering
- Electrical and Control systems

OSD 4 Bootle
- Materials
- Structural integrity
- Navel architecture and marine engineering
- Diving

OSD 5 Aberdeen
- Occupational Health
- Organisational and human factors
- Competency
- Business support
- Special operational projects.

1.1 HSE Correspondence

Official correspondence emanating from the Offshore Safety Division of the Health and Safety Executive may appear in any one of two guises, depending on the nature and urgency of the subject matter.

Distribution will be dependant on the content of the correspondence but will normally include the operators of offshore installations and those companies who subscribe to the HSE publications department. Each item of correspondence is uniquely numbered and retains the format originally introduced by the Department of Energy.

The two levels of communication are:

i) Operations Notice
Operations Notices are intended to advise the offshore industry of new regulatory requirements, contact points within the Offshore Division for communications (eg accident/incident reporting) and of other Government requirements with respect to offshore operations which may have safety implications.

ii) Safety Notice
The Safety Notices provide guidance but following the guidance is not compulsory and equivalent alternative arrangements preferred by the operator may be adopted. Following the advice given in a Safety Notice will normally ensure compliance with the law. Health and safety inspectors seek to secure compliance with the law and may refer to this guidance as illustrating good practice. It is recommended that the Safety Notices are retained on file on the installation for reference purposes.

2.0 UK Offshore Operators Association (UKOOA)

UKOOA was officially established in 1973, having previously existed (from 1964) as an informal organisation. In April 2007 UKOOA became Oil and Gas UK. Every oil company engaged in exploration or production on the UK Continental Shelf (UKCS) is a member of Oil and Gas UK as are a large number of service companies, over 80 in total.

The main objective of Oil and Gas UK is to ensure the long-term health of the offshore oil and gas industry in the United Kingdom. It operates purely in an advisory capacity, working closely with the Government to ensure that legislation obtains the desired results and reflects the requirements of the offshore industry. It also provides a forum for discussion on matters relating to technology, the environment and safety.

As a service to the industry, Oil and Gas UK provide a range of specialist guidance documents which relate to various aspects of the offshore industry, such as Guidelines for the Management of Emergency Response for Offshore Installations, and a wide range of well illustrated literature which is of more general interest.

Contact Details

Oil and Gas UK
6th Floor East
Portland House
Bressenden Place
London
SW1E 5BH

Telephone:	020 7802 2400
Facsimile:	020 7802 2401
email:	info@oiland gasuk.co.uk
web site:	www.oilandgasuk.co.uk

Index

The images in this book are reproduced with the kind permission of:

2.24	Harald M. Valderhaug
2.33	Based on an image from Statoil
2.42	Oilfield Publications Ltd, Homend House, 15 The Homend, Ledbury, Herefordshire, HR8 1BN
2.48	Conoco (UK) Ltd and Chevron (UK) Ltd
3.74	Bettis Actuators and Controls
3.76, 3.78 & 3.79	USCG
3.77	Getty Images
4.1	Based on an image from Kvaerner Cape Surveys
4.24	RGB Stainless Ltd
4.28, 4.30, 4.32, 4.33, 6.18, 6.21, 7.35, 7.39, 7.47, 8.1, 8.6 & 8.8	Cooper Oil Tools, Texas
6.1	Image based on a schematic from Stena Offshore Ltd
6.6	Bibby Offshore
6.8a & 6.11	Hydrovision Ltd
6.8b & 6.12	Slingsby Engineering Ltd
6.19 & 6.22	Konsberg Offshore As
6.25	Based on an original image supplied by ABB Engineering As
6.27	Framo Engineering As
7.11	Baker Service Tools
7.15, 7.27, 7.28 & 7.40	Varco BJ, Orange, California
7.29, 7.31 & 7.32	Hydril, Houston, Texas
7.44 & 7.45	Schlumberger Oilfield Services
8.11 & 8.12	Halliburton Energy Services
9.2	SSR (Int'l) Ltd